SOLVENT EXTRACTION and LIQUID MEMBRANES

Fundamentals and Applications in New Materials

SOLVENT EXTRACTION and LIQUID MEMBRANES

Fundamentals and Applications in New Materials

Edited by
Manuel Aguilar
José Luis Cortina

CRC Press is an imprint of the
Taylor & Francis Group, an **informa** business

CRC Press
Taylor & Francis Group
6000 Broken Sound Parkway NW, Suite 300
Boca Raton, FL 33487-2742

First issued in paperback 2020

ISBN 13: 978-0-367-57751-3 (pbk)
ISBN 13: 978-0-8247-4015-3 (hbk)

Library of Congress Cataloging-in-Publication Data

Solvent extraction and liquid membranes: fundamentals and applications in new materials / editors, Manuel Aguilar, Jose Luis Cortina.
p. cm.
Includes bibliographical references and index.
ISBN 978-0-8247-4015-3 (hardback : alk. paper)
1. Solvent extraction. 2. Liquid membranes. I. Aguilar, Manuel. II. Cortina, Jose Luis. III. Title.

TP156.E8S62 2008
660'.284248--dc22 2007036221

Visit the Taylor & Francis Web site at
http://www.taylorandfrancis.com

and the CRC Press Web site at
http://www.crcpress.com

Dedication

We dedicate this book to our esteemed colleague Abraham Warshawsky who passed away during the preparation of his chapter.

Contents

Foreword

The International Solvent Extraction Conferences (ISECs) have been held every 3 years since 1971 in various countries around the world. The 1999 conference (ISEC99) was held in Barcelona, Spain. It was organized under the concept of being a "good soup," so the best ingredients were combined both to have a good taste and to provide the best healthy nutrition to grow on solvent extraction fundamentals and applications. Under this concept the Organizing Committee decided to cook ISEC99 with a special ingredient by preceding the conference with a Summer School addressing the principles and applications of solvent extraction for the benefit of students and newcomers to this widely used technique in separation science. Although this idea of providing short courses on topics related to the main conference is an established practice elsewhere, this was the first time such a course was offered as part of an ISEC. The aim of the course organizers was to provide a fairly comprehensive overview of the fundamental and practical applications of solvent extraction, so they contacted a number of internationally recognized experts to talk about their particular fields of interest.

As a result, this publication represents a good recipe to feed the present and future newcomers to the solvent extraction community. Thus, the text presents the perspectives of solvent extraction (SX) in the twenty-first century, a new focus of fundamentals of solvent extraction, renewed topics of calculations on extraction equilibria, liquid membranes, and industrial applications. The book also introduces topics of new materials and solid sorbents for solvent extraction including the improved and developing topic of solvent impregnated resins. The content of this publication, presented with a tutorial focus, will benefit the understanding of solvent extraction for today's practice.

Michael Cox
Manuel Valiente

Preface

Following the activities of the first School in Solvent Extraction in Barcelona in 1986, and coincident with the 1999 International Solvent Extraction Conference (ISEC), the second International Solvent Extraction School (ISES) was held in Bellaterra, Spain. The school gathered senior and young scientists and engineers with the main idea of discussing the fundamentals and practice of solvent extraction (SX) and liquid membranes (LM) and to grow the new ideas and trends to contribute to the best understanding of applications of SX and LM in the future. The lecturers were leading experts in solvent extraction and liquid membranes.

One of the conclusions from the International Solvent Extraction Committee was to recommend to the authors of this book and the chairs of the ISES to produce a reference book by converting the lectures into written material.

The different uses of liquid–liquid extraction, liquid membranes, and solvent impregnated materials make the subject important for university students of chemistry, metallurgy, hydrometallurgy, and chemical and mineral processing technology. Some universities offer special courses on separation processes in which those techniques are minor topics in more comprehensive courses. Laboratory experiments on liquid–liquid and liquid membranes are common in chemical and mineral processing engineering curricula. Because of the breadth of the subject, the treatment in such courses is often scarce, and more comprehensive text is difficult to find in a form suitable for use directly with students.

To meet this demand, and to answer the request of the ISEC, we initiated this project to develop a simple text suitable not only for students but also for scientists and engineers in the field. However, no single scientist or engineer can be an expert in all parts of the field. Therefore, it seemed that the best idea was to develop this book as a joint project among many expert authors, each of whom has in many cases years of experience in research, teaching, or industrial development. The result is an international book at a high scientific and technical level.

In this context, we would like to point out that this book represents the effort to bring together the key important topics dealing with thermodynamics, kinetics, interfacial behavior, process and plant applications, including chemical and engineering aspects of new extraction systems based on synthetic materials by the leading experts at the school. The book is directed toward third- to fourth-year undergraduate and postgraduate chemistry and chemical engineering students as well as toward researchers and developers in the chemical industry, the mining and mineral processing industry, and the waste treatment industry. The book is also intended for chemical, metallurgical, mineral processing, and waste treatment engineers who already use this technique but have a desire to understand better or to solve existing process problems. Furthermore, the book should be useful for researchers in solvent extraction who wish to learn about its applications in areas other than their own

So, after an introduction to SX and LM in the perspective of the twenty-first century (Chapter 1), the following two chapters (Chapters 2 and 3) present the physical principles (thermodynamics and kinetics) in SX processes and the tools for computation of the equilibrium and kinetic parameters. They are followed by three chapters of various industrial applications and process experience on SX and LM (Chapters 4 through 6), including optimization and modeling tools. The three final chapters (Chapters 7 through 9) focus on new materials in SX and LM science and technology, including functionalized organic and inorganic solid sorbents and solid impregnated sorbents, indicating the research frontiers and future developments in new materials.

Acknowledgments

We the editors want to express our gratitude to the contributors, who made this book possible through their helpful suggestions and extensive efforts. Professor Michael Cox, Professor Josef Havel, Dr. Karel Jeřábek, Professor Vadim Korovin, Dr. Anil Kumar Pabby, Professor Immaculada Ortiz Uribe, Professor Angel Irabien, Dr. Jun S. Lee, Dr. Sergio Gomez Salazar, Professor Ana María Sastre, Dr. Yuri Shestak, Dr. Kathryn C. Sole, Professor Lawrence L. Tavlarides, Professor Hitoshi Watarai, Dr. Yuri Pogorelov, and Professor Abraham Warshawsky worked hard and successfully on their chapters and provided us much valuable help. The success of this book belongs to our distinguished authors.

We thank Nita Lekhwani and David Russell on the initial editorial staff at Marcel Dekker and to David Fausel and Barbara Glunn and the new staff at Taylor & Francis for their invaluable editorial assistance.

About the Editors

Manuel Aguilar, Ph.D., was born in Spain in 1943. He achieved his licentiate studies in chemistry at the University of La Laguna in Tenerife, Spain with a main focus in analytical chemistry. From 1965 to 1969 he was assistant lecturer at the University de Los Andes in Venezuela. From 1970 through 1976 he was research assistant at the Royal Institute of Technology in Stockholm, Sweden, where he completed his Ph.D. studies under the supervision of Professor E. Hogfeldt. In 1976 he returned to Spain and was assistant professor of inorganic and analytical chemistry at the Universidad Autonoma in Barcelona. Since 1977, he has been professor of chemistry in the Department of Chemical Engineering at the Universitat Politécnica de Catalunya. Dr. Aguilar has been active in chemical research for many years, with a main interest in the field of ionic equilibrium and solvent extraction, and he has published more than 100 papers in this field. He has also been involved in the field of education, in which he has published different books and compendia on ionic equilibrium and has directed two international solvent extraction schools and diverse postgraduate courses in Spain and South America. Dr. Aguilar has been active in equilibrium analysis and computers, participating in the production of various programs for treatment of hydrochemical systems and in the program of virtual laboratory for experimental laboratory teaching for basic chemistry courses.

José Luis Cortina, Ph.D., was born in Ligüerrre de Cinca, Spain, in 1964. He has been professor of chemical engineering at the Universitat Politécnica de Catalunya in Barcelona, Spain, since 2001 and project research technical director at the Water Technology Center (CETaqua) since 2007. Dr. Cortina received B.Sc., D.Sc., and Ph.D. degrees from the University of Barcelona in Spain. He has been a visiting scientist at the Polymer Chemistry Department at the Weizmann Institute of Science in Rehovot, Israel, and at the Center for Process Analytical Chemistry at the University of Washington in Seattle at the Fiber Optical Chemical Laboratory. Dr. Cortina has been active in chemical technology research for many years, with a main interest in the field of treatment and separation processes for industrial and environmental applications using solvent extraction and ion exchange, and he has published more than 70 papers in this field. During the last several years his research has been centered on the treatment technologies for soil and groundwater remediation. He has been a member of the International Committee on Ion Exchange since 2004.

Contributors

Michael Cox
Department of Chemistry
Hertfordshire University
Hatfield, Hertfordshire, United Kingdom

Sergio Gomez-Sakzar
Chemical Engineering and Materials Science Department
Syracuse University
Syracuse, New York

J. Angel Irabien Gulias
Universidad de Cantabria
Santander, Spain

Josef Havel
Department of Analytical Chemistry
Faculty of Science
Masaryk University
Kotlarska, Czech Republic

Karel Jeřábek
Institute of Chemical Process Fundamentals
Academy of Sciences of the Czech Republic
Praha, Czech Republic

Vadim Korovin
Pridneprovsky Scientific Center
Ukrainian Academy of Sciences
Dniepropetrovsk, Ukraine

Jun S. Lee
Chemical Engineering and Materials Science Department
Syracuse University
Syracuse, New York

Anil Kumar Pabby
Bhabha Atomic Research Centre
Nuclear Recycle Group
Tarapur, Maharashtra, India

Yuri Pogorelov
Pridneprovsky Scientific Center
Ukrainian Academy of Sciences
Dniepropetrovsk, Ukraine

Ana María Sastre
Department of Chemical Engineering
Universitat Politécnica de Catalunya
Barcelona, Spain

Yuri Shestak
Pridneprovsky Scientific Center
Ukrainian Academy of Sciences
Dniepropetrovsk, Ukraine

Kathryn C. Sole
Anglo Research, a Division of Anglo Operations, Ltd.
Johannesburg, South Africa

Lawrence L. Tavlarides
Chemical Engineering and Materials Science Department
Syracuse University
Syracuse, New York

Inmaculada Ortiz Uribe
Universidad de Cantabria
Santander, Spain

Abraham Warshawsky
Department of Organic Chemistry
The Weizmann Institute of Science
Rehovot, Israel

Hitoshi Watarai
Department of Chemistry
Graduate School of Science
Osaka University
Osaka, Japan

1 Liquid–Liquid Extraction and Liquid Membranes in the Perspective of the Twenty-First Century

Michael Cox

CONTENTS

1.1 LIQUID–LIQUID EXTRACTION

Liquid–liquid extraction is now very well established, featuring extensively as a selective separation process. Liquid membranes are a more recent development, which as yet have not featured significantly in industrial applications. This chapter focuses on the current and future prospects of these techniques largely in the context of metal extraction, as that is the area in which I have spent most of my time.

It first addresses the subject of liquid–liquid extraction, dealing in turn with extractants, systems, and, lastly, equipment.

1.1.1 EXTRACTANTS

Because metals generally exist in aqueous solution as hydrated ions before the metal can be extracted into a nonpolar organic phase, the water molecules must be replaced and any ionic charge reduced or removed. This can be achieved in different ways by using three types of extractants: acidic, basic, and solvating, which extract metals according to the following equilibrium reactions:

$$\text{Acidic: } M^{z+}_{(aq)} + z\,HA_{(aq,\text{ or }org)} \Leftrightarrow MA_{z\,(org)} + z\,H^{+}_{(aq)} \tag{1.1}$$

$$\text{Basic: } (n\text{-}z)R_4N^{+}_{(aq\text{ or }org)} + MX_n^{(n\text{-}z)-}{}_{(aq)} \Leftrightarrow (n\text{-}z)R_4N^{+}MX_n^{(n\text{-}z)-}{}_{(org)} \tag{1.2}$$

$$\text{Solvating: } MX_{z\,(aq)} + mS_{(org)} \Leftrightarrow MX_z\,S_{m\,(org)} + mH_2O \tag{1.3}$$

Acidic extractants include simple reagents such as carboxylic acids and organophosphorus acids, as well as chelating acids. The latter are often derived from analytical reagents, such as β-diketones, 8-hydroxyquinoline, and hydroxyoximes. Extraction occurs by a change of the acidity in solution; thus, increasing the pH of solution increases extraction, and decreasing the pH promotes stripping or back-extraction. Therefore, by changing the acidity of the system the metal can be cycled to and from the organic phase.

The extent of extraction also depends on the magnitude of the metal-extractant formation constant and concentration of extractant. These factors can be combined with the distribution coefficient in the following equation derived from Equation 1.1 and the law of mass action:

$$\text{Log } D = \log K_f + z\log[HA] + z\text{pH} \tag{1.4}$$

This equation allows the stoichiometry of Equation 1.1 to be confirmed by constructing linear plots of log D versus pH and log D versus log[HA]. Note in situations where the extractant cannot satisfy both the coordination number and ionic charge of the metal ion, then additional nonionized extractant molecules can be added to the complex (Equation 1.5). In this case the linear log D versus log[HA] plot will have a slope of (z + s).

$$M^{z+}_{(aq)} + (z + s)HA_{(aq,\ or\ org)} \Leftrightarrow MA_z\,(HA)_{s(org)} + z\,H^+_{\ (aq)} \tag{1.5}$$

As noted already, the extent of extraction depends on the relevant formation constant of the metal or extractant (K_f). For metals in the first transition series this will generally follow the Irving-Williams series: that is, for divalent metals, V < Cr < Mn < Fe < Co < Ni < Cu > Zn. Thus, unless other factors are involved copper will be extracted at lower pH values than the other elements. In addition it is found that $M^{4+} > M^{3+} > M^{2+} > M^+$, so that thorium(IV) will be extracted before iron(III) in turn before copper(II) before sodium.

Although the majority of acidic extractants is based on oxygen donor groups, some commercial compounds include a sulfur donor atom. These, of course, would be favored when the extraction of a soft metal is required. These generally consist of elements in low oxidation states or the heavier elements in any group in the Periodic Table.

In certain cases, the separation of metals is a result of kinetic rather than thermodynamic considerations, such as extraction of copper at a lower pH than iron(III) by a commercial 8-hydroxyquinoline derivative, Kelex 100 (Cognis, Inc., Tucson, Arizona, formerly Witco Corporation). If contact is restricted to about 5 minutes then copper extraction is preferred, but when the reaction is allowed to progress to equilibrium then the iron is extracted. This is in contrast to the separation of copper and iron(III) with β–hydroxyoximes, such as LIX860 (Cognis, Inc., formerly Henkel Corporation) or M5640 (Avecia, formerly Zeneca Specialties), which is thermodynamically controlled.

Stripping of the metal from the loaded organic phase is just as important as extraction, and here this process is achieved quite simply by a change in acidity of the system. However, it should be remembered that extractants that extract metals at low pH values will require more concentrated acid for stripping. Thus, it is not always a good idea to use the extractant that extracts too well. Also it is important to realize that extraction and separation is only one process in a hydrometallurgical flow sheet, and it is necessary to choose the extractant that will most easily interface with the operations both upstream and downstream.

Basic extractants normally consist of alkylammonium species, although for analytical applications organophosphonium or organoarsonium compounds may be used. Quaternary ammonium species as shown in Equation 1.2 or uncharged alkylammonium compounds may be used as extractants. In the latter case the presence of excess anions in the system allows the formation of charged species as, for example, Equation 1.6:

$$R_3N + HX \Leftrightarrow R_3NH^+X^- \tag{1.6}$$

In the case of alkylammonium extractants it has been observed that the magnitude of extraction follows the order $R_4N^+ > R_3NH^+ > R_2NH_2^+ > RNH_3^+$ with the size of the alkyl group R generally between 8 and 10 carbon atoms. The structure of the alkyl groups is important, and by using straight or branched chain substituents different properties can be found. In commercial practice the trialkylammonium compounds are generally more commonly used in spite of their inferior extraction properties because they are cheaper than the quaternary extractants.

These extractants require the presence of stable anionic metal complexes to form the extractable ion pair complexes, so that only metals that produce such species can be extracted with these compounds. The ease of metal extraction follows the magnitude of the formation coefficient of the anionic complex. Thus, in halide solution, gold (III) > iron (III) > zinc > cobalt > copper >>> nickel.

Stripping of the metal from the loaded organic phase can be achieved by causing the breakdown of either the anionic metal complex with, for example, water, or the cationic extractant by treatment with a base. The latter will, of course, not work in the case of the quaternary tetraalkylammonium ions, R_4N^+.

Solvating extractants operate by replacing the solvating water molecules around the aqueous metal complex, making the resulting species more lipophilic. Such extractants must possess lone pairs of electrons that can be donated to the metal ion, and the most common donor atoms are oxygen and sulfur. Nitrogen and phosphorus can also feature as donors but, for a number of reasons, are not generally used. The types of organic compounds used include those based on carbon—that is, alcohols, ethers, esters, and ketones with compounds such as dibutylcarbitol, nonyl phenol, and iso-butylmethylketone (MIBK). Amides, $RCONR_2$, have also been proposed for specialized applications such as the extraction of actinides and precious metals. However, the most commonly used oxygen-donating solvating extractants are based on the organophosphorus compounds: alkylphosphates, $(RO)_3PO$; alkylphosphonates, $(RO)_2RPO$; alkylphosphinates, $(RO)R_2PO$; and alkylphosphine oxides, R_3PO. The donor properties of the oxygen atom follow the aforementioned trend with the greater the number of C-P bonds, the better the donor; however, for economic reasons trialkylphosphates are the chosen extractants for commercial use, with tri-n-butylphosphate probably the most widely used extractant worldwide.

Sulfur donor extractants are less common, although dialkylsulfides have been used in precious metal extraction, and a trialkylphosphine sulfide, R_3PS (Cyanex 471X, Cytec Inc., New Jersey), is available commercially. Such sulfur donating extractants will need to be considered when extracting and separating soft metals, such as second- and third-row transition metals (e.g., cadmium, mercury, and palladium).

Stripping can be achieved by contact with water (Equation 1.3) or by raising the temperature of the system.

1.1.2 Synergistic Extraction

On occasion, mixtures of two different extractants will enhance the extraction of a metal above that expected from the summation of the performance of the two reagents separately. This gives rise to the synergistic factor (SF), defined as follows:

$$SF = D_{AB}/(D_A + D_B) \tag{1.7}$$

A large number of examples of synergism can be found in the literature, although very few of these have actually been commercialized. The main reason for the lack of industrial interest is probably the difficulty in maintaining the optimum ratio of extractants in the organic phase to provide synergism. Different extractant losses through evaporation, water solubility, and entrainment would provide a difficult control problem.

The most common synergistic system consists of a mixture of an acidic and a solvating extractant acting on a metal ion where the preferred coordination number cannot be satisfied by just the acidic extractant. For example, a divalent metal ion with a preferred coordination number of six would form with a bidentate monobasic acid extractant a complex, $M(L\text{-}L)_2(H_2O)_2$, where HL-L is the extractant. The replacement of the water molecules in the extracted complex with a solvating extractant (S) will make the complex much more lipophilic, $M(L\text{-}L)_2S_2$ and thus will give a higher distribution ratio. It is very unlikely that the solvating extractant will extract the metal ion by itself; therefore, synergism occurs.

$$M^{2+}_{(aq)} + 2H(L\text{-}L)_{(org)} + 2S \Leftrightarrow M(L\text{-}L)_2S_{2\,(org)} + 2H^+_{(aq)} + xH_2O \qquad (1.8)$$

1.1.3 Applications

Currently, approximately 40 extraction reagents are commercially available for the recovery of metals. These represent the main categories of acidic, basic, and solvating reagents (Table 1.1). Some of these extractants like tri-n-butylphosphate (TBP), Versatic acid, and the various amines have been around since the use of liquid–liquid extraction was proposed for the extraction of metals in the 1950s. Others are much more recent introductions. It is interesting to note that although 40 reagents are listed in Table 1.1, only about a dozen are in daily use.

In this section the current status of commercial extractants are considered to set the scene for possible future trends.

1.1.3.1 Nonferrous Metals

1.1.3.1.1 Copper

The introduction of chelating acidic extractants such as the hydroxyoxime mixture LIX64N in the mid 1960s revolutionized the recovery of copper from acidic leach liquors and its separation from iron(III). Since the first introduction of this reagent, several developments have taken place to improve operating performance and to provide subtle modifications to suit particular leach liquors. However, there has not been any development to challenge the position of the salicylaldoximes of the Avecia (Acorga reagents) and Henkel (Cognis, Inc.) (LIX reagents) in this market. The alternative reagent, Kelex 100 (Witco Corporation), based on 8-hydroxyquinoline—which was originally formulated for this process—was not really successful because as noted already the separation was based on extraction kinetic. Also, the reagent was easily protonated by the acidic strip solutions, which then required a water wash so that in the overall process a water balance could not be maintained.

The reagent Acorga CLX-50 was a novel introduction about 10 years ago based on a pyridine dicarboxylic ester. This behaved as a solvating reagent and recovered copper from chloride media as the complex CuL_2Cl_2 but has only been used on a pilot scale.

TABLE 1.1
Liquid–Liquid Extraction Reagents

Type	Examples	Manufacturers	Commercial Uses
		Acid Extractants	
Carboxylic acids	Naphthenic acids Versatic acids	Shell Chemical Co	Cu/Ni separation, Ni extraction Yttrium extraction
Alkyl phosphoric acids	Mono-alkylphosphoric acids	Mobil Chemical Co (MEHPA/ DEHPA mixture) Avecia formerly Zeneca Specialties (Acorga SBX50)	Fe removal Sb,Bi removal from copper electrolytes
	Di-alkylphosphoric acids and sulfur analogs	Daihachi Chem Ind Co Ltd (DP-8R, DP-10R, TR-33, MSP-8) Bayer AG (BaySolvex D2EHPA, D2EHTPA VP Al 4058) Albright and Wilson Americas (DEHPA) Hoechst (PA216, Hoe F 3787)	Uranium extraction Rare earth extraction Cobalt/nickel separation Zinc extraction, etc.
Alkyl phosphonic acids	2-ethylhexylphosphonic acid 2-ethylhexyl ester, and sulfur analog	Daihachi Chem Ind Co Ltd (PC 88A) Albright and Wilson Americas (Ionquest 801) Bayer AG (BaySolvex VP-AC 4050 MOOP)	Cobalt/nickel separation Rare earth separation
Alkyl phosphinic acids	di-alkyl phosphinic acids and sulfur analogs	Daihachi Chem Ind Co Ltd (PIA-8) Cytec Inc (Cyanex 272, 302, 301)	Cobalt/nickel separation Zinc and iron extraction
Aryl sulfonic acids	Dinonylnaphthalene sulfonic acid	King Industries Inc (Synex 1051)	Magnesium extraction
		Chelating Acid Extractants	
Hydroxyoxime derivatives	α-alkarylhydroximes (LIX63) β-alkylarylhydroxyoximess (LIX860) (M5640)	Cognis Inc formerly Henkel Corp (various e.g., LIX860) Avecia formerly Zeneca Specialties (various e.g., M5640) Inspec (MOC reagents)	Copper extraction Nickel extraction
8-hydroxyoxine derivatives	Kelex 100, 120 LIX26	Witco Corp Cognis Inc formerly Henkel Corp	Gallium extraction Proposed for copper extraction
β-diketone derivatives	LIX 54 Hostarex DK16	Cognis Inc formerly Henkel Corp Hoechst	Copper extraction from ammoniacal media

TABLE 1.1 (Continued)
Liquid–Liquid Extraction Reagents

Type	Examples	Manufacturers	Commercial Uses
Alkaryl-sulfonamides	LIX 34	Cognis Inc formerly Henkel Corp	Development reagent
Bis-dithiophos-phoramide derivatives	DS 5968, DS 6001 (Withdrawn)	Avecia formerly Zeneca Specialties	Zinc extraction Cobalt/nickel/manganese separation
Hydroxamic acids	LIX 1104	Cognis Inc formerly Henkel Corp	Nuclear fuel reprocessing, iron extraction Sb, Bi extraction from copper refinery liquors
Basic Extractants			
Type	**Examples**	**Manufacturers**	**Commercial Uses**
Primary amines	Primene JMT, Primene 81R	Rohm and Haaas	No known commercial use
Secondary amines	LA-1, LA-2 Adogen 283	Rohm and Haaas Witco Corp	Uranium extraction Proposed for vanadium, tungsten extraction
Tertiary amines	Alamines (e.g., Alamine 336) Adogens	Cognis Inc formerly Henkel Corp Witco Corp	Uranium extraction Cobalt from chloride media Tungsten, vanadium extraction, etc.
Quarernary amines	Aliquat 336 Adogen 464	Cognis Inc formerly Henkel Corp Witco Corp	Vanadium extraction Possible chromium, tungsten, uranium, etc.
Quaternary amine + nonyl phenol	LIX 7820	Cognis Inc formerly Henkel Corp	Anionic metal cyanide extraction
Momo-N-substituted amide			Iridium separation from rhodium
Trialkylguanidine	LIX79	Cognis Inc formerly Henkel Corp	Gold extraction from cyanide media
Solvating Extractants and Chelating Nonionic Extractants			
Carbon–oxygen donor reagents	Alcohols, (decanol) Ketones (MIBK) Esters Ethers, etc.	Various chemical companies	Niobium/tantalum separation Zirconium/hafnium separation

TABLE 1.1 (Continued)
Liquid–Liquid Extraction Reagents

Type	Examples	Manufacturers	Commercial Uses
Phosphorus–oxygen donor reagents and phosphorus–sulfur donors	Phosphoric esters Phosphonic esters Phosphinic esters Phosphine oxides and sulfur analogs	Albright and Wilson Americas Daihachi Chem Ind Co Ltd Cytec Inc Bayer AG Hoechst (TBP, DBBP, TOPO) (Cyanex 921. 923, 471X) (Hoechst PX324, 320) BaySolvex VP-AC 4046 (DBBP), VP-AC 4014 (DPPP), VP-Al 4059, (DEDP)	U_3O_8 processing Iron extraction Zirconium/hafnium separation Niobium/tantalum separation Rare earth separation Gold extraction
Sulfur–oxygen donors	Sulfoxides sulfides	Daihachi Chem Ind Co Ltd (SFI-6) Hoechst (Hoe F 3440) Others	Palladium extraction in PGM refining
Nitrogen donors	Bi-imidazoles and bi-benzimidazoles Pyridine dicarboxylic ester	Avecia formerly Zeneca Specialties (ZNX 50) Avecia formerly Zeneca Specialties (CLX 50)	Zinc extraction and separation form iron in chloride media Copper extraction form chloride media

Note: This is a historic survey and it is likely that not all of the reagents are currently available or have been superceded and companies may not still be operating in this field.

Another important application of liquid–liquid extraction for the recovery of copper occurs in printed circuit board manufacture. Here ammoniacal etch solutions are contacted with the Henkel (Cognis, Inc.) β-diketone reagent LIX54 to maintain the etchant levels at their optimum concentration and to recover the excess copper by stripping the loaded organic phase and electrowinning. This process is not new but is now very well established as an integral part of etchant line design with more than 100 plants in operation worldwide.

1.1.3.1.2 Cobalt and Nickel

These metals are considered together because the major industrial requirement is the ability to separate these metals efficiently. Once again, one type of reagent dominates the field, and this time it is the Cytec organophosphinic acid reagent Cyanex 272, with about 12 plants currently operating across the world with this reagent separating cobalt from nickel in sulfate solution. A great amount of papers have been published on this system, which is now very well understood.

More recently an interesting reagent, a bis-dithiophosphoramide, DS6001 (Zeneca Specialties), was developed; it was claimed to separate cobalt from nickel and also from manganese, chromium, magnesium, and calcium. This reagent would therefore have an important role in the treatment of leach liquors from nickel laterites, an increasingly important field. However, in comparative test work at the University of Hertfordshire (United Kingdom) and also at SGS Lakefield Research, Ltd. (Ontario, Canada) it was shown that when operating with actual leach liquors, there was a significant tail on the manganese extraction curve that crossed that of nickel. Thus, separation of nickel from manganese was not feasible. The reason for the anomaly between the results from Zeneca and the other works is not known; further development of the reagent has now ceased, and it is has been withdrawn from supply.

As noted previously, a number of projects are currently being developed to treat nickel laterites. Most of these feature acid pressure leaching, which has the advantage of precipitating iron from solution. The resulting leach liquor contains cobalt, nickel, magnesium, and calcium as major components. Separation of these metals can be carried out in a number of ways. Thus, the use of Cyanex 301 (Cytec, Inc.) to extract cobalt and nickel from magnesium and calcium has been chosen by INCO for the New Caledonia Goro project. On the other hand, a very old reagent, Versatic acid, has been chosen for the Bulong project in Western Australia. The University of Hertfordshire also developed a flow sheet using Versatic acid to extract cobalt and nickel from an acidic heap leach liquor arising from Greek laterite ores. Here the main difficulty in the flow sheet was the precipitation of iron, aluminum, and chromium prior to solvent extraction. However, once these elements had been removed the solvent extraction flow sheet worked very well using magnesium Versatate to recover both cobalt and nickel. Any excess magnesium was scrubbed with a nickel solution, and, following acid stripping, the cobalt was extracted with Cyanex 272. There is another slight problem with this circuit in that manganese is also coextracted with the cobalt and nickel. This is still being studied, but it is likely that the manganese will be precipitated as the dioxide.

Although most acidic extractants other than the phosphorus acids extract nickel before cobalt the separation factors are not sufficiently high so it is difficult to carry out an effective separation. Therefore, several synergistic reagent mixtures have been developed at MINTEK to separate nickel from cobalt. These mixtures include monoxime/carboxylic acid, pyridine carboxylate ester/carboxylic acid, and pyridine carboxylate ester/phosphorous acids. However, the industry seems reluctant to use synergistic mixtures, probably because of problems associated with precise control of the ratio of extractants in the organic phase to maintain synergism.

1.1.3.1.3 Zinc

Over the past decade Zeneca (Avecia) has produced two reagents designed to recover zinc selectively from iron. The first of these, ZNX-50, was based on an imidazole structure and was proposed for the extraction and separation of zinc from iron in chloride media. The second, DS5869, is a dithiodialkylphosphoramide, and although it will separate zinc from iron in sulfate liquor it will also strongly extract cadmium, mercury, bismuth, and other soft metals and also irreversibly loads copper. These are obvious disadvantages for commercial application.

1.1.3.2 Precious Metals

The separation and recovery of precious metals continues to create interest. Henkel (Cognis, Inc.) has developed a new reagent, LIX 79, based on a trialkylguanidine for the recovery of gold from cyanide solutions. The reagent behaves as a strong base, and stripping is achieved with alkaline solutions above pH 13.5.

Amides have been of interest for some time in the development of nuclear waste reprocessing because of their ready complete decomposition. However, recently a mono-alkylamide has been proposed for iridium extraction from rhodium and base metals.

A very interesting development that may set the pattern for extractant development in the next century is a paper from Cognis at ISEC'99. In their process a mixture of a quaternary ammonium compound and nonyl phenol, designated LIX 7820, is used. This as expected produces a salt at neutral pH, but on contact with an alkaline solution the phenol is protonated, releasing the quaternary ammonium cation. This will now be free to form an ion pair with any other anion in solution. This extractant has been used to extract complex metal cyanides that can be stripped by lowering the pH below 12.

1.1.3.3 Nuclear Reprocessing

Here the search continues for reagents that can easily be completely decomposed to gaseous products, thus minimizing any residues for ultimate containment. For several years papers have been published on the use of amides, and here several papers feature malondiamides. These also have an advantage of allowing the separation of lanthanides and actinides.

1.1.4 Operating Problems

1.1.4.1 Extractant Losses

It is inevitable that some losses of extractant will occur from evaporation, aqueous solubility, and entrainment of the organic phase after mixing in the aqueous raffinate. However, because of problems caused by the release of such compounds into the environment, it is necessary to minimize these as much as possible. This can be achieved by a combination of chemistry and good equipment design. Process chemistry will generally determine the aqueous solubility, and operating with acidic extractants at high pH values will inevitably increase solubility by the formation of, for example, sodium. This can be controlled in some situations by using other alkali metals, such as magnesium, where the aqueous solubility of the salt will be lower. Calcium is not recommended with sulfate systems because of the precipitation of gypsum. The size and structure of the alkyl chains attached to the reagent also determine solubility losses, but here the size must not be too large; otherwise the molar ratio of reagent to metal becomes unfavorable. Aqueous solubility of the extractant is also reduced by high ionic strength of the aqueous phase. This is generally the situation in hydrometallurgical operations but may not be so common when aqueous effluents are being treated.

Entrainment is determined by the operating conditions of the contactor and in particular the mixing process. Too intense a stirring regime will promote very small droplets that will only settle with difficulty under gravity. Conditions within the settler can also promote a secondary haze that again can cause problems. Mixer–settler design has addressed these problems with a number of ingenious solutions. However, only very recently has a radical redesign of the mixer been published. This is a slow-flow mixer developed by Outokumpu Technology (now Outotec, Finland) that is reported as giving reduced entrainment losses.

1.1.4.2 Solubility Problems—Third-Phase Formation

On occasion, as the organic phase becomes loaded with the metal-extractant complex, the solubility of the latter in the diluent is exceeded and the metal complex comes out of the solution as a third phase. This is obviously a severe disadvantage in processing, so other compounds are added to the organic phase to remove any such tendency for third-phase formation. These compounds are called modifiers and usually are compounds such as nonyl phenol, decanol, or similar compounds capable of forming hydrogen bonds with the metal complex. Again, there may be a problem of differential solubility of the modifier and extractant.

1.1.4.3 Crud

This term is given to the interfacial deposits that occur during the extraction process. The deposits can be caused in a number of ways but often result from the varying solubility of inorganic species such as silica in the aqueous phase. They can cause a lot of problems by upsetting the flow characteristics of the contactor and by causing slow coalescence. Prevention is difficult so the treatment generally consists of periodically removing the interfacial deposit by suction and passing it through a centrifuge to separate the organic phase for recycle.

1.1.5 Solvent Extraction for Environmental Applications

As pressure is maintained in the industry to reduce the toxic metal content of effluents, it will be necessary for them to use selective separation processes. Although solvent extraction has been a very effective unit operation in hydrometallurgy, it has proven to be less attractive in effluent and waste reprocessing. One of the reasons for this is that the demands on the effluent and waste treatment processes are different from primary hydrometallurgy. Thus, the combinations of metals in wastes may be quite different from those arising from ore leach liquors, requiring reagents with different selectivities. Also, demands on the raffinate are much more severe when this has to meet environmental discharge limits rather than recycling within the process. Finally, it can be just as damaging on the environment to release organic pollutants in the form of the extractant as the metals that have been removed. These organic pollutants can be removed with active carbon, but this requires additional plants and costs. It should be noted that the cost of process water is not insignificant so that recycling water within a plant can be economically important, and in this case higher concentrations of impurities may be permissible.

At a previous International Solvent Extraction Conference (ISEC), Gordon Ritcey reviewed the role of solvent extraction in this area. He concluded that if the technology is to play a significant role in effluent treatment, then a number of improvements have to be made that reflect the different demands noted already. These included the following:

- Easily biodegradable reagents and diluents.
- Equipment designed to reduce solvent losses.
- Better solvent removal and recovery systems.
- Improved liquid membrane processes to reduce leakage.

A search of the literature will produce a significant number of processes that have been tested to pilot scale but then did not achieve commercialization. Only about a dozen commercial processes are known to exist to treat metalliferous wastes. Most of these are single operations, and only the recovery of copper from printed circuit board etchants boasts worldwide operation.

In spite of the change in demands on selectivity there does not seem to be a need currently to produce new solvent extraction reagents, with the existing range producing sufficient versatility to remove the metals singly or in combination.

1.1.6 Trends and Future in Solvent Extraction

What trends can we notice in reagent development over the past few years? Most importantly, over the last decade there has been a reduction in the number of new reagents entering the commercial market, and a number of development reagents have been withdrawn. It seems as though there is now a sufficient range of available extractants to carry out most of the required extractions and separations. Reagent manufacturers have developed their products to fulfill their own particular niche to such an extent that it will be very difficult for any new reagent to break into the market at a significant level. There will always be an academic interest in carrying out clever chemistry to produce new reagents with improved performance and unusual selectivities. But clever chemistry does not often come cheap, and for bulk use in liquid–liquid extraction reagent cost will always be an important consideration. Thus, if it is possible to use a compound for extraction that already has another commercial market, this is an advantage because on a tonnage basis, except in a few special cases, solvent extraction is not a large market. It is also very important to consider the commercial use of the reagent and to ensure that the product will fit into the recognized established flow sheet of the process. It is far more difficult to persuade a customer that he should completely redesign his process to fit in with a new extractant. Remember also that solvent extraction competes with other separation processes so that even if the new extractant carries out the required extraction satisfactorily, other reasons may convince the customer to use competing technology. Thus, although the Zeneca reagent Acorga SBX 50 was shown in pilot studies to extract antimony and bismuth from copper electrolytes, in the end a client who piloted the solvent extraction study installed ion exchange. The reasons for such decisions may not be solely based on technical reasons but might include other factors such as expertise in other fields or availability of suitable plant.

So if there are not going to be many new reagents in the future to carry out the desired extractions and separation, we must make the best use of what is available and look for new ways of using the existing compounds. Thus, combination of reagents—although currently not favored by the industry—might provide some answers as shown by the extensive work by John Preston at MINTEK and others. Here maybe it is necessary to look in detail at differential extractant losses and to try to equate these or to provide simple but effective ways of monitoring on line their respective concentrations so that make-up will be easier.

Problems can also arise with the slow kinetics of extraction of some metals, in particular nickel, iron, and aluminum. Is there a way of increasing the rate of extraction while maintaining selectivity by using phase transfer catalysts to move the metal into the organic environment and eliminating the problem of the interface?

Many new types of contactors have been developed over the past 30 years, incorporating some interesting and novel concepts. However, few have really become established, maybe because the science of the process has also developed in parallel. Thus, in the early days of copper extraction with hydroxyoximes a flow sheet would consist of three extraction and two stripping stages. This meant that any device that could improve the contactor performance and reduce contactor size would be welcomed. During this period enhanced coalescence using Knitmesh packs was introduced. However, improvement to reagent design has now made these unnecessary, and the flow sheet consists of two extraction and one strip stages, which immediately gives a cost benefit. The Knitmesh packs also reduced solvent losses, but inserting a filter into the stream could cause many other problems. At any rate, why try to correct a problem if by redesigning it could be prevented—hence the development of the slow-flow mixer.

Although liquid–liquid extraction can be regarded as a mature technology, a number of aspects are still poorly understood.

1. Extractants
   - New ways with old reagents.
   - Synergistic systems, but trying to match the physical properties of the reagents.
   - Degradation of extractants is important because it affects not only the physical operation but also the chemistry. This was recognized a long time ago with the hydrolysis of TBP to give phosphoric acid esters. However, it seems to have been largely ignored recently. Note that degradation will probably be system specific so it needs to be studied for every system.
   - Chemical regeneration of extractants to offset degradation losses during operation.

2. Diluents
   - Role poorly understood and choice is often a matter of what is the least expensive available.
   - Basic work is required to study interactions among extractants, diluents, and modifiers.
   - These reagents can also degrade, and often this information is readily available in the organic chemical literature. For example, the oxidation

of hydrocarbons by cobalt in the presence of oxygen has been known for many years but still is not fully acknowledged by all.

3. Systems
   - For real processes stripping the loaded organic phase is just as important as extraction.
   - Most laboratory studies are carried out with dilute solutions, which do not equate to real life in which ionic strength of the leach liquors can be quite high. This can affect a number of processing parameters.

4. Environment
   - Potential for release of organic compounds into the environment creates a bad image for liquid–liquid extraction.
   - Removal of trace organic compounds from aqueous raffinates.
   - Assess the environmental impact of liquid–liquid extraction on a life-cycle analysis basis compared with competing technologies.
   - Alternative systems, supercritical fluids, membranes, and so forth where environmental impact could be less.

5. Engineering
   - Careful study of rates of mass transfer would allow better design of equipment. This would include not only chemical but also diffusion rates in real systems.
   - Possibility of extraction with large phase ratios.
   - Real solutions are not clean so what affect do the impurities have on the interface?
   - Degradation products affect operating parameters and hence work on systems over a time period long enough to show these problems.
   - Better control of mixing to generate optimum shear.
   - Mixing time based on kinetics to minimize disengagement problems, crud formation, and so forth.
   - Work on real solutions with equipment made of the same material as the eventual plant.

1.2 LIQUID MEMBRANES

1.2.1 Introduction

There are several types of membrane processes used in wastewater treatment; in the majority of these the polymeric membrane operates as a filter retaining various species on the feed side while allowing other smaller species to cross. The size of the membrane pores determines the size of species that may be retained. Thus, the membranes with the smallest size pores, called nanofiltration membranes, can retain large ions such as chromate but not the small monovalent species. Ultrafiltration membranes can retain metal ions when combined with large complexing agents.

However, as these do not involve liquid–liquid extraction, neither needs to be considered here.

The term *liquid membrane* is given to a system in which the membrane that divides the aqueous feed and product phases consists of a thin film of organic reagent. Two types of liquid membrane can be prepared. The first—a surfactant or emulsion liquid membrane, as its name implies—consists of a water-in-oil emulsion formed from droplets of the aqueous strip solution contained in an organic phase. The latter consists of a diluent, an organic extractant to combine with the metal ions, and a surfactant to stabilize the emulsion. This emulsion is then suspended in the aqueous feed solution in a suitable contactor. The metal ions of interest in the feed react with the organic extractant at the aqueous–organic interface and migrate across the organic membrane to the inner aqueous strip interface where they are stripped. The regenerated extractant then migrates back to the feed interface, and the extraction process continues. When the extraction is complete, the loaded emulsion phase is removed from the aqueous raffinate, and the emulsion is broken to separate the aqueous strip and the organic phase. The recovered organic phase is then contacted with fresh strip solution in an emulsifier, and the process repeated.

In supported liquid membranes the organic phase is immobilized in the pores of a porous polymer. The polymeric support usually consists of ultra- or microporous membranes in the form of a sheet or hollow fiber or tube. The polymer is immersed in the organic extractant or diluent that then migrates into the pores to produce the liquid membrane. This then forms a physical barrier between the aqueous feed and strip solutions, with, for example, the strip solution flowing down the lumen and the feed around the outside of a hollow fiber. The mechanism of extraction is the same as in the surfactant liquid membrane except that for the same system the flux is generally slower as the metal complex has to migrate over a longer distance. A modification of the supported liquid membrane exists where the organic phase flows through the lumen of the hollow fiber that is surrounded by the feed solution as usual. The loaded organic phase then is carried to another module where the fiber is bathed in the strip solution or the organic phase can flow from the fiber to be treated in a conventional solvent extraction process.

Emulsion pertraction is a combination of the two types of liquid membrane process where an unstabilized water-in-oil emulsion is fed down the lumen of a hollow fiber that is surrounded by the aqueous feed.

1.2.2 Surfactant Liquid Membranes

These have the advantage of high membrane flux, which results from the very small thickness of the organic membrane. However, there are a number of operational difficulties. The first of these concerns the osmotic transport of water across the membrane as a result of different ionic concentrations in the two aqueous phases. This causes the membrane drops to swell and ultimately to break down, mixing the strip and feed solutions. Another difficulty arises with the ultimate breaking of the emulsion and separation of the two phases that can give rise to entrainment problems. In addition, the overall process is much more complex than that of the supported liquid membrane.

In spite of these disadvantages emulsion liquid membranes have been widely studied in terms of combinations of extractant and emulsifier. The process has also been piloted for the recovery of zinc from rayon spinning waste liquor.

1.2.3 Supported Liquid Membranes

These have the advantage of a very simple flow sheet requiring only circulation of the two aqueous feeds through, for example, a shell and tube hollow fiber module, which are already available for membrane filtration. Concerns have been expressed over the stability of the membrane and the possible short-circuiting of the feed and strip solutions following loss of organic phase from the pores and the overall membrane lifetime. Several modifications have been made to reduce the loss of organic phase, including the construction of a thin polymer over the membrane pores. This considerably reduces extractant losses overall without a significant loss of performance.

1.2.4 Applications

The main application of this technology for metal extraction will probably be in the treatment of effluents and wastewaters as shown by the many research papers that have been published. This is particularly true of the supported liquid membrane because of the many modules required to treat significant volumes of feed solution. This then creates a large capital expenditure, and, although lifetimes of polymeric membranes have now been increased considerably, the initial outlay will probably be too great for the value of any benefits or metal recovered. However, if this process can be used for high-value products, then the expenditure can be more easily justified—hence, the potential use of such systems for recovery of pharmaceutical compounds.

1.2.4.1 Recovery of Metals

The systems that have been studied show that most, if not all, of the liquid–liquid extraction processes can operate successfully in membranes, often with better overall performance. The reason for this is that these membrane processes are dynamic and the extraction is not limited by loading of the organic phase. Also, the chemistry allows the metal to be transported uphill against the concentration gradient. Thus, it is possible to reach very, very low concentrations of metals in the raffinate (parts per billion)—well below the legal discharge limits for most metals. Currently, to reach such levels does take some time (i.e., days), but this is where future work could help.

Considering the current system, it is surprising that supported liquid membranes work so well because of the limitations from both the extractants and the membrane. Thus, neither of these has been optimized for liquid membranes. Consider the extractant—these are generally commercial compounds that have been optimized for solvent extraction equipment with its requirements for good loading and adequate kinetics, with a two- or three-minute mixing time. Reagents for liquid membranes require the reverse of these parameters of adequate loading and fast kinetics. Also, in solvent extraction to improve metal extraction the reagent concentration can be increased with little difficulty. However, in liquid membranes this is not so easy because as the reagent concentration in the diluent increases, the viscosity of the organic phase also increases, which slows down the rate of membrane diffusion. Thus, there is an optimum concentration where maximum flux occurs.

Similarly with the membrane, this has been optimized for filtration with a smooth surface on the product side and a rough surface with crevices on the feed side. This is ideal to trap particles but not to retain an organic phase, as flow of the aqueous phase on the rough side will set up turbulence, which will cause the physical removal of the organic phase.

Thus, the optimum properties for supported liquid membranes would be as follows:

1. An extractant with ideally the following properties:
   - Fast metal complex kinetics.
   - Retention of metal selectivity.
   - High intrinsic capacity for metal ions (high ratio of donor atoms).
   - Minimum (zero) aqueous solubility.
   - Low viscosity (however, as only low volumes are required reagent cost is not so important).

2. Membranes:
   - Smooth outer surfaces with uniform pore size.
   - No internal barrier to transport.

These requirements are obviously not possible, so perhaps the configuration of the flowing membrane where the organic phase flows across the feed side of the membrane is the best compromise.

BIBLIOGRAPHY

As this is a general review of the subject, detailed references are not given. However, a wide range of books and conference proceedings provide both good coverage of the theory and practice of the technology.

Books

L. Alders, *Liquid–Liquid Extraction*, Elsevier, New York (1955).

E. Hecker, *Verteilungsverfahren in Laboratorium*, Verlag Chemie GmbH, Weinheim (1955).

G. H. Morrison and H. Freiser, *Solvent Extraction in Analytical Chemistry*, John Wiley and Sons, New York (1957).

R. E. Treybal, *Liquid Extraction*, McGraw-Hill Book Co., New York (1963).

A. W. Francis, *Liquid–Liquid Equilibriums*, Wiley-Interscience, New York (1963).

J. Stary, *The Solvent Extraction of Metal Chelates*, Pergamon Press, New York (1964).

Y. Marcus and S. Kertes, *Ion Exchange and Solvent Extraction of Metal Complexes*, Wiley-Interscience, New York (1969).

Yu A. Zolotov, *Extraction of Chelate Compounds* (translated from Russian ed., 1968), Humprey, Ann Arbor, MI (1970).

A. K. De, S. M. Khopkar, and R. A. Chalmers, *Solvent Extraction of Metals*, Van Nostrand Reinhold, New York (1970).

C. Hanson (Ed.), *Recent Advances in Liquid–Liquid Extractions*, Pergamon Press, New York (1971).
V. S. Schmidt, *Amine Extraction,* Israeli Programme for Scientific Translations (1973).
T. Sekine and Y. Hasegawa, *Solvent Extraction Chemistry*, Marcel Dekker, New York (1977).
T. C. Lo, M. H. I. Baird, and C. Hanson, *Handbook of Solvent Extraction*, Wiley-Interscience, New York (1983).
G. Ritcey and A. W. Ashbrook, *Solvent Extraction: Part I and II*, Elsevier, Amsterdam (1984).
W. W. Schultz and J. D. Navratil (Eds.), *Science and Technology of Tributylphosphate,* 2 vols., CRC Press, Boca Raton, FL (1987).
S. Alegret (Ed), *Developments in Solvent Extraction,* Ellis Horwood, Chichester, U.K. (1988).
J. Rydberg, C. Musikas, and C. R. Choppin, *Principles and Practices of Solvent Extraction,* Marcel Dekker, New York (1992).
J. Thornton (Ed.), *Science and Practice of Liquid–Liquid Extraction,* 2 vols., Oxford Science Publications, Oxford, U.K. (1992).
J. Szymanowski, *Hydroxyoximes and Copper Hydrometallurgy,* CRC Press, Boca Raton, FL (1993).
J. C. Godfrey and M. J. Slater (Eds.), *Liquid–Liquid Extraction Equipment,* John Wiley, Chichester, UK (1994).
K. Schügerl, *Solvent Extraction in Biotechnology,* Springer, Berlin (1994).
R. A. Bartsch and J. Douglas Way (Eds.), *Chemical Separations with Liquid Membranes,* American Chemical Society, Washington, D.C. (1996).
R. Hatti-Kaul (Ed.), *Aqueous Two-Phase Systems: Methods and Protocols (Methods in Biotechnology Series Volume 11),* Humana Press, New Jersey (2000).
J. Rydberg, M. Cox, C. Musikas, and G. Choppin, *Solvent Extraction Principles and Practice* (2nd edition), Marcel Dekker, Inc., New York (2004).

Book Series

J. A. Marinsky and Y. Marcus (Eds.), *Ion Exchange and Solvent Extraction, A Series of Advances,* Vols. 1–13, Marcel Dekker, New York.

In addition to these citations, chapters on technology can be found in books on separation science and processes and hydrometallurgy.

Conference Proceedings

Solvent Extraction Chemistry of Metals (H. A. C. McKay, T. V. Healy, I. L. Jenkins, and A. Naylor, eds.), Macmillan, London (1965).
Solvent Extraction Chemistry (D. Dyrssen, J. O. Liljenzin, and J. Rydberg, eds.), North-Holland, Amsterdam (1967).
Solvent Extraction Research (A. S. Kertes and Y. Marcus, eds.), John Wiley and Sons, New York (1969).
Solvent Extraction, Proceedings of ISEC '71, The Hague (J. G. Gregory, B. Evans, and P. C. Weston, eds.), Society of Chemical Industry, London, 3 vols. (1971).
Proceedings of ISEC '74, Lyon (J. D. Thornton, A. Naylor, H. A. C. McKay, and G. V. Jeffries, eds.), Society of Chemical Industry, London, 3 vols. (1974).
Proceedings of ISEC '77, Toronto, CIM Special Volume 21 (G. Ritcey, B. H. Lucas, and H. W. Smith, eds.), Canadian Institute of Mining and Metallurgy, Montreal, 2 vols. (1979).
Proceedings of ISEC '80, Liege, Association d'Ingeneur de l'Université de Liège, Liège, 3 vols. (1980).

Selected Papers of ISEC '83, Denver, American Institute of Chemical Engineers Symposium Series No. 238, Vol. 80, American Institute of Chemical Engineering, New York (1984).

Preprints of ISEC '86, Munich, DECHEMA, Frankfurt-am Main, 3 vols. (1986).

Proceedings of ISEC '88, Moscow, Vernadsky Institute of Geochemistry and Analytical Chemistry of the USSR Academy of Sciences, Moscow, 4 vols. (1988).

Solvent Extraction, Proceedings ISEC '90, Kyoto (T. Sekine and S. Kusakabe, eds.), Elsevier Science, 2 vols. (1992).

Solvent Extraction in the Process Industries, Proceedings ISEC '93, York (D. H. Logsdail and M. J. Slater, eds.), Society of Chemical Industry, London, 2 vols. (1993).

Value Adding through Solvent Extraction, Proceedings ISEC '96, Melbourne, (D. C. Shallcross, R. Paimin, and L. M. Prvcic, eds.), University of Melbourne, 3 vols. (1996).

Solvent Extraction for the 21st Century, Proceedings ISEC '99 Barcelona (M. Cox, M. Hildalgo, and M. Valiente, eds.), Society of Chemical Industry, London, 2 vols. (2000).

Proceedings of ISEC2002, Cape Town (K. C. Sole, P.M. Cole, J. S. Preston, and D. J. Robinson, eds.), South African Institute of Mining and Metallurgy, Johannesburg, 2 vols. (2002).

Solvent Extraction for Sustainable Development, Proceedings of ISEC'2005, Beijing (available on CD) ISBN 7-900692-02-9 (2005).

In addition to these specialized conferences, solvent extraction papers can also be found in conferences on hydrometallurgy and specialized conferences concerned with the recovery of metals from primary and secondary sources. Extraction of organic compounds are also featured in conferences on biotechnology and pharmaceutical products.

2 Fundamentals in Solvent Extraction Processes
Thermodynamic, Kinetic, and Interfacial Aspects

Hitoshi Watarai

CONTENTS

2.1 INTRODUCTION—WHAT IS THE DRIVING FORCE OF SOLVENT EXTRACTION?

Solvent extraction is a separation method that uses the difference in solubilities of some chemical species dissolved in the two-phase system. Solubilities are governed by the solvation free energy when the species is dissolved into an organic phase or aqueous phase. Therefore, the driving force of the solvent extraction is the difference or gradient of chemical potential of a given species between an organic phase and an aqueous phase.

2.1.1 Nernst Distribution Isotherm

Chemical potentials of species A in organic and aqueous phases are represented by

$$\mu_o = \mu_o{}^0 + RT\ \ln[A]_o \quad (2.1)$$

$$\mu_o = \mu^0 + RT\ \ln[A] \quad (2.2)$$

When the distribution equilibrium is attained, $\mu_o = \mu$ is hold, and the next relation is derived as

$$[A]_o/[A] = \exp\{(\mu^0 - \mu_o{}^0)/RT\} = K_D \quad (2.3)$$

where K_D is the distribution constant or the partition coefficient. A schematic presentation of the relationship between μ_o and μ is shown in Figure 2.1.

2.1.2 Classical Extraction Mechanism of Metal Chelate

Considering only the reactions in two-bulk phases, the formation constants of metal complexes, β_n, and the distribution constants of the ligand and the complex, K_D and

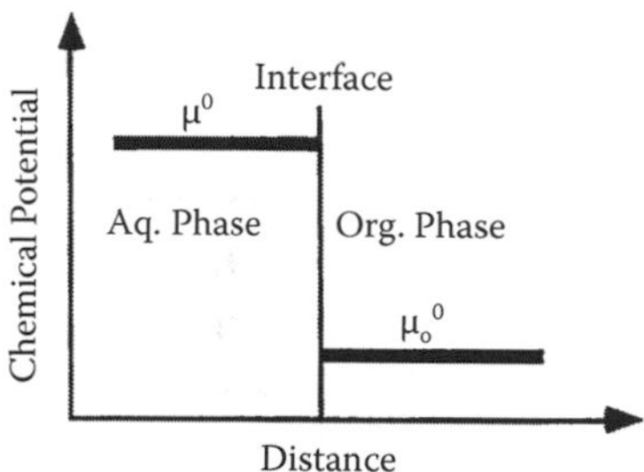

FIGURE 2.1 Conceptual drawing of the driving force of the distribution in terms of the difference of standard chemical potentials between the aqueous phase and organic phase (see Equation 2.3).

K_{DM}, are primary factors governing the extraction constant, K_{ex}, as written by the equation

$$M^{n+} + n\,HL_o \rightleftharpoons ML_{n,o} + nH^+ \tag{2.4}$$

provided that the metal ion exists as a hydrated ion in the aqueous phase.

Extraction constant is defined by

$$K_{ex} = [ML_n]_o[H^+]/[M^{n+}][HL]_o^{\,n} = K_{DM}\beta_n(K_a/K_D)^n \tag{2.5}$$

where $\beta_n = [ML_n]/[M^{n+}][L^-]^n$ and $K_{DM} = [ML_n]_o/[ML_n]$.

Distribution ratio of metal element M is represented by

$$D = \frac{[ML_n]_o}{[M]+[ML]+[ML_2]+\ldots} = \frac{K_{DM}\beta_n[L^-]^n}{1+\beta_n[L^-]+\beta_2[L^-]^2+\beta_3[L^-]^3+\ldots} \tag{2.6}$$

When $[M] >> [ML] + [ML_2] + \ldots$, Equation (2.6) is simplified as

$$D = K_{DM}\beta_n[L^-]^n = K_{ex}([HL]_o/[H^+])^n$$

or

$$\log D = \log K_{ex} + n \log [HL]_o + n\,pH \tag{2.7}$$

On the other hand, when all metal ion in the aqueous is converted to the neutral chelate of ML_n, the next relation is obtained since $[ML_n] >> [M] + [ML] + \ldots$,

$$D = K_{DM} \text{ or } \log D = \log K_{DM} \tag{2.8}$$

Thus, the logarithmic value of the distribution ratio depends on the pH of the equilibrated aqueous phase as shown in Figure 2.2.

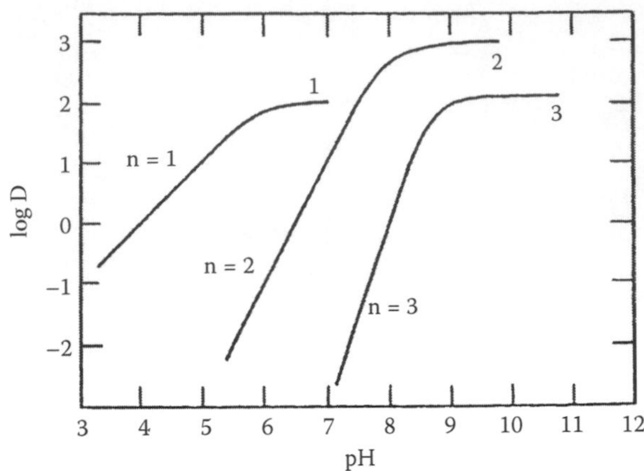

FIGURE 2.2 Typical distribution curves of ML_n for n = 1, 2, and 3.

2.1.3 Modern Extraction Mechanism of Metal Chelate

Recent research on the solvent extraction mechanism revealed that the role of liquid–liquid interface has to be taken into account for understanding the extraction mechanism, especially in the extraction kinetic mechanism. Breakthrough research has been attained with the invention of a high-speed stirring (HSS) method that made it possible to measure the interfacial concentration of extractant (Lewis base) and extraction rate of the complex. Thus, this technique could determine a rate law for the interfacial reaction for the first time [1].

Concentration at the interface is represented in the unit of mol/cm^2 and correlated to the bulk phase concentration by the Langmuir isotherm as discussed later.

2.2 THERMODYNAMICS OF SOLVENT EXTRACTION EQUILIBRIA

Gibbs free energy or chemical potential governs the whole equilibrium reaction in extraction systems as well as in individual reactions.

2.2.1 Structure-Free Energy Relationship

Distribution constants of homologous series of compounds can be reconstructed empirically by the group distribution constants. For example, the distribution constants of keto and enol forms of n-alkyl substituted β-diketones in CCl_4/H_2O were linearly increased with the carbon numbers of the n-alkyl substituent as shown in Figure 2.3. The increment of 0.64 refers to the distribution constant of $-CH_2-$ [2].

2.2.2 Theoretical Prediction of Distribution Constant

Some approaches have been reported that could predict or understand the value of distribution constant from a theoretical point of view.

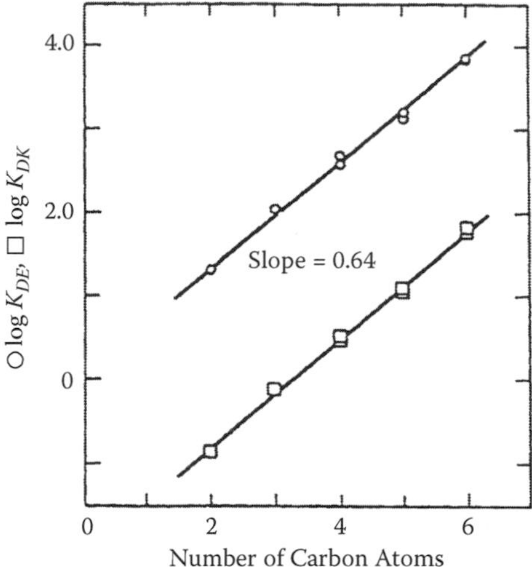

FIGURE 2.3 Linear free energy relationship between the distribution constant of enol form (K_{DE}) or keto form (K_{DK}) of n-alkyl substituted β-diketones and the carbon numbers of the alkyl group.

2.2.2.1 Application of Regular Solution Theory—Solvent Effect on the Extraction Equilibria

The activity of solute 1 in solvent 2 is represented by the next equation in the frame of regular solution theory [3].

$$\ln a_1 = \ln \varphi_1 + \varphi_2\{1 - (V_1/V_2)\} + V_1\varphi_2^2(\delta_1 - \delta_2)^2/RT \tag{2.9}$$

where φ, V, and δ refer to the volume fraction, molar volume, and solubility parameter, respectively. At the distribution equilibrium, $\ln a_{1,o} = \ln a_{1,a}$, where the subscripts o and a refer to an organic phase and an aqueous phase, respectively. Therefore, the distribution constant can be written as [4]

$$\ln K_D = \ln(\varphi_{1,o}/\varphi_{1,a}) = (V_1/RT)\{(\delta_o - \delta_a)^2 - (\delta_1 - \delta_o)^2\} + V_1\{(1/V_o) - (1/V_a)\} \tag{2.10}$$

or

$$\frac{\log K_D}{(\delta_a - \delta_o)} = \frac{V_1}{2.303RT}(\delta_a - \delta_o' - 2\delta_1) \tag{2.11}$$

where

$$\delta_o' = \delta_o + RT(V_a - V_o)/V_aV_o(\delta_a - \delta_o) \tag{2.12}$$

Figure 2.4 demonstrates the linear correlation between $\log K_D/(\delta_a - \delta_o)$ and δ_o' for the distribution of some β-diketone. Advantages of the regular solution theory are that there is no adjustable parameter in the calculation and that the effect of organic solvents is fairly explained by the solubility parameter that is experimentally determined independently from the vaporization energy and molar volume (Table 2.1).

TABLE 2.1
Solubility Parameters of Typical Solvents Used in Solvent Extraction

Number	Solvent	δ_o
2	n-hexane	7.3
8	carbontetrachloride	8.6
9	toluene	8.9
12	benzene	9.15
13	chloroform	9.3
15	chlorobenzene	9.5
16	1,1,2,2-tetrachloroethane	9.7
22	m-xylene	8.8
25	dichloromethane	9.7
26	o-dichlorobenzene	10.0
34	bromoethane	8.9
	aqueous phase	17.55

Note: δ (water) = 23.5 $(cal/cm^3)^{1/2}$.

2.2.2.2 Application of Scaled Particle Theory (SPT)—Concept of Cavity Formation Energy

The distribution equilibria of a neutral solute is governed by the difference in the solution free energies in aqueous and organic phases:

$$RT \ln K_D = \Delta G_{s,aq} - \Delta G_{s,org} \tag{2.13}$$

The Gibbs free energy of solution, ΔG_s, can be expressed by the sum of the Gibbs free energy for the cavity formation, G_c, and the interaction free energy between solute and solvent molecules, G_i:

$$\Delta G_s = G_c + G_i + RT \ln (RT/10^3) \tag{2.14}$$

where the Gibbs free energies are defined on molarity unit.

A simple but effective theory for calculating the cavity formation energy is the scaled particle theory (SPT) [6]. According to the scaled particle theory, the cavity formation energy is expressed as

$$G_c = -RT \ln (1 - y) + RT(3y/(1 - y))(\sigma_2/\sigma_1)+$$

$$RT[3y/(1 - y) + (9/2)(y/(1 - y)^2)(\sigma_2/\sigma_1)^2 + (NyP/\rho)(\sigma_2/\sigma_1)^3 \tag{2.15}$$

where y is the compactness factor, $y = \pi\rho\sigma^1{}_3/6$, and ρ is the number density of the solvent, σ, is the molecular diameter and subscripts 1 and 2 refer to solvent and solute, respectively.

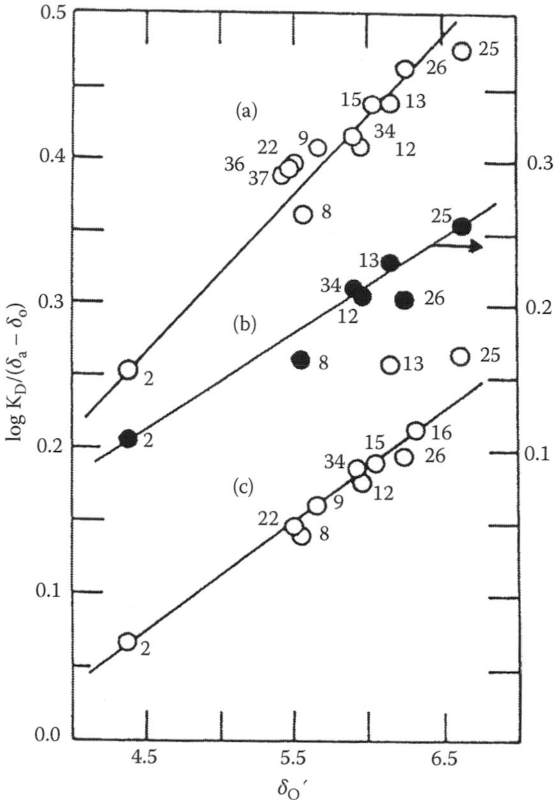

FIGURE 2.4 Linear correlations between $\log K_D/(\delta_a - \delta_o)$ and δ_o' for the distribution of 2-thenoyltrifluoroacetone, V = 155 cm^3/mol, δ = 10.1; (b) trifluoroacetylacetone, V = 134 cm^3/mol, δ = 9.6; (c) acetylacetone, V = 102 cm^3/mol, δ = 10.6. The numbers correspond to the solvents listed in Table 2.1.

The interaction energy, provided that there are no strong interactions such as hydrogen bonding and charge-transfer interaction, is expressed as the sum of the dispersion, Gdis, the inductive, Gind, and dipole-dipole, G_{dip}, energy terms:

$$G_i = G_{dis} + G_{ind} + G_{dip}$$

$$= -N(16/3)(4\varepsilon_{12}\sigma_{12}^6)(\pi\rho/6\sigma_{12}^3)$$

$$-8N(\mu_1^2\alpha_2 + \mu_2^2\alpha_1 + 2\mu_1^2\mu_2^2/3kT)\pi\rho/6\sigma_{12}^3 \quad (2.16)$$

where

$$\sigma_{12} = (\sigma_1 + \sigma_2)/2 \quad (2.17)$$

$$\varepsilon_{12} = (\varepsilon_1\varepsilon_2)^{1/2}$$

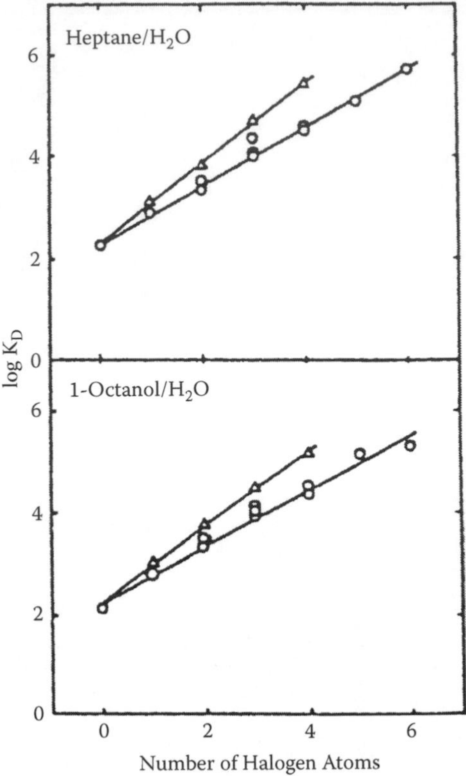

FIGURE 2.5 Linear correlations between the logarithmic distribution constants of chlorobenzenes and bromobenzenes and the number of halogen atoms in heptane/water and 1-octanol/water systems.

and ε, μ, and α stand for the Lennard-Jones parameter, the dipole moment, and the molecular polarizability.

To demonstrate the utility of this approach, we will deal with the distribution of simple molecules. Figure 2.5 shows the distribution constants of chlorobenzenes and bromobenzenes in heptane/water and 1-octanol/water, in which the logarithmic values of K_D are proportional to the number of halogen atoms in the molecule. The theoretical calculation of RT ln K_D was only made for the distribution of benzene, since all of the parameters were required for the calculation. The results are shown in Table 2.2, in which the agreement of the observed values and the calculated values are fairly good. We can find that the primary factor governing the magnitude of ΔGs is the difference in the cavity formation energies, G_c, between aqueous phase and the organic phase, since the difference in the interaction energies, G_i, is not significant. More precisely, the larger positive value of the cavity formation energy in aqueous phase than that in organic phase is predominant factor. Therefore, only the cavity formation energy was compared with the observed RT ln K_D in Figure 2.6 and Figure 2.7. It is well represented that the distribution is governed by the cavity formation

TABLE 2.2
Distribution Constants of Benzene in Heptane/Water and 1-Octanol/Water

Solvent	G_c/kcal mol^{-1}	G_{dis}/kcal mol^{-1}	G_{ind}/kcal mol^{-1}	($RT \ln K_D$)/kcal mol^{-1} calcd	obsd
Water	9.05	−9.42	−1.21		
Heptane	4.36[a], 5.80[b]	−9.22	0	3.28[a], 1.84[b]	3.12
1-octanol	6.51	−10.14	−0.28	2.33	2.90

[a] σ (Heptane) = 5.996 × 10^{-8} cm.
[b] σ (Heptane) = 6.268 × 10^{-8} cm.

energy. The situation is same for the distribution of metal complexes. In Figure 2.8 and Figure 2.9, log K_D values of bis, tris, and tetrakis acetylacetonato-metal complexes are plotted against the molecular diameter of the solute complex. In both systems of dodecane/water and benzene/water, a significant contribution of cavity formation energy is recognized, in which the discrepancy of log K_D from the predicted line from the SPT indicates the stronger interaction between water and the complex. We can note specific hydration effects in the distribution of $Cr(acac)_3$ in dodecane/water and $Cu(acac)_2$ and $VO(acac)_2$ in benzene/water as understandable from their structure [7].

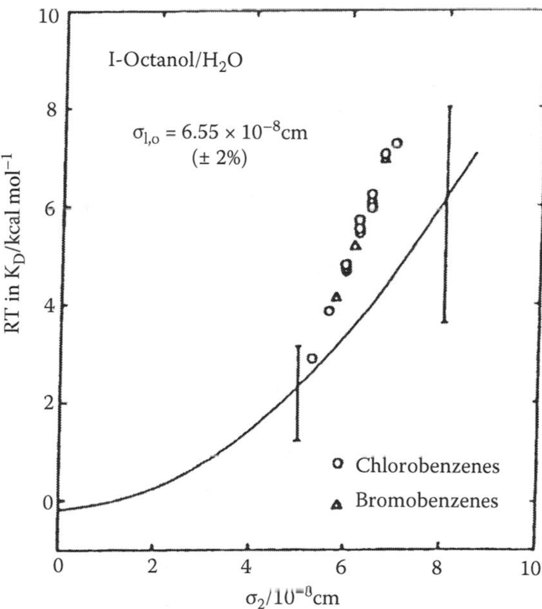

FIGURE 2.6 Comparison between the observed distribution constants of halobenzenes and those predicted from the scaled particle theory (solid line) in 1-octanol/water system.

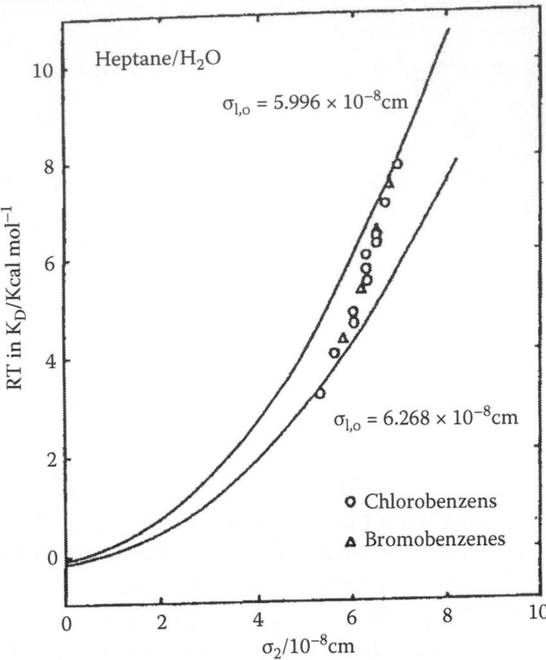

FIGURE 2.7 Comparison between the observed distribution constants of halobenzenes and those predicted from the scaled particle theory (solid line) in a heptane/water system.

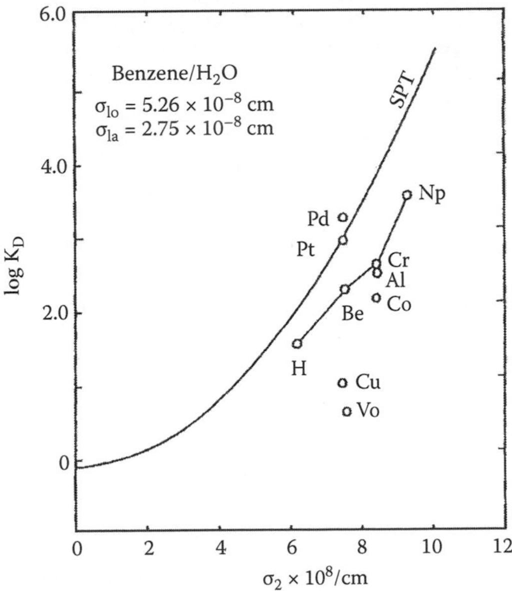

FIGURE 2.8 Distribution constants of metal (II, III, or IV)-acetylacetone complexes in a benzene/water system. The solid line denoted SPT is the predicted K_D from the scaled particle theory.

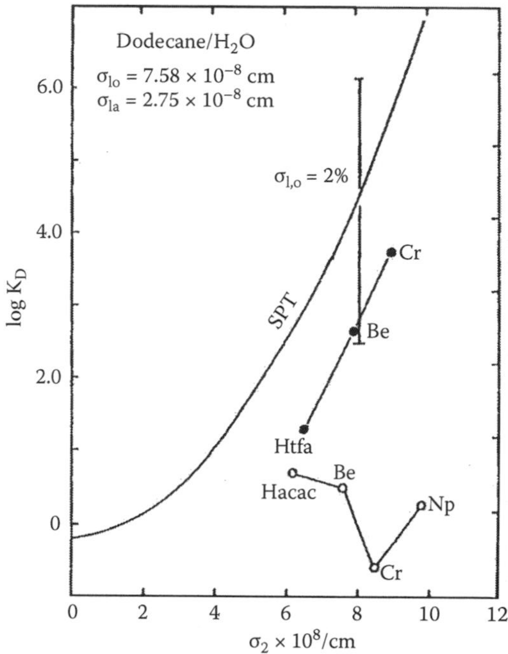

FIGURE 2.9 Distribution constants of metal(II, III or IV)-acetylacetone complexes in dodecane/water system. The solid line denoted SPT is the predicted KD from the scaled particle theory. Closed circles are data of trifluoroacetylacetone complexes.

2.3 ADSORPTION AT LIQUID–LIQUID INTERFACE

2.3.1 Structure of Liquid–Liquid Interface

The liquid–liquid interfaces are generated whenever two mutually immiscible solvents are put together in a vessel. There have been many reports on the interfacial free energy in terms of interfacial tension; on the other hand, there is little information concerning the structure because of the difficulty of the measurement of the interface. Therefore, the molecular simulation technique is a very helpful approach to understanding the thickness of the interfacial region.

The liquid–liquid interfaces of heptane/water and toluene/water were simulated by the use of the DREIDING force field [8] on the software of Cerius2 Dynamics and Minimizer modules (MSI, San Diego, CA).

In the frame of the molecular mechanics and molecular dynamics according to the DREIDING force field, the internal energy, E, of the system is calculated by the combination of E(valence interactions) and E(nonbonded interaction).

Each of them are represented by

$$E(\text{valence interactions}) = 1/2k_e(R - R_e)^2 \quad \text{bond stretch}$$
$$+ 1/2C_{IJK}(\cos\theta_{IJK} - \cos\theta_J) \quad \text{bond-angle bend}$$
$$+ 1/2V_{JK}\{1 - \cos[n_{JK}(\varphi - \varphi_{JKo})\}] \quad \text{dihedral angle torsion}$$
$$+ 1/2K_{inv}(\psi - \psi_0)^2 \quad \text{inversion}$$

$$E(\text{nonbonded interaction}) = AR^{-12} - BR^{-6} \quad \text{van der Waals (dispersion)}$$
$$+ (322.0637)q_iq_j/\varepsilon R_{ij} \quad \text{electrostatic}$$
$$D_{hb}\,[5(R_{hb}/R_{DA})^{12} - 6(R_{hb}/R_{DA})^{10}]\cos4(\theta_{DHA}) \quad \text{hydrogen bonds}$$

The most dominant interaction is the electrostatic interaction between partial charges of neighboring atoms.

The two-phase systems were constructed by 62 heptane molecules and 500 water molecules or 100 toluene molecules and 500 water molecules in a quadratic prism cell. Each bulk phase was optimized for 500 ps at 300 K under NPT ensemble in advance. The periodic boundary conditions were applied along all three directions. The calculations of the two-phase system were run under NVT ensemble. The dimensions of the cells in the final calculations were 23.5Å × 22.6Å × 52.4Å for heptane/water system and 24.5Å × 24.3Å × 55.2Å for toluene/water system. The time-step was 1 fs in all cases, and the simulation almost reached equilibrium after 50 ps. The density versus distance profile showed a clear interface with a thickness of about 10Å in both systems as shown in Figure 2.10. Interfacial adsorption of an extractant can be simulated by the similar procedure after the introduction of the extractant molecule at the position from where the dynamics will be started [9].

2.3.2 How to Measure the Interfacial Adsorption

2.3.2.1 Interfacial Tension Measurements

Interfacial tension reflects the interfacial free energy of the interface. One of the simplest methods to measure the interfacial tension is the drop volume method, which just measures the volume of the detaching aqueous drop from the tip of a capillary immersed in an organic solution. A reversed situation is also constructed, when the density of the organic solution is higher than the aqueous solution as shown in Figure 2.11. Interfacial tension γ is analyzed by the Gibbs isotherm

$$\Gamma = -(1/RT)(\partial\gamma/\partial\ln C) \tag{2.18}$$

where Γ is the interfacial excess and C the bulk concentration at the equilibrium. Furthermore, by the combination with the Langmuir isotherm

$$C_i = aK'C/(a + K'C) \tag{2.19}$$

one can derive

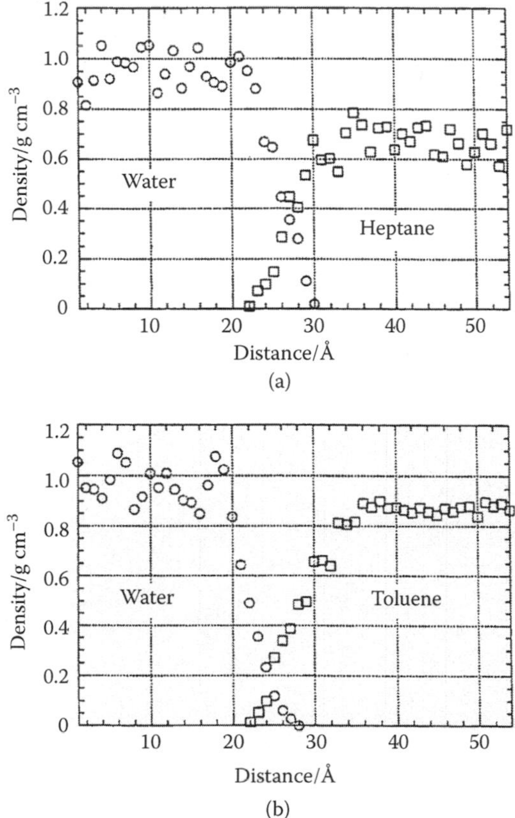

FIGURE 2.10 MD simulation of 2-hydroxy-5-nonylbenzophenone oxime (LIX65N) at heptane/water interface after 120 ps. Heptane molecules are not shown.

$$\gamma = \gamma_0 - aRTln\ (K'/a) - aRT\ ln\ C \quad (2.20)$$

where C_i refers to the interfacial concentration, a the saturated interfacial concentration, and K' the adsorption constant defined at the infinite dilution. Equation (2.20) has the same meaning as the Szyskowski equation.

As an example, the interfacial adsorption of 1,10-phenanthroline(phen) complexes in carbontetrachloride/water and chloroform/water systems are shown in Figure 2.12 [10]. The cationic ion of $Fe(phen)_3^{2+}$ is adsorbed at the interface from the aqueous phase side. Analysis of the results according to Equation (2.20) afforded the parameters of C*, which is the bulk concentration at the interface began to be saturated, as well as a and K'. Stronger adsorptivity in $CHCl_3/H_2O$ system is attributable to the higher K_D value in the system, as shown in Table 2.3.

2.3.2.2 High-Speed Stirring Spectrometry

A simple but effective method to measure the interfacial amount in highly agitated systems is the high-speed stirring method. The principle of this method is that when

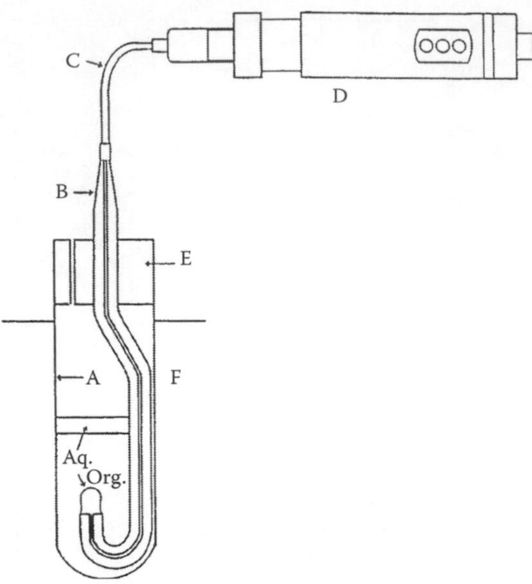

FIGURE 2.11 A simple device for the drop volume method to measure interfacial tension between water and another phase, which is heavier than water.

a solute in the organic phase is interfacially adsorbable, the bulk concentration in the organic phase has to be decreased by the interfacial adsorption. A Teflon phase separator can separate only the organic phase from the agitated system, and the concentration under the stirring can be determined photometrically. Figure 2.13 shows the schematic diagram of the apparatus. Usually, 50 ml of aqueous phase and the same volume of organic phase were employed for the measurement. Table 2.4 shows the result of the interfacial adsorption of β-diketones in vigorously stirred heptane/aqueous phase systems where Ai is the total interfacial area under the stirring and K_L' is the adsorption constant of the dissociation form from the aqueous phase to the

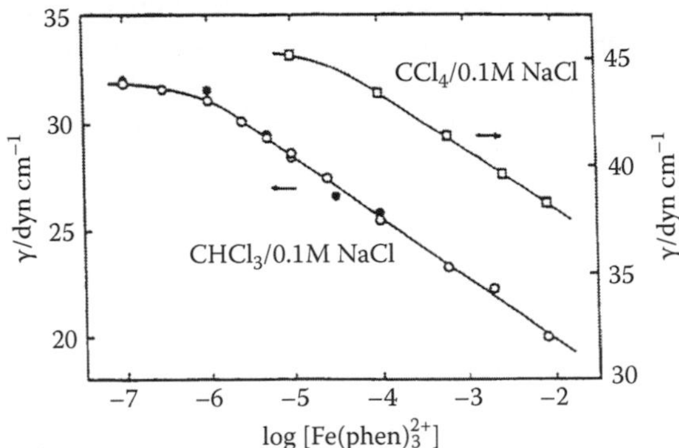

FIGURE 2.12 The interfacial adsorption of 1,10-phenanthroline(phen) complexes in carbon tetrachloride/water and chloroform/water systems.

TABLE 2.3
The Parameters for the Interfacial Adsorption of $Fe(phen)_3^{2+}$

Parameter	Organic Solvent	
	CCl_4	$CHCl_3$
a/mol cm^{-2}	4.80×10^{-11}	4.98×10^{-11}
	(346)[a]	(333)
C^*/mol l^{-1}	4.50×10^{-5}	5.40×10^{-7}
K'/l cm^{-2}	1.07×10^{-6}	9.23×10^{-5}
K_D (phen)	1.14 ± 0.07	723 ± 18

[a] Values in parentheses refer to an area (Å2) per molecule.

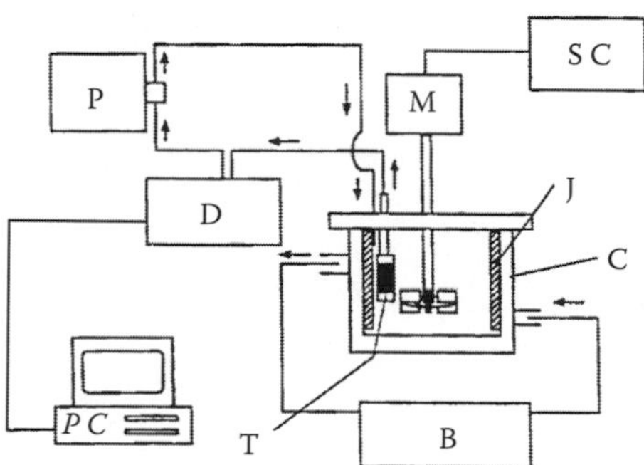

FIGURE 2.13 Schematic drawing of the high-speed stirring (HSS) apparatus. An organic phase and an aqueous phase were stirred in a vessel C with baffles (J) by a motor (M) with a controller (SC), thermostated by a circulating water bath (B). Through the Teflon phase separator (T), the organic phase is continuously circulated through the photodiode array detector (D) by a pump (P).

interface [11]. Percent adsorptivity for some β-diketones under the given concentrations are shown in Figure 2.14 as a function of pH. Significant adsorptivity is found in highly hydrophobic extractants.

2.3.2.3 Reflection Spectrometry

As for the measurement of the interfacial concentration directly, the total internal reflection fluorescence (TIRF) measurements is an effective method. Total internal reflection of condition is attained by adjusting the incident angle of the excitation beam so as to exceed the critical angle determined from the refractive indices of both phases. The penetration depth, d, is determined by

TABLE 2.4
The Interfacial Adsorption of β-Diketones in Hepaten/Water Systems

β-Diketone	pK_a	$\log K_D$	$\log aA_1$	$\log K_L'A_1$
TTA	6.17	0.57	−5.98	−2.74
BFA	6.31	1.70	−6.05	−1.78
PTA	6.99	1.85	−5.69	−2.72
NFA	6.27	3.10	−6.02	−0.58
LIX51	6.18	8.9	−5.87	4.39
BA	8.70	1.99	—	—
LIX54	9.64	4.73	−6.14	0.33

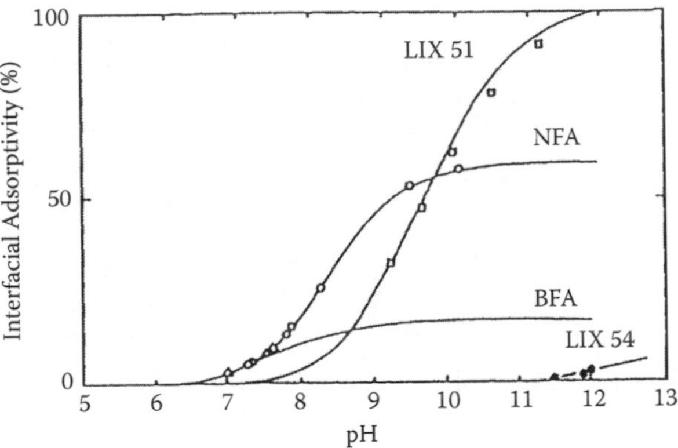

FIGURE 2.14 Interfacial adsorptivity for some β-diketones as a function of pH.

$$d = \frac{\lambda}{2\pi(n_1^2 \sin 2\theta - n_2^2)^{1/2}} \tag{2.21}$$

When the incident angle is 72.0° of the 420 nm beam, d is 136 nm. Simple optical modification of a fluorescence cell as shown in Figure 2.15 is enough for the measurement on a conventional fluorospectrometer. Interfacial ion-association adsorption of water-soluble porphyrins at toluene/water interface was measured by this method as shown in Figure 2.16 [12]. Cationic(TMPyP) or anionic(TPPS) porphyrin can be adsorbed with the surface active counter ions. The maxima in Figure 2.16 correspond to the interfacial saturation.

2.4 FACTORS DETERMINING THE INTERFACIAL ADSORPTION

2.4.1 Capacity of Liquid–Liquid Interface

The capacity of the liquid–liquid interface to accommodate the adsorption of a solute depends on the specific interfacial area defined by the ratio of the total interfacial

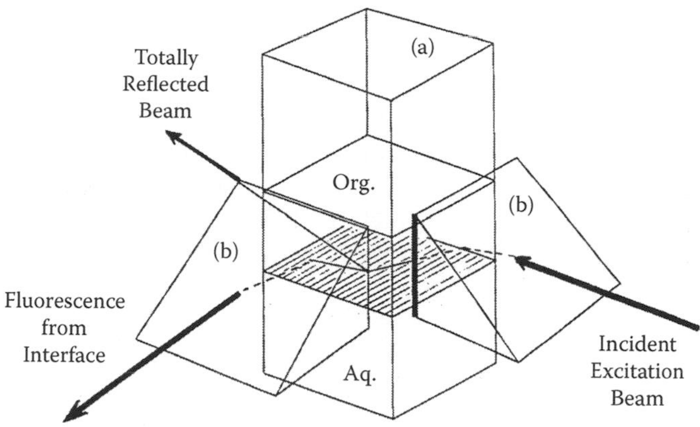

FIGURE 2.15 Schematic drawing of the simple optical modification of a fluorescence cell for the measurement of the interfacial concentration using reflection spectrometry.

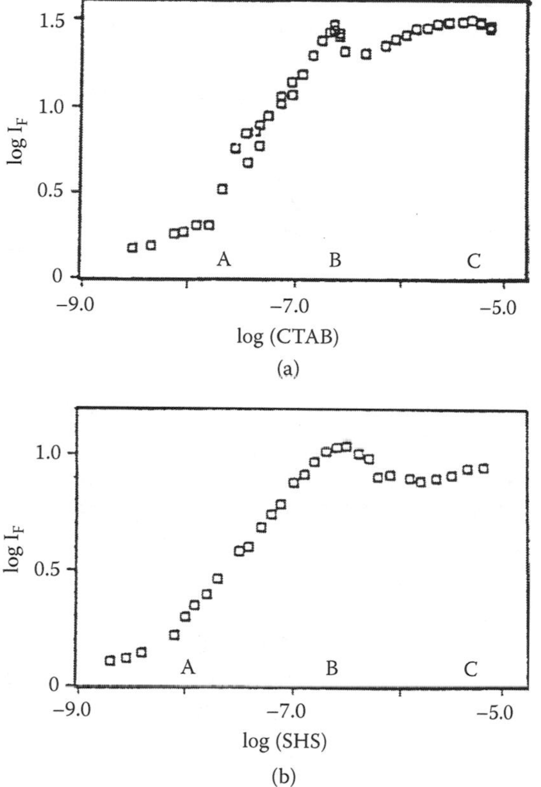

FIGURE 2.16 Interfacial fluorescence intensity of water soluble anionic [TPPS, (a)] and cationic [TMPyP, (b)] porphyrins as a function of the concentration of the surface active counterions of cetyltrimethylammonium bromide (CTAB) and sodium hexadecylsulfate (SHS), respectively.

TABLE 2.5
Capacity of the Liquid–Liquid Interface to Accomodate the Adsorption of a Solute on the Specific Interfacial Area of the Bulk Phase

Diameter of Droplet (mm)	Interfacial Area (cm^2)	Specific Interfacial Area (cm^{-1})	Maximum Interfacial Amount (mol)	Concentration Change in 50 ml (M)
—	ca. 30	0.6	3×10^{-9}	6×10^{-8}
1.0	3×10^{3}	60	3×10^{-7}	6×10^{-6}
0.1	3×10^{4}	600	3×10^{-6}	6×10^{-5}
0.01	3×10^{5}	6000	3×10^{-5}	6×10^{-4}
0.001	3×10^{6}	60000	3×10^{-4}	6×10^{-3}

Note: Maximum interfacial concentration is assumed as 10^{-10} mol/cm^2.

area to the volume of the bulk phase. Table 2.5 lists the probable situations in which a two-phase system was changed to a dispersed system by shaking or stirring. With the decrease of the droplet diameter, the specific interfacial area increases. When the diameter is 10 μm, the capacity or the maximum amount of a strongly adsorbable solute will be 3×10^{-5} mol, provided that the maximum interfacial concentration is assumed as 10^{-10} mol/cm^2. This amount corresponds to the bulk concentration of 6×10^{-4} M. This is concentrated enough to be measurable by a standard analytical technique. Thus, we can find that the liquid–liquid interface has a rather high capacity for the adsorption.

2.4.2 Correlation between Adsorption Constant and Distribution Constant

As shown in Tables 2.3 and 2.4, a highly hydrophobic ions tend to adsorb with high adsorption constant from the aqueous phase. So, we will examine the correlation between the hydrophobicity and the adsorptivity.

The distribution or transfer process of a solute, HL, from an aqueous phase to an organic phase, $\Delta G_{tr}(HL)$, is divided into the two transfer processes: adsorption into the interface, $\Delta G_{ad}(HL)$, and dehydration from the interface, $\Delta G_{deh}(HL)$, as schematically drawn in Figure 2.17.

$$\Delta G_{ad}(HL) = \Delta G_{tr}(HL) - \Delta G_{deh}(HL) \tag{2.22}$$

The adsorption process of the neutral species is the same with the dissociated one because the adsorption is the solvation of only the hydrophobic part of the solute molecule

$$\Delta G_{ad}(HL) = \Delta G_{ad}(L^-) = - RT \ln K_L' \tag{2.23}$$

Since $\Delta G_{tr}(HL) = -RT \ln K_D$, we can derive the next equation:

$$\log K_L' = \log K_D + \Delta G_{deh}(HL)/2.303RT \tag{2.24}$$

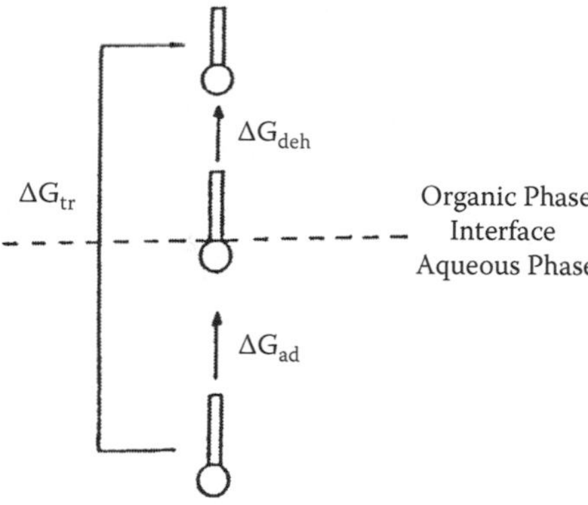

FIGURE 2.17 Schematic diagram for the distribution process of a solute, HL, from an aqueous phase to an organic phase, $\Delta G_{tr}(HL)$, is divided into the two transfer processes; adsorption into the interface, $\Delta G_{ad}(HL)$, and dehydration from the interface, $\Delta G_{deh}(HL)$.

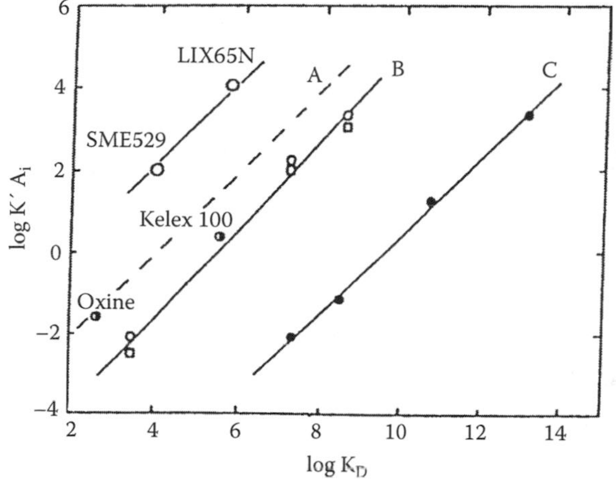

FIGURE 2.18 Linear correlation between log K_L' and log K_D for a family of extractants.

In Figure 2.18, the expected linear correlations between log K_L' and log K_D are confirmed for the adsorption of β-diketone, 1,10-phenanthroline and n-alkyl dithizone. The intercept of this figure should correspond to $\Delta G_{deh}(HL)/2.303RT$. These empirical relationships will be very helpful for the prediction of the adsorptivity of any solute with known distribution constants.

2.4.3 Prediction of Adsorptivity by Computational Simulations

The prediction of the interfacial adsorptivity can be made from the calculation of solvation energies in a bulk phase and at the interface. A modern method to calculate the solvation energy is the application of molecular mechanics (MM) and molecular dynamics (MD). In Figure 2.19, the procedure of the calculation is illustrated [13]. First, an MD simulation is carried out to obtain feasible configurations of the

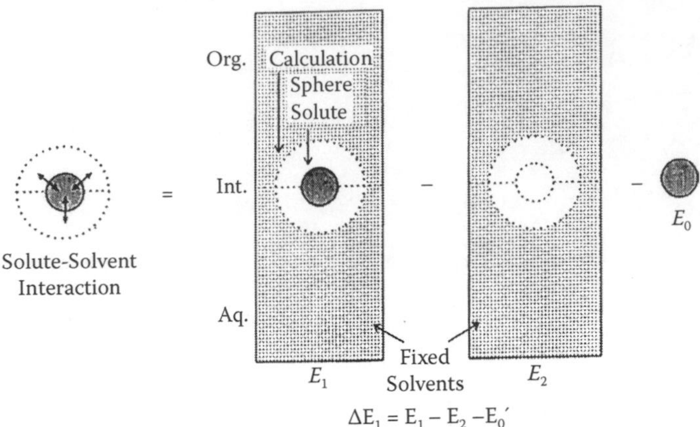

FIGURE 2.19 Procedure for the prediction of the interfacial adsorptivity and solvation energies in a bulk phase and at the interface by application of molecular mechanics(MM) and molecular dynamics(MD).

extractant and solvent molecules at the equilibrium state. Second, a spherical region with the radius of 8.5 Å around the center of mass of the extractant molecule is defined, and then the internal energy of the spherical region (E_1), which contained all bonding and nonbonding interactions of atoms, was calculated by MM calculation. The surrounding of the spherical region is fixed during the MM calculation. Third, removing the extractant molecule from the spherical region, the energy of the residual solvent molecules in the spherical region (E_2) and the internal energy of the removed extractant molecule in a vacuum (E_0') were calculated. Finally, by the relation of

$$\Delta E_i = E_1 - E_2 - E_0' \tag{2.25}$$

$$\Delta\Delta E_i = \Delta E_i\,(\text{int}) - \Delta E_i\,(\text{org}) \tag{2.26}$$

the solute–solvent interaction energy, ΔE_i, and the adsorption energy of the extractant from a bulk organic phase to the interface, $\Delta\Delta E_i$, were calculated. Figure 2.20 clearly shows that LIX65N molecule is more stable at the heptane than in bulk heptane [14]. Table 2.6 shows the calculated results of the interfacial adsorption energy for the three 2-hydroxy oximes. In all systems dealt with in the present simulation, they showed negative value and strongly suggested the preferential adsorptivity.

2.5 SOLVENT EXTRACTION KINETICS AND INTERFACIAL PHENOMENA

Solvent extraction is intrinsically dependent on the mass transfer to or across the interface and the chemical reaction at the interfacial region. The researchers of solvent extraction especially in the field of analytical chemistry and hydrometallurgy

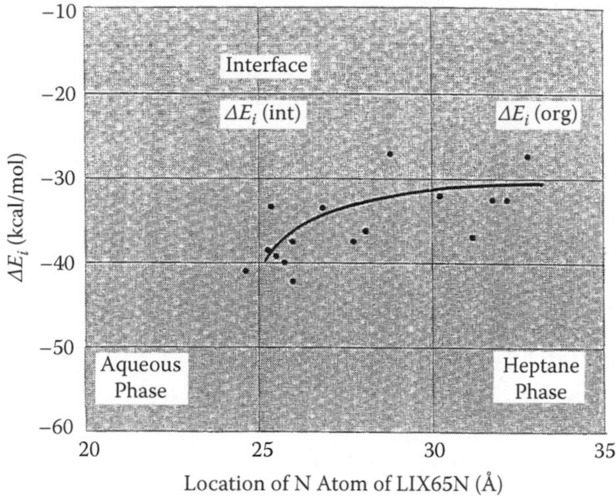

FIGURE 2.20 Solute-solvent interaction energy, ΔE_i of the extractant from a bulk organic phase to the interface as a function of the N atom location of LIX65N molecule.

TABLE 2.6
Calculated Results of the Interfacial Adsorption Energy for Three 2-Hydroxy Oximes

2-Hydroxy Oxime	pK_a	log K_D	log K'[a] (cm)	log k_i[b] (1/Ms)	$\Delta\Delta E_i$[c] (kcal/mol)
5-Nonylsalicylaldoxime (P50)[5]	9.00	3.36	−3.34	4.57	−9.3
2-Hydroxy-5-nonylaceto-phenone oxime (SME529)[6]	9.79	3.99	−3.29	5.11	−12.1
2-Hydroxy-5-nonylbenzo-phenonc oxime (LIX65N)[7]	8.70	5.69	−2.97	5.14	−6.6

[a] Adsorption constant from organic phase to interface.
[b] Rate constant for the interfacial complexation.
[c] Adsorption energy from organic phase to interface.

used to observe frequently that interfacial phenomena happened in the solvent extraction systems and had strong motivation to measure what happened at the interface.

From the beginning of 1980s, some effective experimental approaches based on new principles have been invented for the studies of interfacial reactions in solvent extraction systems, which included the HSS method [15,16], the two-phase stopped flow method [17], the capillary plate method [18], the reflection spectrometry [19], and the centrifugal liquid membrane (CLM) method [20].

2.5.1 Modern Techniques for the Measurement of Solvent Extraction Kinetics

2.5.1.1 High-Speed Stirring (HSS) Method

There are many classical methods to investigate extraction kinetics, including a Lewis cell, a single drop method, and rotating disc method [21,22]. All of these methods, however, could not measure the extraction rate and interfacial amount of extractant simultaneously. The HSS method made it possible for the first time by a relatively simple principle: When the two-phase system is highly stirred or agitated, the interfacial area is extremely enlarged. The interfacial area can easily be increased 500 times larger than that in a standing condition. The ratio of the interfacial area to the volume of the bulk phase, termed a *specific interfacial area,* is a good measure of the contribution of the interfacial phenomena in a whole system. The HSS condition can produce a specific interfacial area up to 400 cm^{-1}. In this situation, an interfacially active solute in bulk phases can adsorb at the interface. The maximum interfacial concentration of an ordinary compound is the order of 10^{-10} mol/cm^2. Therefore, the maximum amount of the solute adsorbable at the interface can be estimated as 10^{-4} M in the bulk phase concentration. This means that the bulk phase concentration of an adsorbable solute is decreased accompanied by the interfacial adsorption at the interface generated by the high-speed stirring. The HSS method succeeded to measure the concentration depression spectrophotometrically using a Teflon phase separator and photodiode array spectrometer. The extraction rate can also be measured by this method simultaneously [23]. A schematic drawing of the apparatus is shown in Figure 2.13.

2.5.1.2 Centrifugal Liquid Membrane Method

A high specific interfacial area and a direct spectroscopic observation of the interface were attained by the CLM method shown in Figure 2.21. The two-phase system of about 100 μL in each volume is introduced into a cylindrical glass cell with a diameter of 19 mm. The cell is rotated at the speed of 5,000 to 10,000 rpm. By this procedure, two-phase liquid membrane with the thickness of 50 to 100 μm is produced inside the cell wall, which attains the specific interfacial area over 100 cm^{-1}. UV/Vis spectrometry, spectrofluorometry, and other spectroscopic methods can be used for the measurement of the interfacial species and its concentration as well as those in the thin bulk phases. This method can be excellently applied for the measurement of interfacial reaction rate as fast as the order of seconds.

2.5.1.3 Two-Phase Stopped Flow Method

Stopped flow mixing of organic and aqueous phases is an excellent way to produce dispersion within a few milliseconds. The specific interfacial area of the dispersion becomes as high as 700 cm^{-1} and the interfacial reaction in the dispersed system can be measured by photodiode array spectrophotometer. A drawback of this method is the limitation of a measurable time, although it depends on the viscosity. After approximately 200 ms, the dispersion system starts to separate, even in a rather

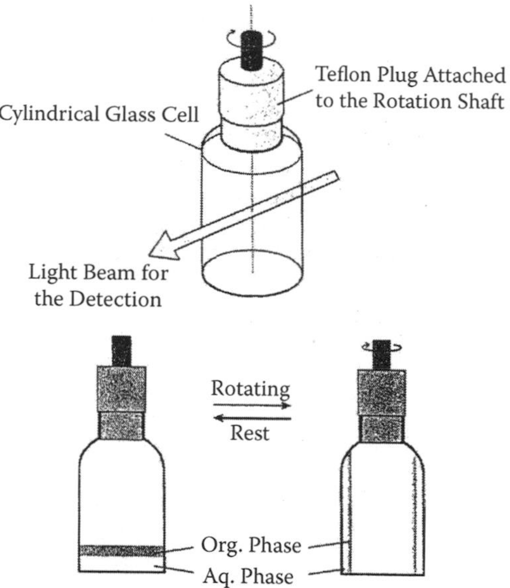

FIGURE 2.21 Schematic drawing of the centrifugal liquid membrane (CLM) cell.

viscous solvent like a dodecane. Therefore, rather fast interfacial reactions such as diffusion rate limiting reactions are preferable systems to be measured.

2.5.1.4 Time-Resolved Laser Spectrometry

By the total internal reflection condition at the liquid–liquid interface, one can observe interfacial reaction in the evanescent layer, a very thin layer of approximately 100 nm in thickness. Fluorometry is an effective method for a sensitive detection of interfacial species and its dynamics [24]. Time-resolved laser spectrofluorometry is a powerful tool for the elucidation of rapid dynamic phenomena at the interface [25]. Time-resolved total reflection fluorometry can be used for the evaluation of rotational relaxation time and the viscosity of the interface [26]. Laser excitation can produce excited states of adsorbed compound. Thus, the triplet–triplet absorption of interfacial species was observed at the interface [27].

2.5.2 Kinetic Extraction Mechanism and Interfacial Reactions

2.5.2.1 Chelate Extraction Systems

2.5.2.1.1 n-Alkyl Substituted Dithizone

The extraction system that was measured by the HSS method for the first time was the extraction kinetics of Ni(II) and Zn(II) with n-alkyl substituted dithizone (HL) [28]. The observed extraction rate constants linearly depended on both concentrations of the metal ion $[M^{2+}]$ and the dissociated form of the ligand $[L^-]$. This seemed to suggest that the rate-determining reaction was the aqueous phase complexation that formed 1:1 complex. However, the observed extraction rate constant k′ was not decreased with the distribution constant K_D of the ligands as expected from the aqueous phase mechanism.

When the extraction kinetics is governed by the aqueous phase reaction

$$M^{2+} + L^- \rightarrow ML^+ k_1;\text{ formation rate constant} \quad (2.27)$$

the rate law for the extraction is obtained as

$$-\frac{d[M^{2+}]}{dt} = k' \frac{[M^{2+}][HL]_o}{[H^+]} \quad (2.28)$$

Thus, the extraction rate constant k′ is related to k_1 and a dissociation constant K_a by

$$k' = k_1 K_a / K_D \quad (2.29)$$

The extraction rate constants for Ni(II) and Zn(II) with n-alkyl substituted dithizone did not obey the aforementioned relationship (Figure 2.22). In addition, the calculated value for k_1 in the extraction of Zn(II) with dihexyldithizone gave an impossibly large value of 10^{13} $M^{-1}s^{-1}$. These experimental results strongly suggested a significant contribution of other process than the aqueous reaction. On this issue, the HSS method revealed that the dissociated form of the n-alkyl-dithizone did adsorb at the interface generated by the vigorous agitation [14]. Therefore, the extraction rate was analyzed by introducing the interfacial reaction

$$M^{2+} + L^-_i \rightarrow ML^+_i\ k_i;\text{ interfacial formation rate constant} \quad (2.30)$$

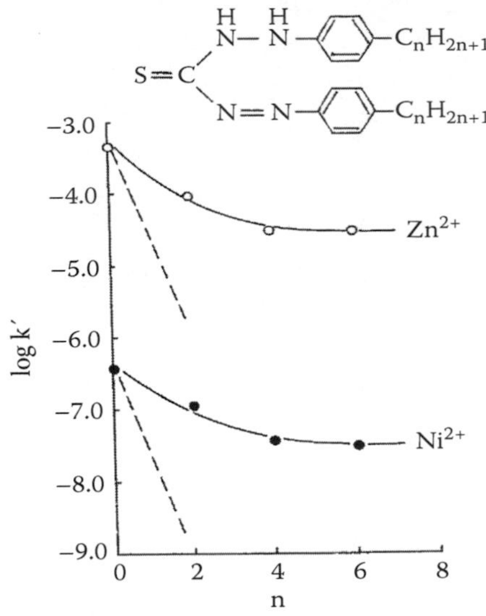

FIGURE 2.22 Dependence of the extraction rate constants, k′, of Zn(II) and Ni(II) on the number of carbon atoms in the substitution of n-alkyl dithizones. The solid lines trace the observed values and the broken lines the values predicted from the aqueous mechanism.

The extraction rate constant in this circumstance is represented by

$$k' = (k_1 + k_i K_L' A_i/V) K_a / K_D \quad (2.31)$$

where Ai/V refers to the specific interfacial area. Under the conditions of $k_1 \ll k_i K_L' A_i/V$, the previous equation is reduced to

$$\log k' K_D / K_a = \log k_i + \log K_L' A_i/V \quad (2.32)$$

The value of $K_L'A_i$ was determined by HSS method. Linear relationship of the previous equation was determined experimentally and gave $\log k_i = 8.08$ for Zn^{2+} system and $\log k_i = 5.13$ for Ni^{2+} system, respectively [15].

From these results, one can understand that the liquid–liquid interface can assist effectively the interfacial reaction through the adsorption of extractants like a solid catalyst. Whole extraction scheme of chelate extraction system is represented in Scheme 2.I.

2.5.2.1.2 2-Hydroxy Oxime

The second example is the extraction of Ni(II) with 2-hydroxy oxime. 2-Hydroxy oxime—including 5-nonylsalicylaldoxime (P50) [29], 2-hydroxy-5-nonylacetophenone oxime (SME529) [30], and 2-hydroxy-5-nonylbenzophenone oxime (LIX65N) [31]—are widely used as commercial extractants of Ni(II), Cu(II), and Co(II) in hydrometallurgy. These extractants are adsorbed at the interface even in their neutral forms following the Langmuir isotherm

$$[HL]_i = aK'[HL]_o/(a + K'[HL]_o) \quad (2.33)$$

where $[HL]_i$ and $[HL]_o$ refer to the concentrations at the interface and organic phase, respectively, a is the saturated interfacial concentration, and K′ is the interfacial adsorption constant defined by $K' = [HL]_i/[HL]_o$ under the condition of $[HL]_o \rightarrow 0$. The reaction between the dissociated form, L^-, and metal ions at the interface governed the extraction rate. We determined the adsorption constants of the three 2-hydroxyoximes and the complexation rate constants with Ni(II) ion at the interface of heptane/water system by means of the HSS method. The measurements were carried out by employing 50 ml for each phase at the stirring speed of 5000 rpm. A linear proportionality between the initial extraction rate and the interfacial concentration of the extractant was confirmed experimentally. The adsorption constants (K′) of the

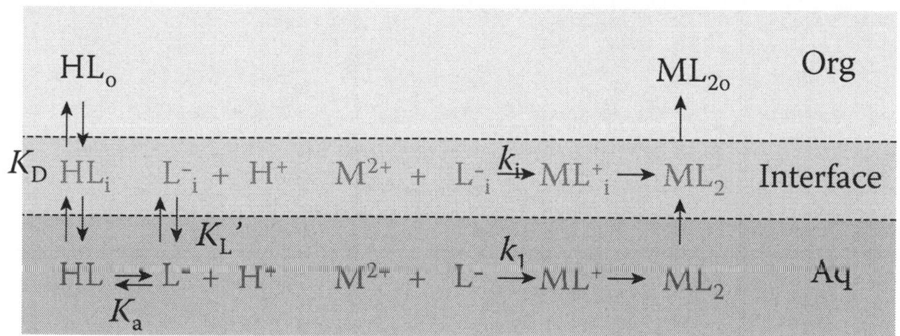

Scheme 2.I

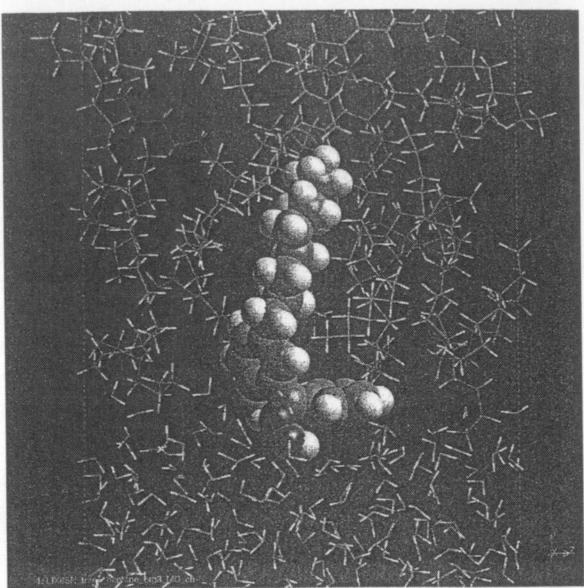

FIGURE 2.23 MD simulation of 2-hydroxy-5-nonylbenzophenone oxime (LIX65N) at heptane/water interface after 120ps.

neutral forms were all in the order of 10^{-3} cm and the complexation rate constants of the dissociated form with Ni(II) ion at the interface were in the order of 10^5 $M^{-1}s^{-1}$ for the three extractants, regardless of the large difference in the distribution constants (log K_D) as listed in Table 2.6. The complexation rate constants at the interface of heptane/water did not seriously differ from those in bulk aqueous solutions.

The adsorptivity and the reactivity of the 2-hydroxy oxime were well simulated by the MD simulations as shown in Figure 2.23, where the polar groups of –OH and =N–OH of the adsorbed LIX65N molecule are accommodated in the aqueous phase so as to react with Ni(II) ion in the aqueous phase [13]. This is the reason why the reaction rate constants of Ni(II) at the interface have almost same magnitude with those in the aqueous phase.

2.5.2.1.3 Pyridylazo-Ligand

Pyridylazo-ligands have been widely been used in colorimetric analyses of various metal ions. For example, 1-(2-pyridylazo)-2-naphthol (Hpan) is one of the most well-known reagents for the colorimetric determination and complexometric titration of metal ions. However, it showed slow extraction rate for some metal ions including Ni(II) and Pd(II). 2-(5-Bromo-2-pyridylazo)-5 diethylaminophenol (5-Br-PADAP) is a relatively new reagent that is more sensitive than Hpan for Cu(II), Ni(II), Co(II), and Zn(II), giving the metal complexes of molar absorptivities in the order of 105 $M^{-1}cm^{-1}$. 5-Br-PADAP and its complex can readily be extracted into organic solvents.

5-Br-PADAP showed a significant adsorption at the interface of heptane/water under HSS conditions at 5000 rpm. At the toluene/water interface, the adsorptivity was significantly reduced. Hpan did not adsorb at the toluene/water interface at all.

TABLE 2.7
Adsorption Constants of 5-Br-PADAP(HL) at the Heptane/Water and Toluene/Water Interfaces

Solvent	log K_D	$K'A_i/L$[a]	log ($k'/M^{-1}s^{-1}$)
Heptane	3.22	4.4×10^{-2}	3.8×10^2
Toluene	4.78	4.3×10^{-4}	1.8×10

[a] $K' = [HL]i/[HL]_o$.

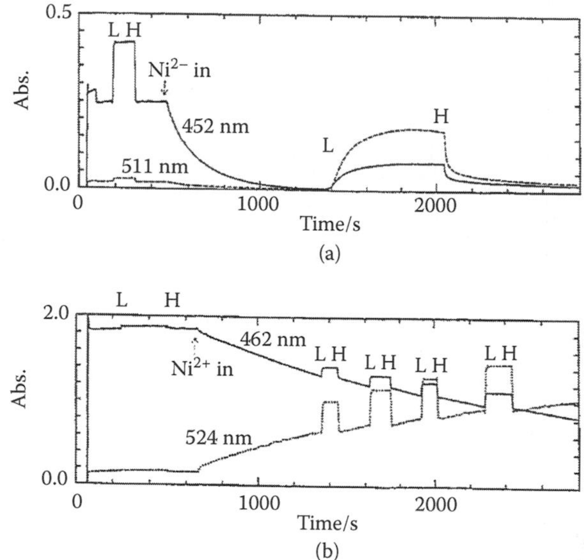

FIGURE 2.24 Measurements of the interfacial adsorptivity and the extraction rate from the absorbance changes of the organic phase by means of the high-speed stirring method, (a) heptane/water system containing 7.9×10^{-6} M 5-Br-PADAP and 1.0×10^{-4} M Ni^{2+} at pH = 5.95 observed at 452 nm for the ligand and 511 nm for the complex, and (b) toluene/water system containing 4.8×10^{-5} M 5-Br-PADAP and 2.0×10^{-3} M Ni^{2+} at pH = 5.61 observed at 462 nm for the ligand and 524 nm for the complex. The L and H in the Figure correspond to the low speed stirring of 200 rpm and the high speed stirring of 5000 rpm, respectively.

By the use of Langmuir isotherm, the adsorption constants of 5-Br-PADAP (HL) at the heptane/water and toluene/water interfaces were obtained as listed in Table 2.7.

When a heptane solution of 5-Br-PADAP and an aqueous solution of Ni2+ were stirred, the ligand in the organic phase was continuously consumed according to the complexation, but there was no extraction of the complex. The complex formed was completely adsorbed at the interface. On the other hand, in toluene system, the complex was extracted very slowly (Figure 2.24). The complexation mechanism in the two solvent systems could be analyzed by taking into account the interfacial adsorption of the ligand. The next equation was derived for the initial rate of the consumption of HL_o in the heptane system:

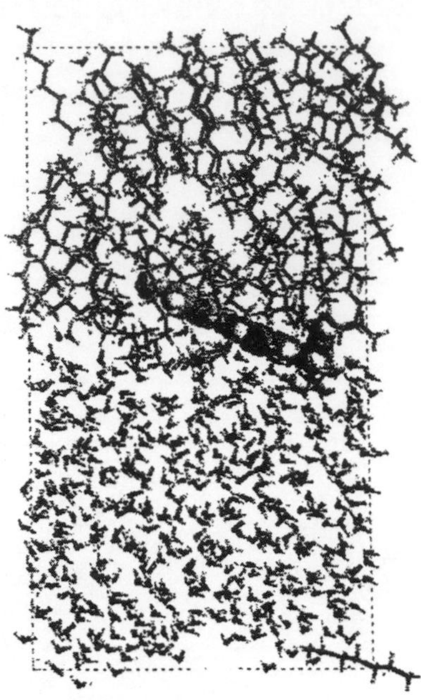

FIGURE 2.25 MD simulation of the adsorption of 5-Br-PADAP at heptane/water interface after 100 ps.

$$r_o = -d[HL]_o/dt = \frac{V}{V+K'A_i}\left(\frac{k}{K_D}+\frac{k'K'A_i}{V}\right)[Ni^{2+}][HL]_o \quad (2.34)$$

where V refers to the organic phase volume and k and k' to the 1:1 formation rate constants in a bulk aqueous phase and at the interface, respectively. The value of k = 5.3×10^2 $M^{-1}s^{-1}$ was determined by a stopped flow spectrometry in the region where the formation rate was independent of pH. The experimentally obtained kinetic results were well explained by Equation (2.34). The values of k' = 3.8×10^2 $M^{-1}s^{-1}$ for heptane/water and k' = 1.8×10 $M^{-1}s^{-1}$ for toluene/water were determined. These results mean that the percentage of the interfacial reaction rate over the total reaction rate was 99.9% in heptane/water and 94.5% in toluene/water in the HSS condition.

The molecular dynamics simulation of the adsorptivities of 5-Br-PADAP in heptane/water and toluene/water interfaces explained the solvent effect on the interfacial complexation. The MD simulation was started by putting one molecule of 5-Br-PADAP at the center of the interface. Figure 2.25 and Figure 2.26 show the averaged location of the ligand after 100 ps in the two liquid–liquid systems. In heptane/water system (Figure 2.25), the interfacial region is not filled with solvent molecules because of poor interaction between water and heptane molecules. Therefore, 5-Br PADAP stayed at the interfacial region almost parallel to the interface during the simulation. The diethylamino group was interacting with water molecules, and this moiety was preferentially attracted to the interface (or was repelled from the heptane phase). This must be the reason why it could react with Ni^{2+} ion with almost the same rate constant in the aqueous phase. On the other hand, the interfacial region

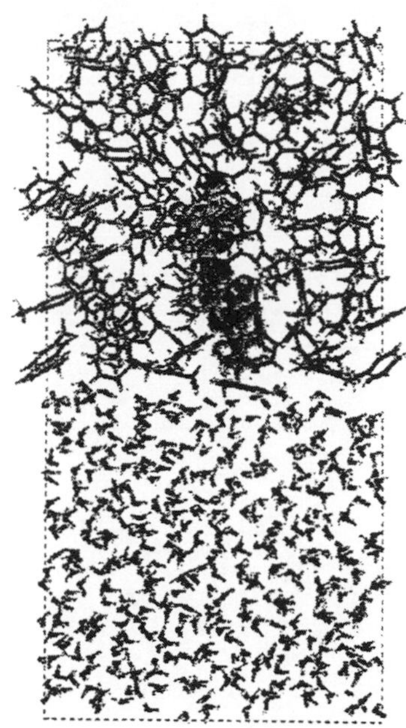

FIGURE 2.26 MD simulation of the adsorption of 5-Br-PADAP at toluene/water interface after 100 ps.

in toluene/water system is composed of a mixture of water and toluene molecules that resulted from the attractive interaction. 5-Br-PADAP and solvent molecules moved around the interface during the simulation. In Figure 2.26, 5-Br-PADAP is located near the interface, probably by the attractive interaction with water phase, although it is surrounded by toluene molecules. It can be expected that the 5-Br-PADAP molecule at the interface can react with Ni^{2+} ion in the aqueous phase, but the probability is less than that in the heptane/water interface because of the solvation by toluene molecule. Thus, the MD simulation can explain the experimentally observed solvent effects in the interfacial formation rate of Ni(II)-5-Br-PADAP complex as well as the interfacial adsorptivity.

2.5.2.2 Ion-Association Extraction Systems

In the ion-association extraction systems, hydrophobic and interfacially adsorbable ions are encountered very often. Complexes of Fe(II), Cu(II), and Zn(II) with 1,10-phenanthroline (phen) and its hydrophobic derivatives exhibited remarkable interfacial adsorptivity, although the ligands themselves can hardly adsorb at the interface, except for the protonated species [32–34]. Solvent extraction photometry of Fe(II) with phen is widely used for the determination of a trace amount of iron(II). The extraction rate profiles of Fe(II) with phen and its dimethyl (DMP) and diphenyl (DPP) derivatives into chloroform were shown in Figure 2.27. In the presence of 0.1 M $NaClO_4$, the interfacial adsorption of phen complex is most remarkable.

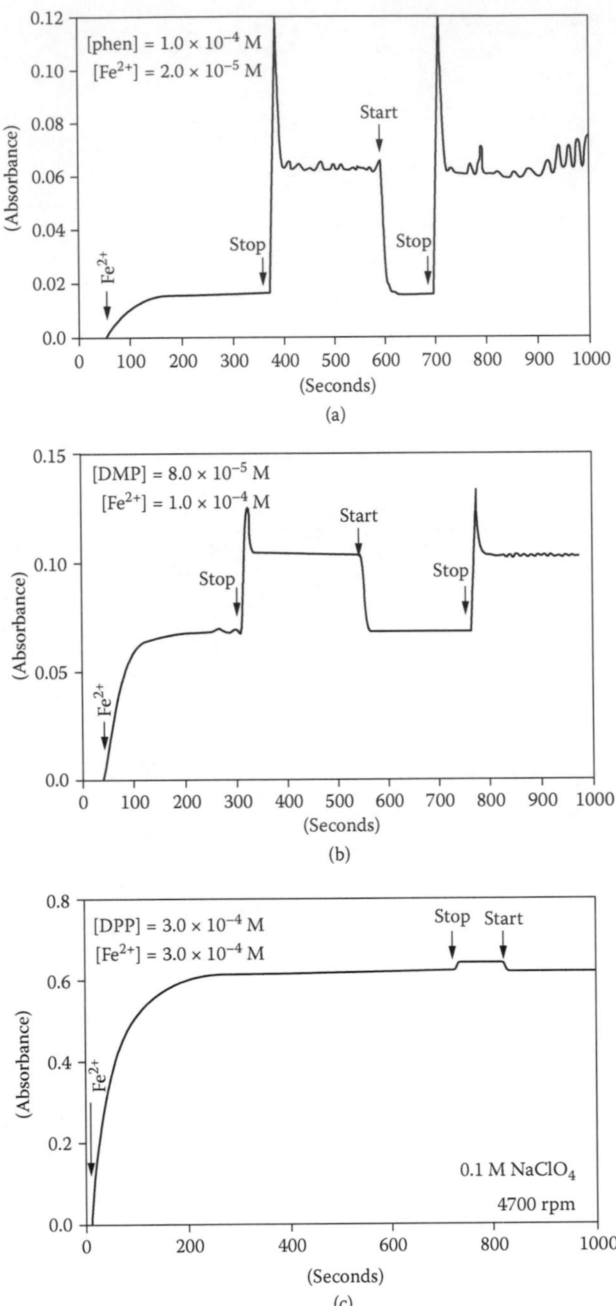

FIGURE 2.27 Extraction rate profiles of the ion-association extraction of Fe(II) with (a) phen and its (b) dimethyl (DMP) and (c) diphenyl (DPP) derivatives into chloroform in the presence of 0.1 M $NaClO_4$. Effect of stirring (4700 rpm) indicates the interfacial adsorption of the complexes.

Therefore, we have to take into account the adsorption of the extractable complex in the analysis of the extraction kinetic mechanism of these systems. The observed initial rate r_o showed the relation

$$r_o = k_{obs}[Fe^{2+}][L]_o \tag{2.35}$$

where k_{obs} refers to the observed extraction rate constant. The initial extraction rate can be described by

$$(d[FeL_3X_2]/dt)_{t=0} = k_1[Fe^{2+}][L] + k_1'\,[Fe^{2+}][L]_iA_i/V_o \tag{2.36}$$

where k_1 and k_1' stand for the 1:1 formation rate constant of FeL_2^+ in the aqueous phase and at the interface, respectively. Thus, k_{obs} is written as

$$k_{obs} = \frac{1}{1+K_C'A_i/V_o}\left(\frac{K_1}{K_D}+kiK_L'\frac{Ai}{V_o}\right) \tag{2.37}$$

where K_L' and K_C' refer to the adsorption constants of L and FeL_3X_2 from organic phase to the interface, respectively. The values of parameters are listed in Table 2.8. The results showed the following:

1. The rate determining step was the 1:1 complex formation in aqueous phase and interface.
2. The extraction rate was lowered by the adsorption of FeL_3X_2.
3. The adsorption of ligand accelerates the extraction (a positive catalytic effect) but the adsorption of the complex apparently suppresses the extraction rate (a negative catalytic effect).
4. The effects of anion and solvent on the extraction rate can be explained through the change of the adsorption constant K_C' of FeL_3X_2.

TABLE 2.8
Extraction Rate of the Ion-Association Extraction of Fe(II) with Phen and Its Dimethyl (DMP) and Diphenyl (DPP) Derivatives into Chloroform/Water Mixtures

Ligand/$CHCl_3$	Anion (0.1 M)	log K_D	$K_C'A_i$/dm³	log k_1 ($M^{-1}s^{-1}$)	log k_1' ($M^{-1}s^{-1}$)
Phen	ClO_4^-	2.85	0.368	5.65	nd
Dimethylphen	ClO_4^-	3.62	0.278	5.70	6.36
Diphenylphen	ClO_4^-	7.45	0.00452	5.52	5.00
	Cl^-	7.13	0.473	—	4.30
	SO_4^{2-}	7.14	>1	5.32	5.19

2.5.2.3 Synergistic Extraction Systems

When an auxiliary ligand added to a chelate extraction system enhances an extractability and extraction rate, such phenomena is termed *synergism.*

A typical synergism has been reported for the Ni(II)-dithizone(Hdz)-phen-chloroform system [35]. The extraction equilibrium constant ($\log K_{ex}$) was enhanced from –0.7 to 5.3 in the extraction reaction

$$Ni^{2+} + 2HDz_o + phen \rightleftharpoons NiDz_2phen_o + 2H^+ \tag{2.38}$$

At the same time, the extraction rate was accelerated by the reactions in aqueous phase [36]

$$Ni^{2+} + phen \rightarrow Niphen^{2+} \quad k = 6.8 \times 10^3\ M^{-1}s^{-1} \tag{2.39}$$

$$Niphen^{2+} + Dz^- \rightarrow NiDzphen^+ \quad k = 9.2 \times 10^4\ M^{-1}s^{-1} \tag{2.40}$$

and at the interface NiDzphen $^{2+}$ was produced during the extraction

$$Niphen^{2+} + phen \rightarrow Niphen_2^{2+} \quad k = 7.4 \times 10^3\ M^{-1}s^{-1} \tag{2.41}$$

$$Niphen_2^{2+} + Dz^- \rightarrow NiDzphen^{2+} \tag{2.42}$$

An interfacially trapped complex ion, most probably NiDzphen^{2+}, showed an interfacial disproportionation reaction:

$$2NiDzphen^{2+}{}_i \rightarrow NiDz2phen_o + Niphen_3^{2+} \quad k = 3.1 \times 10^{-3}s^{-1} \tag{2.43}$$

The synergistic effect of DPP on the extraction of Ni(II) with dithizone was also studied and confirmed the formation of an interfacial complex [37].

2.5.2.4 Interfacial Catalysis in the Solvent Extraction of Metal Ions

From fundamental knowledge concerning the interfacial complexation mechanism obtained from the kinetic studies on the chelate extraction, ion-association extraction, and synergistic extraction, one can design the interfacial catalysis. The main strategy is to raise the concentration of reactant or intermediate at the interface.

The extraction rate of Ni(pan)$_2$ into toluene is known to be very slow even under an HSS condition, whereas the addition of PADA even at the diluted concentration of 10^{-5} M could accelerate the extraction rate about ten times. At this time, only Ni(pan)$_2$ was extracted without any consumption in PADA after the extraction, except a significant decrease in the organic phase concentration of PADA during the stirring, as shown in Figure 2.28. Analyses of the experimental results suggested the interfacial adsorption of Ni(pada)$^{2+}$ complex, which was rapidly formed in the

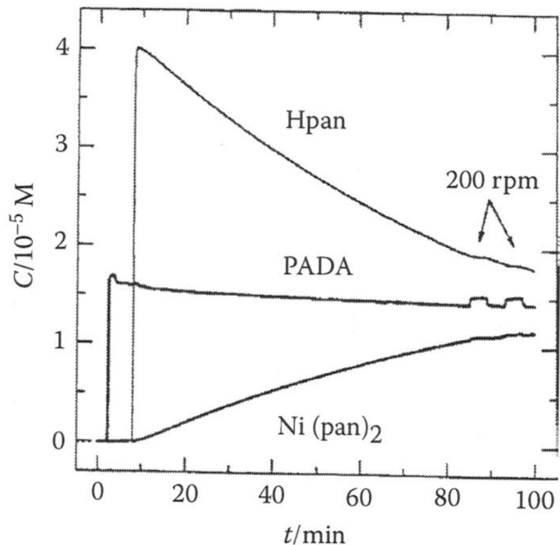

FIGURE 2.28 Concentration change of Hpan and PADA in the catalytic extraction of $Ni(pan)_2$ into the toluene phase. Initial concentration are $[Ni^{2+}] = 1.0 \times 10^{-5}M$, $[Hpan] = 4.0 \times 10^{-5}M$, and $[PADA] = 1.7 \times 10^{-5}M$ in pH = 5.6.

aqueous phase. Thus, the catalytic role of PADA could be ascribed to the interfacial ligand-exchange mechanism

$$Ni(pada)^{2+}{}_i + Hpan_o \xrightarrow{k'} Ni(pan)(pada)_i{}^+ + H^+ \quad (2.44)$$

$$Ni(pan)(pada)_i{}^+ + Hpan_o \rightarrow Ni(pan)_{2,o} + pada_o + H^+ \quad (2.45)$$

The value of k′ was obtained as 90 $M^{-1}s^{-1}$ [38].

The key process of the catalytic extraction of Ni(II)-Hpan-PADA system is the raising of the interfacial concentration of Ni^{2+} by the facilitated adsorption of $Ni(pada)^{2+}$. MD calculations could simulate the high adsorptivity of the complex ion, affording the adsorption energy from aqueous phase to the interface as –7.4 kcal/mol.

In general, acidic condition is not favored for the extraction rate of metal ion, because the dissociation of extractant is suppressed or a coordinating atom in the extractant is protonated. However, an interesting phenomenon was found in the extraction of Pd(II) with 5-Br-PADAP: Lowering the pH accelerated the extraction rate [39]. This phenomenon was ascribed to the interfacial adsorption of the protonated 5-Br-PADAP, which accelerated the interfacial complexation. The protonation of 5-Br-PADAP was made at the nitrogen atom of diethylamine which is not coordinating atom. Therefore, the interfacial protonation in this system worked so as to increase the interfacial concentration of reactable ligand.

Interfacial reactions in this system were represented by the equations

$$PdCl_2 + H_2L^+{}_i \rightarrow PdLCl_o + Cl^- + 2H^+ k_{i1} = 1.0 \times 10^3\ M^{-1}s^{-1} \quad (2.46)$$

$$PdCl_3{}^- + H_2L^+{}_i \rightarrow PdLC_{lo} + 2Cl^- + 2H^+ k_{i2} = 3.3 \times 10^2\ M^{-1}s^{-1} \quad (2.47)$$

$$PdCl_4^{-2} + H_2L^+_i \rightarrow PdLCl_o + 3Cl^- + 2H^+ k_{i3} = 3.2 \times 10^2\ M^{-1}s^{-1} \quad (2.48)$$

This finding of a new type of catalyst will provide a useful hint for the design of molecular structure of interfacially adsorbable and strongly reactive ligand for a specific metal ion.

2.5.3 Fast Dynamic Process at Liquid–Liquid Interface

2.5.3.1 Analysis of Interfacial Complex by a Time-Resolved Fluorescence Spectrometry

In the mechanism of an interfacial catalysis, the structure and reactivity of the interfacial complex is very important as well as those of the ligand itself. Recently, a powerful technique to measure the dynamic property of the interfacial complex was developed: a time-resolved total reflection fluorometry. This technique was applied for the detection of the interfacial complex of Eu(III), which was formed at the evanescent region of the interface when bathophenanthroline sulphate (bps) was added to the extraction system of Eu(III) with 2-thenoyltrifuluoroacetone(Htta) [25]. Double component luminescence decay profile experimentally observed showed the presence of dinuclear complex at the interface as illustrated in Scheme V. The lifetime $\tau = 31$ μs of the dinuclear complex was much shorter than the lifetime $\tau = 98$ μs for an aqua-Eu(III) ion, which has nine coordinating water molecules, because of a charge transfer deactivation.

Rotational dynamics of a fluorescent dye adsorbed at the interface provides useful information concerning the rigidity of the microenvironment of liquid–liquid interface in terms of the interfacial viscosity. The rotational relaxation time of the rhodamine B dye was studied by the time-resolved total internal reflection fluorescent anisotropy. In-plane rotational relaxation time of octadecylrhodamine B cation was evaluated under the presence or absence of a surfactant [26]. Table 2.8 shows that by adding a surfactant, the relaxation time and the interfacial viscosity increased. Anionic surfactants SDS and HDHP (hydrogen dihexadecylphosphate) were more effective in reducing the rotational motion, because of the electrostatic interaction. HDHP with double long chains hindered the interfacial rotation more [40].

2.5.3.2 Diffusion Controlled Interfacial Reaction

Interfacial reaction is accompanied by a diffusion process that provides reactants from the bulk phase to the interface. When the reaction rate is faster than the diffusion rate, the overall reaction has to be governed by the diffusion rate of the reactant. Diffusion, in general, takes place in the stagnant layers that are the region of both sides of the interface and are not disturbed even under the stirring of a bulk region.

A typical example is the protonation of tetraphenylporphirin (TPP) at the dodecane/acid solution interface. The interfacial protonation rate was measured by a two-phase stopped flow method [6] and a CLM method [9].

In the former method, the stagnant layer of 1.4 μm still existed under the highly dispersed system. In the CLM method, the liquid membrane phase of 50 to 100 μm thickness behaved as a stagnant layer where the TPP molecule has to migrate according to its self-diffusion rate.

2.6 SOLVENT EXTRACTION DATABASE (SEDATA)

Since the 1940s, a great number of papers have been reported that are relevant to solvent extraction. The best way to use these data is to make a database for solvent extraction. The database for the solvent extraction of metal ions now available through the Internet is Solvent Extraction DATAbase (SEDATA) (http://sedatant.chem.sci.osaka-u.ac.jp/) [41], which was originally constructed by Suzuki et al.

In this database two kinds of data are included: (1) equilibrium constant data and (2) a set of raw data. The raw data include a set of points on one extraction curve in a plot (e.g., %E versus pH, log D versus pH) in a literature, which was converted to numerical values so as to be reconstructed as a plotted graph. The equilibrium constant data are accompanied by a set of data corresponding to items such as classification, element, valence of metal ion, organic solvent, extractant, acid, salt, buffer, ionic strength, volume of organic and aqueous phases, temperature, shaking time, extraction equation, adduct formation equation, extracted species, value of constant, and referenced paper ID. A most useful advantage of SEDATA is that one can draw a log D versus pH plot as a figure from the constant values for interesting elements by using a selected reagent and a solvent, and thus one can predict the conditions or make a plan for the separation of a desired element.

REFERENCES

1. H. Watarai and H. Freiser, Role of the interface in the extraction kinetics of zinc and nickel ions with alkyl-substituted dithizones, *J. Am. Chem. Soc.*, 105(2), 189–190 (1983).
2. H. Watarai and N. Suzuki, Partition kinetic studies of n-alkyl substituted β-diketones, *J. Inorg. Nucl. Chem.*, 40, 1909–1912 (1978).
3. J. H. Hildebrand, J. M. Prausnitz, and R. L. Scott, *Regular and Related Solutions*, New York: Van Nostrand Reinhold Co. (1970).
4. T. Wakabayashi, S. Oki, T. Omori, and N. Suzuki, Some applications of the regular solution theory to solvent extraction—I: Distribution of β-diketones, *J. Inorg. Nuclear Chem.*, 26, 2255–2264 (1964).
5. H. Watarai, M. Tanaka, and N. Suzuki, Determination of partition coefficients of halobenzenes in heptane/water and 1-octanal/water systems and comparison with the Scaled Particle calculation, *Anal. Chem.*, 54(4), 702–705 (1982).
6. R. Pierotti, A Scaled Particle Theory of aqueous and nonaqueous solutions, *Chem. Rev.*, 76, 717–726 (1976).
7. H. Watarai, H. Oshima, and N. Suzuki, Regularities of the partition coefficients of bis, tris, and tetrakis(acetylacetonato)metal (II, III and IV) complexes, *Quant. Struct.-Act. Relat.*, 3, 17–22 (1983).
8. S. L. Mayo, B. D. Olafson, and W. A. Goddard, III, DREIDING: A generic force field for molecular simulations, *J. Phys. Chem.*, 94, 8897–8909 (1990).
9. H. Watarai, M. Gotoh, and N. Gotoh, Interfacial mechanism in the extraction kinetics of Ni(II) with 2-(5-bromo-2-pyridylazo)-5-diethylaminophenol and molecular dynamics simulation of interfacial reactivity of the ligand, *Bull. Chem. Soc. Jpn.*, 70, 957–964 (1997).

10. H. Watarai, Y. Horii, and M. Fujishima, Interfacial adsorption of 1,10-phenanthroline complexes in solvent extraction systems, *Bull. Chem. Soc. Jpn.*, 61(4), 1159–1162 (1988).
11. H. Watarai, K. Kamada, and S. Yokoyama, Interfacial adsorption of β-diketones in vigorously stirred heptane/aqueous phase systems, *Solv. Extr. Ion Exch.*, 7(2), 361–376 (1989).
12. H. Watarai and Y. Saitoh, Total internal reflection fluorescence measurements of ion-association adsorption of water-soluble porphyrins at liquid/liquid interface, *Chem. Lett.*, 283–284 (1995).
13. Y. Onoe and H. Watarai, Evaluation of the interfacial adsorptivity of 2-hydroxy-5-nonylbenzophenone oxime by a molecular dynamics simulation, *Anal. Sci.*, 14, 237–239 (1998).
14. H. Watarai and Y. Onoe, Molecular dynamics simulation of interfacial adsorption of 2-hydroxy oxime at heptane/water interface, *Solv. Extr. Ion Exch.*, 19(1), 155–166 (2001).
15. H. Watarai and H. Freiser, Effect of stirring on the distribution equilibria of n-alkyl-substituted dithizones, *J. Am. Chem. Soc.*, 105, 191–194 (1983).
16. H. Watarai and K. Sasabuchi, Interfacial adsorption of 2-hydroxy-5-nonylbenzophenone oxime in static and vigorously stirred distribution systems, *Solv. Extn. Ion Exch.*, 3(6), 881–893 (1985).
17. H. Nagatani and H. Watarai, Two-phase stopped flow measurement of the protonation of tetraphenylporphyrin at the liquid–liquid interface, *Anal. Chem.*, 68(7), 1250–1253 (1996).
18. H. Watarai and Y. Chida, Simple devices for the measurements of absorption spectra at liquid–liquid interfaces, *Anal. Sci.*, 10, 105–107 (1994).
19. Y. Saitoh and H. Watarai, Total internal reflection fluorometric study on the ion-association adsorption of ionic porphyrins at liquid–liquid interface, *Bull. Chem. Soc. Jpn.*, 70, 351–358 (1997).
20. H. Nagatani and H. Watarai, Direct spectrophotometric measurement of demetallation kinetics of tetraphenylporphyrinato zinc(II) at the liquid–liquid interface by a centrifugal liquid membrane method, *Anal. Chem.*, 70, 2860–2865 (1998).
21. P. R. Danesi, In J. Rydberg, C. Musikas, G. R. Choppin, eds., *Principles and Practices of Solvent Extraction*, New York: Marcel Dekker, 157–207 (1992).
22. G. J. Hanna and R. D. Noble, Measurement of liquid–liquid interfacial kinetics, *Chem. Rev.*, 85, 583–598 (1985).
23. H. Watarai, What's happening at the liquid–liquid interface in solvent extraction chemistry? *Trends in Anal. Chem.*, 12, 313–318 (1993).
24. H. Watarai and F. Funaki, Total internal reflection fluorescence measurements of protonation equilibria of rhodamine B and octadecylrhodamine B at toluene/water interface, *Langmuir*, 12, 6717–6720 (1996).
25. M. Fujiwara, S. Tsukahara, and H. Watarai, Time-resolved total internal reflection fluorometry of ternary europium(III) complexes formed at the liquid–liquid interface, *Phys. Chem. Chem. Phys.*, 1, 2949–2951 (1999).
26. S. Tsukahara, Y. Yamada, T. Hinoue, and H. Watarai, Measurement of the rotational relaxation of octadecylrhodamine B adsorbed at a liquid–liquid interface by time-resolved fluorescence anisotropy under the total internal-reflection condition, *Bunseki-kagaku*, 47, 945–952 (1998).
27. S. Tsukahara and H. Watarai, Transient attenuated total internal reflection spectroscopy to measure the relaxation kinetics of triplet state of tetra(N-methylpyridinium-4-yl)porphine at liquid–liquid interface, *Chem. Lett.*, 89–90 (1999).
28. H. Watarai, L. Cunningham, and H. Freiser, Automated system for solvent extraction kinetic studies, *Anal. Chem.*, 54, 2390–2392 (1982).

29. H. Watarai and K. Satoh, Kinetics of the interfacial mechanism in the extraction of nickel(II) with 5-nonylsalicylaldoxime, *Langmuir*, 10, 3913–3915 (1994).
30. H. Watarai, M. Takahashi, and K. Shibata, Interfacial phenomena in the extraction kinetics of nickel(II) with 2′-hydroxy-5′-nonyl-acetophenone oxime, *Bull. Chem. Soc. Jpn.*, 59, 3469–3473 (1986).
31. H. Watarai and M. Endo, Interfacial kinetics in the extraction of copper(II) and nickel(II) with 2′-hydroxy-5′-nonylbenzophenone oxime, *Anal. Sci.*, 7, 137–140 (1991).
32. H. Watarai, Interfacial adsorption of 1,10-phenanthrolines in vigorously stirred solvent extraction systems, *J. Phys. Chem.*, 89, 384–387 (1985).
33. H. Watarai, Effect of stirring on the ion-association extraction of copper and zinc 4,7-diphenyl-1,10-phenanthroline complexes, *Talanta*, 32, 817–820 (1985).
34. H. Watarai and Y. Shibuya, Interfacial adsorption of iron(II)-4,7-diphenyl-1,10-phenanthroline complex in ion-association extraction systems, *Bull. Chem. Soc. Jpn.*, 62, 3446–3450 (1989).
35. B. Freiser and H. Freiser, Analytical applications of mixed ligand extraction equilibria Nickel-ditbizone-phenanthroline complex, *Talanta*, 17, 540–543 (1970).
36. H. Watarai, K. Sasaki, K. Takahashi, and J. Murakami, Interfacial reaction in the synergistic extraction rate of Ni(II) with dithizone and 1,10-phenanthroline, *Talanta*, 42, 1691–1700 (1995).
37. H. Watarai, K. Takahashi, and J. Murakami, Synergistic extraction rate of Ni(II) with dithizone and 4,7-diphenyl-1,10-phenanthroline, *Solv. Extr. Research Develop. Jpn.*, 3, 109–116 (1996).
38. Y. Onoe, S. Tsukahara, and H. Watarai, Catalytic effect of N,N-dimethyl-4-(2-pyridylazo)aniline on the extraction rate of Ni(II) with 1-(2-pyridylazo)-2-naphthol: Ligand-substitution mechanism at the liquid–liquid interface, *Bull. Chem. Soc. Jpn.*, 71, 603–608 (1998).
39. A. Ohashi, S. Tsukahara, and H. Watarai, Acid-catalyzed interfacial complexation in the extraction kinetics of palladium(II) with 2-(5-bromo-2-pyridylazo)-5-diethylaminophenol, *Anal. Chim. Acta.*, 364, 53–62 (1998).
40. S. Tsukahara, Y. Yamada, and H. Watarai, Effect of surfactants on in-plane and out-of-plane rotational dynamics of octadecylrhodamine B at toluene-water interface, *Langmuir*, 16, 6787–6794 (2000).
41. H. Watarai, M. Fujiwara, S. Tsukahara, N. Suzuki, K. Akiba, K. Saitoh, et al., Construction of an internet compatible database for solvent extraction of metal ions, *Solv. Extr. Research Develop. Jpn.*, 7, 197–205 (2000).

3 Computation of Extraction Equilibria

Josef Havel

CONTENTS

3.1 INTRODUCTION

The knowledge of chemical equilibria in solvent extraction and other chemical technologies, environmental chemistry, and all branches of chemistry is very important. The quantitative knowledge of chemical equilibria—that is, knowledge of species composition, their formation constants, and distribution of species in solution as a function of pH or component concentration—is of a great importance in chemistry.

In spite of the fact that computers revolutionized the calculation of stability constants from all types of equilibrium data and that any new data can be handled with relative ease, there are still some unresolved problems and difficulties.

The purpose of this chapter is to discuss the task of chemical model determination (CMD) not only with respect to solvent extraction (SX) and liquid membranes (LM) but also to give overview of the fundamentals and also to elucidate the problems of extraction equilibria computation more deeply. There is no intention of providing an exhaustive description of computation in extraction systems due to the limited space and because it is not necessary. Also no exhaustive review of the computer programs will be given here but will be limited to the most important questions.

On the other hand, the latest developments and approaches applying modern methods in chemical equilibrium analysis using, for example, factor analysis or so-called soft modeling, is discussed here as well. The intention is also to demonstrate that the problem of CMD in solvent extraction but also in other instrumental techniques is to date not completely solved and is open to further development. It is worth stressing that a rather new chemical discipline, chemometrics, might be of great help here. A generally accepted definition of *chemometrics* is a chemical discipline that uses mathematical and other methods employing formal logic (1) to design or select optimal measurement procedures and experiments and (2) to provide maximum relevant information by analyzing chemical data. Such a discipline is needed and is useful in SX as well.

3.2 CHEMICAL MODEL DETERMINATION

The task of solving extraction equilibria means finding out what the complexes and stability and extraction constants are. More generally, it means finding out the chemical model that agrees best with experimental data.

Let us consider a chemical system containing a metal ion M, ligand L, another ligand X, and OH^- or H^+ ions, where formation of complex species $M_pL_qX_r(OH)_s$ (charges omitted) might occur according to

$$p\,M + q\,L + r\,X + s\,... = M_pL_qX_r(OH)_s\,...\,(\beta_{pqrs})$$

where equilibrium (stability) constants β_{pqrs} of the species are given by relation

$$\beta_{pqrs} = [M_pL_qX_r(OH)_s]/([M]^p[L]^q[X]^r[OH]^s) \tag{3.1}$$

When studying extraction equilibria the aim is to find or determine the correct chemical model. CMD consists of determining the following:

1. The number of $M_pL_qX_r(OH)_s$ species formed.
2. Composition of the products (i.e., the determination of $p,q,r,$ ... indices).
3. Equilibria/extraction (stability) constants β_{pqrs}.

4. Free concentrations (distribution) of [M], [L], [X], [OH], and of M_pL_q $X_r(OH)_s$ species.

However, there are some questions we should ask:

- How to fulfill the task?
- Can the chemical model be determined uniquely?
- Are the values of the parameters of the reaction scheme unique?
- Can some models be indistinguishable?

Although the answer to most of these questions can be found in [1,2], a brief overview is given here. Though a classical approach is mostly using general regression and a "trial-and-error" approach, there is a possibility of (I) direct estimation of a number of species in solution by principal component analysis, and via (II) simultaneous regression estimation of stoichiometry and equilibria constants. Such a novel scheme of the CMD based on (I) and (II) was first suggested in [1] and used later in several studies.

The tasks of CMD can be fulfilled by various means. Formerly, graphical methods were used [3,4], and later computer methods were developed to a high degree of sophistication; presently, many algorithms are available. There is almost a jungle to penetrate. A review of computer methods used in equilibria studies is offered by Gaizer [5], whereas critical comments and a background of various mathematical methods used for the purpose of solution equilibria studies is given by Gans [6]. From other monographs on computation of solution equilibria we have to mention, among others, Burgess et al. [7], a monograph of Smith and Missen [8] dealing with equilibrium composition computation, a book edited by Leggett [9], and two monographs by Meloun and Havel [10,11].

At this moment graphical methods are almost forgotten, and the philosophy of CMD is based mainly on the use of computers. However, the computational approach of CMD should be always supported by a sound chemical experience and then on the "intelligent" computational methods, even if the stage of data measurement the automatic experimentation applying online computers helps to increase both the precision and the economy of the experimental data collection. Sophisticated experimental devices have been developed for the purpose of data acquisition.

3.2.1 General Regression

CMD is mostly performed by applying a general regression method. The procedure is the following.

First, the primary chemical model is constructed, which consists of a certain number of species, their stoichiometry, and first estimates of equilibria and extraction constants. Chemical knowledge and experience with analogous systems play an important role in this stage.

Second, a suitable minimizing program is then applied to minimize the residual square sum, U, as a function of parameters (e.g., β_{pqrs}) to obtain the best fit of theoretically calculated curves (from the parameters) with the experimental ones. The quality of the fit is examined, and if it is not found satisfactory, the primary chemical

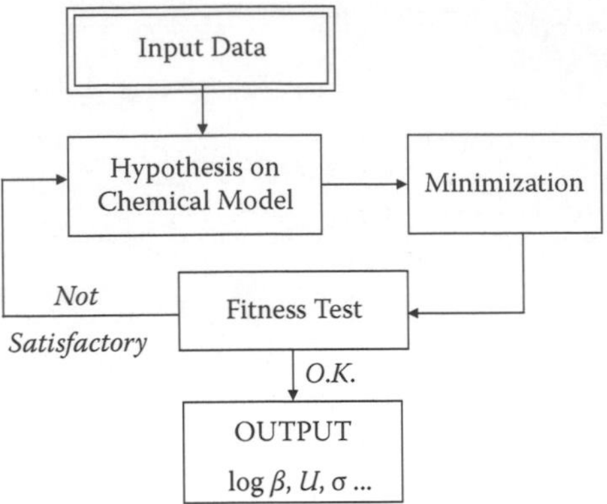

FIGURE 3.1 Scheme of classic trial-and-error approach toward the chemical model determination using general regression minimization methods.

model should be changed—that is, some of the former species originally suggested are excluded, or rejected, or another species is added. Then the minimization is repeated until the satisfactory fit is obtained. A scheme of this classical approach as previously described is given in Figure 3.1, and an example of the use of such a computational approach is given in Table 3.1 for case of Neodymium(III) extraction.

However, what are the criteria of goodness of fit? As the best fit we consider such a fit that gives the minimum value for the sum of squares of deviations (residuals):

$$U = \Sigma\, w_i\, (D_{exp} - D_{calc})^2 \rightarrow \min \tag{3.2}$$

where D_{exp} and D_{calc} are experimental or calculated values of the distribution ratio, D (dependent variable), and w_i is the statistical weights, often taken equal to unity, whereas the summation is taken over all experimental points. In the same time, the standard deviation of the dependent variable, $\sigma(D)$, should also reach the minimum value, as near as possible to the value of the experimental error.

$$\sigma(D) \rightarrow min \tag{3.3}$$

As for the stability constant, the species is accepted if at the end of the minimization process the equilibrium constant comes out with a physically meaningful value and low value of standard deviation. Sillén [12] suggested that a species is rejected if the ratio of the constant and their standard deviation is less than certain rejection factor F_σ (*Sigfak*). The value of *Sigfak* equal to 3 or 2 (or even 1) is used dependent on the data accuracy. Most often, the value equal to 3 is recommended and used.

Another criterion applied as a measure of the agreement between experimental and calculated data is Hamilton's *R*-factor, used, for example, in the MINIQUAD program [13].

It is defined by

$$R = \surd\, (\Sigma\, w_i\, (D_{exp,i} - D_{calc,i})^2 / \Sigma\, w_i\, (D_{exp,i})^2) \tag{3.4}$$

TABLE 3.1
Review of Some Extraction Agents Oligomerization as Calculated with the CPLET Program

System (Ligand)	Solvent	Species	Log β	Reference
Di-(2-ethylhexyl) phosphoric acid (HDEHP)	Benzene	L_2	$\log \beta_2 = 3.67$	[55]
DINA .HCl	Benzene	L_2, L_3	$\log \beta_2 = 3.46$	[47]
			$\log \beta_3 = 5.63$	
			$\log \beta_2 = 3.24$	[50]
			$\log \beta_3 = 5.68$	
Tri laurylamine + n-octanol	Benzene	[TLA, ol]	$\log \beta_{11} = 0.56$	[56]
mono-(2-ethyl-hexyl)ester 2-ethylhexylphosphonic acid	Toluene	L_2	$\log \beta_2 = 2.81 \pm 0.09$	[50]
Tri-n-octylamine Benzene	Benzene	L_2, L_5	$\log \beta_2 = 1.64$	[57]
			$\log \beta_5 = 4.91$	
			$\log \beta_2 = 1.63 \pm 0.04$	[51]
			$\log \beta_5 = 4.89 \pm 0.05$	
Di-(2-ethylhexyl)monothio-phosphoric acid (DEHTPA)	Toluene	L_2, L_{12}	$\log \beta_2 = 1.76 \pm 0.03$ $\log \beta_{12} = 14.84 \pm 0.12$	[51]
Trioctylmethyl ammonium chloride (TOMACl)	Toluene	L_3, L_4	$\log \beta_3 = 4.69 \pm 0.09$ $\log \beta_4 = 6.67 \pm 0.07$	[51]

where $D_{exp,i}$ are i-th experimental and $D_{calc,i}$ calculated values of dependent variable and w_i the weights. If errors in experimental quantities measured can be estimated, then using rules for propagation of error limiting value R_{lim} can be obtained as

$$R_{lim} = \sqrt{(\Sigma\, w_i\, \varepsilon_i^2 / \Sigma\, w_i\, (D_{calc,i})^2)} \tag{3.5}$$

where ε_i is the residual calculated from pessimistic estimates of the errors. This quantity can be regarded as the significance limit, and if $R <= R_{lim}$, we can conclude that the fit of the data is satisfactory.

These values can be applied for R-factor ratio test (i.e., the Hamilton test) performance [14]. If for hypothesis H_o a minimum value of R factor, R_o and for alternative hypothesis H_i the value of R-factor equal to R_i is obtained, the H_i hypothesis can be rejected at the significance level α, if

$$R_i/R_o > R_{p,\,(N\text{-}p),}\ \alpha \tag{3.6}$$

where p is the number of unknown parameters to be calculated and (N-p) is the number of degrees of freedom. Significant values of R for different values of p and (N-p) and α have been reported. On the other hand, if (N-p) is large, this quantity may also be calculated from Chi-quadrat value [14]. Details on the problem of species selection using the technique described already can be found in [2].

Selection of the species can be even automated by a computer. Such a device, called the species selector, or STYRE, was first introduced by Sillén [12]. From the list given one species after another is added to the starting chemical model, and a new species is accepted if $\beta/\sigma(\beta) < F_\sigma$ and rejected if $\beta/\sigma(\beta) < F_\sigma$ or if β has no physical meaning (e.g., if $\beta < 0$). A scheme of such selector work is given in Figure 3.2.

However, even with a computer it is not an easy task to examine all possible species or species combinations. Let us consider, for example, a system of $M_pL_qX_r(OH)_s$ complexes with p from zero up to three, q from zero up to three, and r from zero up to five. There are 92 possible species (including oligomers of the components) to be considered. But the number of all possible combinations of two species is 4,186 and 125,580 of three, 2,794,155 of four, and so forth. The number of possible combinations of 10 species is approximately 10^{13}. Thus, to examine all possible combinations it would be difficult even with use of the present largest supercomputers. Of course, if a chemist takes into account his chemical experience and knowledge, the number of species to be examined could be radically lowered. For example, if p equals 1 (only mononuclear complexes are assumed), the total number of species is 24.

In practice, not more than a few tens of species combinations are usually tested, such as in the course of uranium (VI)-maltol-H^+ equilibria study [15]. As an exception we can mention a paper by Varga et al. [16], in which up to 128 different models in a SX study were examined. Recently, in a potentiometric study of tungsten (VI) complex formation with tartrate up to 40,000 various combinations were examined [17].

However, the examination of all possible combinations with present computers is still not possible. Therefore, we have concentrated our aims to find out some more realistic means, how to find out the best combination of species, and how to speed up the CMD. Principally there are two possibilities:

1. To estimate directly the number of species in solution by factor analysis.
2. To estimate species composition computing stoichiometric indices, called the estimation of stoichiometric indices (ESI) approach.

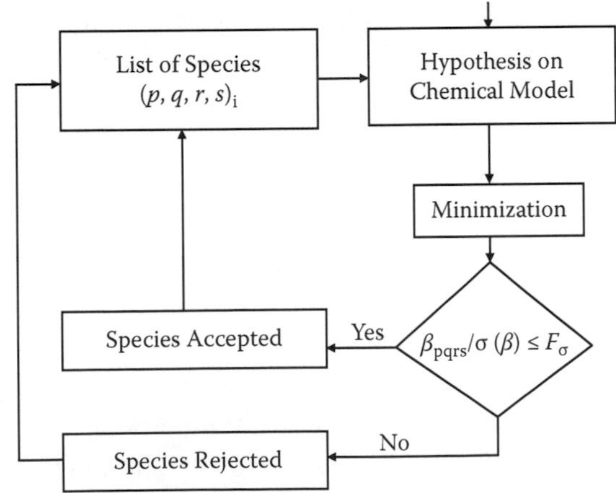

FIGURE 3.2 Scheme of species selector STYRE [12].

3.2.2 Application of Factor Analysis to Extraction Data

3.2.2.1 Factor Analysis

The number of absorbing species in solution in spectrophotometry can be estimated by factor analysis or principle component method (PCM) especially. Sometimes, the term *eigenvalues analysis* is also used. The method is traditionally used in spectrophotometry. Because of the simplicity of spectrophotometry, a detailed description is given here even though this method is not frequently used in extraction studies.

If absorbance is measured for N_s solutions and N_w wavelengths, the absorbance for i-th solution and j-th lambda is given by the sum of contributions of individual N_c species:

$$A_{ji} = \sum_{k=1}^{Nc} \varepsilon_{j,k} c_{k,i} \tag{3.7}$$

where $\varepsilon_{j,k}$ are molar absorptivities for k-th species and j-th wavelength and $c_{k,i}$ are concentrations of k-th species in i-th solution. Equation (3.7) can be written in matrix notation as

$$\mathbf{A} = \mathbf{E}\ \mathbf{C} \tag{3.8}$$

where **A** is $N_w \times N_s$, **E** is $N_w \times N_c$ and **C** is $N_c \times N_s$ matrix.

It follows from the theory of matrices that

$$\text{rank}\ (\mathbf{A}) = \min\ (\text{rank}\ (\mathbf{E}), \text{rank}\ (\mathbf{C}) = \min\ (N_w, N_c, N_s) \tag{3.9}$$

Provided that N_w and N_s are both >= N_c, then rank (**A**) <= N_c. Thus, the rank of the absorbance matrix directly gives the number of species. Survey of the computational methods for the rank determination can be found in the original literature [18–29] or is reviewed and discussed in [1,10]. In case the spectra of some species are the same or are linear combination of others, such species cannot be distinguished. In spite of this limitation the method is very useful [1,2].

3.2.2.2 Factor Analysis in Extraction

Extension of PCM to potentiometry and extraction has been reported [30,31]. The number of species formed in the organic phase in extraction may be found by use of normalized graphs [32] or suitable computer programs such as LETAGROP-DISTR [33] and EXLET [34], whereas even minor species can be searched in this way [35]. The limitations of this approach were previously discussed.

In a system with formation of polynuclear complexes the program MESAK [36] computes the average *p, q* values for formation of A_pB_q complexes [12]. This program is of great help in defining the part of the *p, q* area in which to search for possible species. However, neither MESAK nor computer programs give direct information about the possible number of species formed. The number of species involved in extraction equilibria may be estimated by principal component analysis of extraction data by finding the rank of the normalized or transformed data matrix. The method was already applied to two-phase titrations and to distribution data [30]. Two-phase electromotoric force (e.m.f.) titrations and distribution data are discussed later.

3.2.2.3 Two-Phase Titrations

In this method the titration is carried out in an aqueous phase that is in equilibrium with an organic phase containing the extractant [37]. Consider for simplicity binary system described by the following reaction:

$$p\ A_{aq} + q\ B_{org} = A_pB_{q\ (org)}$$

The average number of A extracted per B, Z, is defined by

$$Z = [A]_{org}/B = \Sigma p\ [A_pB_q]/(b + \Sigma q\ [A_pB_q]) = \Sigma p\ \beta_{pq}\ a^p\ b^q\ /(b + \Sigma q\ \beta_{pq}\ a^p\ b^q) \quad (3.10)$$

B is an extractant assumed to be practically insoluble in the aqueous phase. Free concentrations of A and B are a (in the aqueous phase) and b (in the organic phase). The equilibrium constant of Equation (3.10) is given by β_{pq}, and the summation is made over all complexes in the organic phase, N_c.

Let us assume that a number of curves $Z(\log a)_B$ are collected. Each titration contains N_s points, whereas the number of B concentrations used is N_b. For each log a, N_b of Z-values are measured or interpolated on the curves $Z(\log a)_B$. This gives the $(N_s \times N_b)$ matrix **Z**. Equation (3.10) can be rewritten as

$$Z = e_i\ c_i \quad (3.11)$$

where $e_i = p_i/B$; $c_i = [A_pB_q]$ and i refers to i-th complex. Equation (3.11) can further be written in matrix notation as

$$\mathbf{Z} = \mathbf{E}\ \mathbf{C} \quad (3.12)$$

The dimensions of the matrices are $\mathbf{Z} = (N_s \times N_b)$, $\mathbf{E} = (N_b \times N_c)$, and $\mathbf{C} = (N_c \times N_s)$, and from matrix algebra it follows that the rank of **Z** (the number of either independent rows or columns or nonzero eigenvalues) is given by

$$\text{rank}(\mathbf{Z}) = \min[\text{rank}\ (\mathbf{E}), \text{rank}\ (\mathbf{C})] = \min(N_{b,}, N_{c,}, N_s) \quad (3.13)$$

Let the experiments be arranged such that $N_b > N_s > N_c$, or $N_s > N_b > N_c$); then

$$\text{rank}(\mathbf{Z}) <= N_c \quad (3.14)$$

Thus, the rank of the matrix is less than or equal to the number of species formed in the organic phase. This treatment is the same as factor analysis for one-phase potentiometric data [38]. This is due to the fact that Equation (3.10) is the same in both cases—only that b in the present case belongs to the organic phase.

3.2.2.3.1 Distribution Data

For distribution ratio D for the metal B concerning reaction p Aq + q B_{org} = ApBq_{org}, $D = B_{org}/B_{aq}$ becomes

$$D = [B]_{org}/[B]_{aq} = \Sigma\, q\, [A_pB_q]/(B - \Sigma q\, [A_pB_q]) = \Sigma\, q\, \beta_{pq}\, a^p\, b^q\, /(B - \Sigma q\, \beta_{pq}\, a^p\, b^q) \tag{3.15}$$

and B = [B]org + [B]aq, assuming the phase ratio $f = V_{org}/V_{aq} = 1$. If the phase ratio differs from unity it can be included in D, and the formalism outlined following can be applied as well.

Similarly as before for two-phase titration data, Equation (3.15) can be transformed and formulated in matrix notation as

$$\mathbf{D}_l = \mathbf{E}_l\, \mathbf{C} \tag{3.16}$$

where **C** is the concentration matrix. The similar equation can be derived if the distribution ratio of the ligand (A) is measured; the details can be found elsewhere [31].

The rank of $\mathbf{D}_1$ is obtained from a family of curves with D plotted against log a for each B value studied and is equal to the number of species with different q in Equation (3.15). Similarly, the rank of the matrix obtained from a family of curves plotted against log b for different A values studied is equal to the number of species in the organic phase with different p values.

For mononuclear complexes (all $q_j = 1$) the extraction curves are independent of B; that is, curves $D = f(\log a)_{\mathbf{B}}$ coincide giving rank (**D**) = 1. In such a case a family of curves $D = f(\mathrm{pH})_A$ can be measured and analyzed in the manner just outlined.

3.2.2.3.2 Determination of Rank

The procedure described earlier [27] and proposed originally by Kankare [28] was used.

The second moment **M** of a matrix **D** (or **Z**) is defined by

$$\mathbf{M} = (1/N_s)\, \mathbf{D} \times \mathbf{D}^T \tag{3.17}$$

assuming the eigenvalues of **M** are r_i for k independent species in the system. The residual standard deviation $s_k(\mathbf{D})$ is given by

$$s_k(\mathbf{D}) = [(\mathrm{tr}(\mathbf{M}) - r_i/(\mathrm{Nb} - k)]^{1/2} \tag{3.18}$$

where tr(**M**) is the trace of matrix **M**. If $s_{inst}(\mathbf{D})$ is the experiment uncertainty of the elements of **D**, then those eigenvalues for which $s_k(\mathbf{D}) = s_{inst}(\mathbf{D})$ should be neglected. This means that the first k, for which this condition is fulfilled, indicates the value of the rank. Values of relative variance and cumulative relative variance can also be calculated. Eigenvalues with variance of 0.01% are considered to be negligible.

Determination of the rank of matrix has been applied to several systems [31], and the results are briefly summarized in Table 3.2. Only the first example is discussed in detail; for others we refer to literature [31]. Högfeldt and Fredlund [39] studied the

TABLE 3.2
Estimation of the Number of Species for Selected Extraction Systems

	Chemical System	Chemical Species	Rank	References
1.	TLA-*m*-xylene-HNO_3-H_2O	(TLA HNO3)$_n$, n = 1,2,3 n = 12 uncertain	3	[39]
2.	TLAHCl-*o*-xylene-Fe(III)-{Li, H)Cl	$FeCl_3(TLAHCl)_q$ q = 1, 2 ,3	3	[40,41]
3.	TLA-toluene-H_2SO_4-H_2O	$(H_2SO_4)_p(TLA)_q$ (p, q) = (1,2) and (2,4)	2	[42]
4.	HDBP-TBP-hexane-UO_2^{2+}-H_2SO_4	MA_4H_2, MA_4H_2B, MA_3HB, MA_2B_2 and $MB_2(SO_4)$ (three different indices for A)	3	[43]
5.	HDBP-TBP-hexane-H_2SO_4	HAB, HAB_2, and H_2A_2B	2–3	[43,44]

extraction of nitric acid by trilaurylamine (TLA) dissolved in m-xylene. They found species $(TLAHNO_3)_n$ with n = 1, 2, 3 and possibly 12, whereas the composition of the larger species was uncertain. LETAGROP treatment gave three species with n = 1, 2, and 12. The **Z** matrix for a number of log {[H^+] [$NO^-{}_3$]} values and nine amine concentrations was obtained by interpolation in Figure 3.4 of [39]. The estimated uncertainty in Z was $s_k(Z)$ ~ 0.01 and the rank equals to 3. It can be concluded that at least three species are formed in the organic phase in agreement with the model suggested.

The method gives the results in fair agreement with models originally suggested. However, only the number of species with different p or q values is obtained. Even if the method estimates only the number of recognizable species formed, it can be of some help for the search of the chemical model.

3.2.3 Oligomerization of Extraction Agents in the Organic Phase—Vapor Pressure Osmometry

Extraction reagents are often used in organic diluents in a high concentration, and they often aggregate in the organic solvent. This process complicates determination of chemical model and evaluation of extraction equilibria, and the knowledge of the aggregation is important. Potentiometry cannot be usually used, and because the extraction reagents are often strongly absorbing the spectrophotometry cannot be used either. On the other hand, vapor pressure osmometry (VPO) has some advantages, such as studies on aggregation of long-chain amines in organic solvents [45–47]. Measuring colligative property data, as particularly collected in VPO, the problem can be solved. The method is similar to cryscopy and ebullioscopy, and the data can be described as

$$Y = K_x \times S \tag{3.19}$$

where Y is a measured experimental quantity (e.g., vapor pressure lowering), S is the sum of the species' concentrations, and K_x is a proportionality factor that may often be considered constant for a given solvent. Equation (3.19) can be more generally written as

$$Y = K_{x1} \times S + K_{x2} \times S^2 + K_{x3} \times S^3 \tag{3.20}$$

where Y is, for example, vapor pressure lowering, freezing point depression, or resistance difference ΔR in VPO, S is the sum of the concentrations of all solute species, and K_{x1} is a proportionality factor that can be usually considered as constant for a given solvent. In some cases a second and third term can also be applied to correct for nonideal behavior [48].

For thermodynamic methods like freezing point depression, this value can be calculated from enthalpy data, whereas for the VPO method it is determined experimentally using solutions with known S (usually a monomer). The method can be applied studying aggregation equilibria of solutes A, B, C, and D:

$$pA + qB + rC + sD = A_pB_qC_rD_s$$

$$\beta_{pqrs} = [A_pB_qC_rD_s]\,[A]^{-p}\,[B]^{-q}[C]^{-r}\,[D]^{-s} \tag{3.21}$$

by VPO, where R corresponds to Y; a, b, c, and d are free; and A, B, C, and D are total (analytical) concentrations of the components. Then,

$$S = a + b + c + d + \Sigma\, \beta_{pqrs}\, a^p\, b^q\, c^r\, d^s \tag{3.22}$$

The usual total mass balance equations can be written but are not given here. Applying the generalized least squares method like in equilibrium analysis, a set of unknown parameters (β_{pqrs} and K_{xi}) can be found searching for the minimum of residuals square sum, U,

$$U = \Sigma\, w_i(Y(\text{calc}) - Y(\text{exp}))^2 \rightarrow minimum \tag{3.23}$$

where w_i are statistical weights, usually taken equal to unity, and Y(calc) or Y(exp) are calculated and experimental Y values. For analysis of such data the program LETAGROP-SUMPA has been written [49]. This program, similar to most of the other LETAGROP programs, searches the chemical model by a trial-and-error method; that is, various combinations of species composition are tried until a satisfactory fit is obtained.

However, SUMPA has limited statistical evaluation of the quality of fitness and no possibility of simulation. A new program, CPLET, developed by Havel et al. [50,51], includes (1) the possibility of simulating the data, (2) the complete statistical analysis of residuals, (3) a semigraphical form of output, and (4) direct evaluation of the stoichiometry when applying the so-called ESI approach [30,52], consisting of simultaneous regression estimation of equilibrium constants and species stoichiometry. Thus, the composition (indices p, q, r, s) can be calculated.

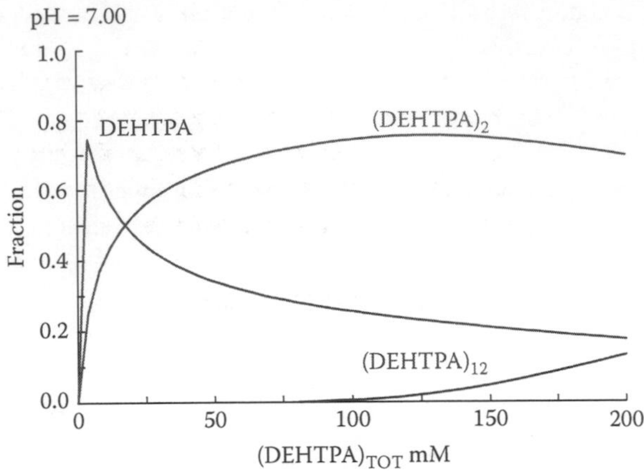

FIGURE 3.3 Distribution diagram concerning di-(2-ethylhexyl)monothio phosphoric acid (DEHTPA) oligomerization in toluene (log $\beta_2 = 1.76$, log $\beta_{12} = 14.84$).

3.2.3.1 Description of the Program

The program uses several minimizing routines that can easily be chosen, such as the pit-mapping method developed by Sillén, but also Simplex, Davidon-Fletcher-Powell, and Monte Carlo methods of MINUIT package [53]. As for ΔR, up to third-degree equations can be used:

$$\Delta R = K_{x1} \times S + K_{x2} \times S^2 + K_{x3} \times S^3 \tag{3.24}$$

The program can evaluate not only the equilibrium constants β_{pqrs} but also K_{x1}, K_{x2}, and K_{x3} constants. If specially required, selected values of *p, q, r,* and *s* can also be searched for by simultaneous regression estimation. The highest number of parameters should not exceed 16. Apart from this, a simulation can be performed while theoretical data calculated from assumed values of parameters β_{pqrs} *(p, q, r, s)*, and K_{xi} are loaded with a random noise having the normal (Gaussian) distribution of errors. An arbitrary level of noise can be chosen by the user.

The input data organization is described, and the program description is available in original papers [50,51] and in a manual [54]. Examples of some results obtained with the CPLET program are given in Table 3.2, and an example of the distribution diagram for this system is shown in Figure 3.3. In conclusion, VPO is a suitable technique for the determination of oligomerization; simultaneously, calibration constants K_{xi} are also calculated, and species stoichiometry can be searched directly by applying the ESI approach.

3.3 EVALUATION OF EXTRACTION DATA WITH LETAGROP-DISTR

3.3.1 Introduction

The LETAGROP-DISTR program was published after the death of L. G. Sillén in 1970 by Liem [58] and new types of input later on by Liem and Ekelund [59]. Since that time the program has been successfully used in the study of all types of liquid–liquid distribution data by a wide variety of laboratories all around the world. The program has the general characteristics similar for all members of the LETAGROP series and for many workers nowadays represents an old but still powerful tool. On the other hand, for a newcomer there are difficulties of giving the chemical and other information to the program. Therefore, in this chapter a detailed description of use together with input and some practical explanations and strategies are given.

3.3.2 Description of the Program

In a two-phase system, consisting of phase 1 (e.g., an organic solvent) and phase 0 (e.g., an aqueous solution) we might have up to a four-component system with species $A_pB_qC_rL_t$. Usually B represents metal ions, A hydrogen ions, and C and L various ligands. Given the values of the equilibrium constants, pH (or log a), $[B]_{tot}$, $[C]_{tot}$, and $[L]_{tot}$, the program solves unknown ln b ln c and ln l and calculates D_{calc}. The distribution ratio $D_{exp} = [B]_1/[B]_0$ estimated experimentally is then used by the program and similarly as all other members of the LETAGROP family, the LETAGROP-DISTR works with the scheme shown in Figure 3.2.

Conveniently, in the event that B can be labeled with some radioactive isotope, the distribution ratio D can be determined radiometrically as the ratio of the amount of radioactive disintegrations of equivalent samples of phase 1 and phase 0. Correction to D should be made for the different absorption or quenching in the two liquids (factor λ) and for the dead time of the counter (factor τ), whereas the corrected distribution ratio can be written as

$$D = [B]_1/[B]_0 = \lambda(I_1 + \tau I_1^2)/(I_0 + \tau I_0^2) \quad (3.24)$$

The program searches the best values for a series of unknown parameters $K1$, $K2$, … Kn by minimizing the square sum U, which in the majority of the cases is

$$U = \Sigma w_i (\log D_{calc} - \log D_{exp})^2 \quad (3.25)$$

where the summation is taken over all Np points, w_i represents a weight factor, D_{exp} is obtained from the experimental data, and using the information and conditions given in the input D_{calc} is calculated. The standard deviation σ(Y) (or, e.g., σ(log D)) is defined as

$$\sigma^2(Y) = U_0/(Np - n) \quad (3.26)$$

where n is the number of parameters being determined, U_0 is the value of U in the minimum, and Np the number of experimental points. Each $A_pB_qC_rL_t$ species is characterized by stoichiometric coefficients that may assume positive, zero, or negative values, and additional coefficient f_i (called "fas[i]" in the program) must be added that can have the value 0 or 1, depending on whether the species exists in phase 0 or in phase 1. Consequently, even the reaction components B, C, and L must be listed among the Nk complexes after Rurik = 7, with formation constant K = 1. The volume ratio, V, of phase 1 to phase 0 is assumed to be known for each experimental point.

3.3.3 General Rules for Use of the Program

General input data for LETAGROP are described in Table 3.3. Table 3.4 shows how the data are defined for different control types (number *Typ*). We present here also general information about the Ruriks, as these instructions are hardly available [Input for Letagrop, The Royal Institute of Technology, Oorg. Kemi, mimeographed edition].

In everyday practice, it is usual to treat to the different sets of data and to choose the model that gives the lowest values for *U* and σ(Y). Nevertheless, care must be taken because results reached for the best minimum should be evaluated from the point of chemical reality and physicochemical meaning. An example is given following to demonstrate the data and process of how the extraction data are solved.

3.3.4 Example: Use of LETAGROP-DISTR to Search Chemical Model in Nd (III) Extraction with di-(2-Ethylhexyl) Phosphoric Acid in Hexane

Increasing demand for rare earth elements (REE) and their applications in industry motivated Sanchez et al. [61], who performed a detailed study of the extraction of Nd (III) at a constant ionic strength 0.1 M (Na^+, H^+) NO_3 with di-(2-Ethylhexyl) phosphoric acid (D2EHPA) in hexane. The experiments were done as a two-phase potentiometric titration, whereas pH in the aqueous phase was measured with a glass electrode and Nd (III) extracted to the organic phase was determined by a FIA system, applying derivatization with Arsenazo III. The distribution data were evaluated by both graphical and numerical analysis using LETAGROP-DISTR. The paper is a nice example of how the chemical model was searched for by a computational approach applying the species selector STYRE.

There were nine models examined (see Table 3.5). As a criterion for the selection of the model the value of σ(log *D*) and *U* was used. It is evident that as the best model number 9, assuming NdA_3 3HA, NdA_3, $NdA_3 \cdot NO_3$ and NdA $(NO_3)_2$ species, was obtained with σ(log *D*) = 0.063 and *U* = 0.216. The value of σ(log *D*) reaches the value near to the experimental noise, and thus the model obtained can be considered as being in a very good agreement with the experiment. Data are given in Table 3.6; graphs of the experimental data are demonstrated in Figure 3.4.

3.4 OTHER POSSIBILITIES AND METHODS

3.4.1 Estimation of Stoichiometric Indices from Extraction Data

It has been shown in previous chapters that the search of the chemical model of extraction process is based on a trial-and-error method. The best model is that one that gives the lowest *U* and σ. However, statistical analysis of the LETAGROP-DISTR program is rather limited. Other programs for extraction [2] were written after this program; for example, EXLET [34] is trying to improve the quality of analysis by introduction of given statistical packages as sometimes the different quality of the experimental data makes it difficult to establish the equilibrium models. In this sense, any improvement of the search of the chemical model is appreciated.

TABLE 3.3
General Rules and Input Values for LETAGROP Programs

Rurik = 1 U is calculated for Rs1 through Rs2 using available data and parameters. Rs means the individual groups of data (e.g., the individual (separated) titrations).

Rurik = 2 U is calculated, and for each point quantities of interest (e.g., input data, errors) are printed (Uttag, OUTPUT). U = sum of squares of residuals (Y(exp)-Y(calcd)).

Rurik = 3, N, (ivar[i], w)$_N$ Preparation for shot, while varying only common parameters. The N parameters k[ik] with the numbers ik = ivar[i], i = 1 ->N are to be varied. If w is positive, it gives the step, stek[i] := w; if w is any negative number, stek[i] := stekfak x abs(dark[ik].

Rurik = 4, stekfak If a value for the standard deviation (in twisted space) dark[ik] or darks[Rs, ik] of the parameters has been calculated, and the quantity is positive, then the step stek[i] in the next variation is the product of dark or darks with stekfak.

Stekfak is automatically set := 0.5 in the beginning of the calculations, but if the pit deviates strongly from a pure second-degree function, it may be advisable to set a lower value such as 0.2 or 0.1. The input is then, for example, 4,0.1. The new stekfak then remains until changed.

***Rurik* = 4, - tol U** If stekfak < 0 => tol U = stekfak

Rurik = 5 A shot is made, with the instructions given after Rurik =3, 19, or 20.

Rurik = 6 , Ns, Nag, Nas, Nap, (ag)$_{Nag}$, (Np, (as)$_{Nas}$,((ap)$_{Nap}$)$_{Np}$)

The data are given: Nag common quantities ag; for each of the Ns groups (satser) Nas quantities as; and for each of the Np[Rs] points of a certain group, the Nap quantities ap. (DATA).

Rurik = 7, (common), (group), (twist)

Information on adjustable parameters (LÄSK). There are alternatives for the three parts:

(common) = (a) **Nk, Nk, Nak, (k, (ak)$_{Nak}$)$_{Nk}$**; means new problem

(b) **Nk, Nbyk, (ik, k, (ak)$_{Nak}$)$_{Nbyk}$**; partial change, or addition of Nbyk k values

(c) **Nk,0**; no change

(group) = (a) **Nks, Nks, ((ks)$_{Nks}$)$_{Ns}$** ; new problem, same if Nks=1 or 0.If Nks =0, only 0,0 is given.

(b) **Nks,1,j,(ks)$_{Ns}$** ; ks[Rs, j] is exchanged in all sets.

(c) **Nks,-1, m, (j)$_m$** ; ks[Rs,j])$_m$ are set :=0 for all the j values given.

(d) **Nks, 0, m, (j)$_m$, ((ks[Rs,j])$_m$)$_{ms}$** ; all ks = 0 except m of them in each group.

(e) **Nks, 0, 0** ; no change.

(twist) = (a) **skin,(ik, jk, sk)$_{skin}$** ; values given for some twist matrix terms

(b) **0** ; no change

Rurik =7, Nk,-1,Negk,(ik)$_{Negk}$ The posk protection is removed from Negk of the k parameters, which are thus allowed to be negative.

Rurik = 8, Nok, stegbyt, (start, tol)$_{Nok}$ The equations for each point in UBBE contains Nok unknown quantities to be determined, and start[i] and [tol[i] are the starting values and tolerances to be used in solving them. The order of the components B,(C, L,) A if BDTV is used. (See also BDTV and the input for special programs).

Stegbyt is the value for the step in Kålle (or Kille) where one switches from binary approach to chord shooting. Since the steps take the values 2^{-n}, any value between 0.50 and 0.25 (e.g., 0.45, 0.4, 0.3) gives the same effect. For equations of high degree n_{deg}, a stegbyt value around 1/ n_{deg} seems appropriate, but the time does not seem very sensitive to the choice of stegbyt.

TABLE 3.3 (Continued)
General Rules and Input Values for LETAGROP Programs

Rurik = 9, Typ The number Typ tells the type of problem and picks out the right labels in the switches in PUTS and UBBE.

Rurik = 10, vmax. Gives the number of loops in Kålle after which x and y values are printed, as an indication of bad values for tol and start after Rurik = 8.

Rurik = 11, Rs1, Rs2 Gives the numbers of the first and last in the series of groups to be treated.

Rurik = 12, Skrikut Skrikut is a number that may be used to suppress certain types of output. At present the values 1, 0, –1, and –2 are used. The output stated following is printed for the first Skrikut values but suppressed for those in parentheses.

Variation of k, n o t Tage

kv and U values during shot, Minskasteg, Stegupp, Komner, s [ik, jk] values: 1 0 (1 2) if not Koks, 1 0 1 (2) if Koks

Minusgrop, kbom, darr, sigy, U at PROVA, Gamla konstanter (Old parameters), slumpskott, Umin+kmin in slumpshott, MIKO partial results: all

Variation of Ks; Tage and Koks

kv and U during shot: 1 (0 1 2)

Minska steg, Stegupp, kbom, darr, sigy, s[ik, jk] values, U at PROVA, MIKO

Gamla konstanter, partial results: 1 0 (1 2)

Slumpskott, Umin+kmin in Slumpskott: 1 0 1 (2)

Minusgrop: all

Rurik = 13 Print available best values for k and ks, with their darr (standard deviation as calculated) (proc. SKRIK).

Rurik = 14 /text on the next card or on the next line/

Text on following card/line/ is printed. Can be repeated: 14 /text/ 14/text/ ...

Rurik = 15 The information on the k [ik] and their standard deviations is transformed to decadic logarithms. (Procedure Logkik). The limits given correspond approximately to log (k +– 3 σ (k)); if σ (k) > 0.2 × k , only the best value for log k and the maximum value, log (k + 3 σ (k)) are given.

Rurik = 16 New page in the output.

Rurik = 17, 1(Styr), sigfak, Nskytt, Nvar, $(ik)_{Nvar}$, Nin (k, ak)$_{Nak}$, dark)$_{Nin}$

This starts one cycle of the species selector. Sigfak is the rejection factor F(sigma) so that a protected parameter is rejected if, after Nskytt shots, it comes out as k < $F_{\square}$ × σ (k). Out of the initial Nk k values, Nvar are to be varied, and their numbers are given. Nin new parameters will be added in time and tried: their ak values, and first guesses for suitable k and stek values are given. Warning: Nk + Nin must not exceed 20, or the number of k's allowed in the program.

Rurik = 17, 5(Styr), sigfak, Nskytt

Another cycle of the species selector, giving previously rejected parameters (species) one more chance, is started.

Rurik = 18, val

Val tells which deviation fel [val] is to be used in calculating U. Val is automatically set :=1 after Rurik = 9 (new Type).

Rurik = 19,N, $(ik,w)_N$, Nskott, Nvaks, $(ik,w)_{Nvaks}$

TABLE 3.3 (Continued)
General Rules and Input Values for LETAGROP Programs

Preparation for shot with Koks true. N of the common parameters are to be varied (information as for Rurik = 3) on an upper level. For each set of the common k on the upper level, a certain number of ks parameters are to be adjusted to give for each group the lowest possible U value.

With Rurik = 19, a certain set of Nvaks group parameters ks are to be varied, the same ones in each group. The numbers are given, together with w: w is the stek if positive, else stekfak × abs (darks[Rs, ik]) is used for stek. For each set of k parameters during a shot on the upper level, Nskott shots are to be made varying the ks in each group.

Rurik = 20, N, $(\mathbf{ik,w})_N$, Nskott, Nvaks, $(\mathbf{Rs, ik, w})_{\mathbf{Nvaks}}$

Like for Rurik = 19 except that the ks [Rs, ik] parameters to be adjusted are handpicked: so, Nvaks is the total number of ks to be adjusted.

After the preparation by Rurik = 3, 19, or 20, the shot is ordered by Rurik = 5, three shots by "5,5,5" and so forth (Rurik = 4 need not be stated every time as earlier).

For varying only ks parameters and no common k (earlier Rurik = 10) one may use Rurik = 19 with N = 0.

Order of Input

The input will be represented here by the underlined Rurik numbers.

Begin with descriptive text and data, such as 14, 9, 6.

Uttag(OUTPUT) (2) must be preceded by 7, 8 if equations are to be solved or by 11 if not all groups will be treated. Note that if 2 is given, N is set :=0 so that 3,or 19, or 20 must again be given if a new 5 is wanted.

A shot (5) or a cycle of the species selector (17) must be preceded by 7,8,11, and 3 (or 19, or 20), and by 18 (if val # 1).

The run is finished by –1, or any other Rurik < 1 or > 20. This can happen by accident if there are misprints in the input data.

Note: This general manual describes information that should follow after each RURIK number. Please observe that here the Ruriks are given not in the order they should be given in input data but in order corresponding to their values.

The initial chemical stoichiometry is given to the computer (in the form of a set of species), whereas the stoichiometric coefficients of the species either are previously determined graphically [61] or average p, q, r values are found by means of a computer program such as MESAK [36,62]. Many nice examples of numerous applications of LETAGROP-DISTR program can be found in literature, such as in [15,57–63]. However, the search of many combinations of species might be a tedious and time-consuming process.

It is possible to apply a simultaneous regression estimation of stoichiometric coefficients and stability constants (i.e., ESI) [30,64], in which both stoichiometric coefficients and extraction constants are given as adjustable parameters and the program searches for the best model changing also stoichiometric indices as real numbers. This approach introduced by Havel et al. [30,64] has been implemented in the program POLET [65] and used for the treatment of potentiometric [66], spectrophotometric [67], and kinetic data [68]. The method has also been applied to

TABLE 3.4
Specific Input Data for LETAGROP-DISTR; Details of the Calculations and Type Selection

Typ = 1 or 2

ag (information related to the system as a whole) = λ (-1 if λ is given as ap), τ

as (information related to the different series of experiments) = *V*, +ln *V*

ap (information related to the different points of each experiment) = log *a*, *Btot*, *Ctot*, (*Ltot*), I_0, I_1, (λ)(if ag[1] = −1), +D_{exp},+ln *a*, + ln *b*, +ln *c*, (+ln *l*)

k = β

ak (information related to different equilibrium constant) = *pot*, *p*, *q*, *r*, (*t*), *fas*

ks = none

Data: 14(*Rurik*), text, 9(*Rurik*), *Typ* (1 or 2), 6(*Rurik*), *Ns*, 2(*Nag*), 1(*Nas*),(5 or 6(*Nap*), λ) or (6 or 7(*Nap*), −1, if λ is given as *ap*), τ (*Np*, *V*,(loga, *Btot*, *Ctot*, *Ltot*, I_o, I_1, (λ, if given as *ap*))$_{Np}$)$_{Ns}$

Day order follows.

Typ = 3 or 4

ag, *as* = none

ap = log *a*, *Btot*, *Ctot*, (*Ltot*), *V*, D_{exp}, +ln *a*, +ln *b*, +ln *c*, (+ln *l*), + ln *V*

k, *ak*, *ks* = the same as Typ = 1, 2

Data: 14, text, 9, *Typ* (3 or 4), 6, *Ns*, 0(*Nag*), 0(*Nas*), (5 or 6 (*Nap*)), (*Np*, (log*a*, *Btot*, *Ctot*, *Ltot*, *V*, D_{exp},)$_{Np}$)$_{Ns}$

Day order follows.

Typ = 5 or 6

Liquid–liquid distribution data for which the volume ration *V* is given as one of the ap values—that is, *V* is given for each experimental point. For Typ = 1 or 2 the volume ratio is given a constant value for each group of data. These input data are suitable for distribution data collected by the AKUFVE technique [63,64].

Data: 14, text, 9, *Typ*(5 or 6), 6, *Ns*, 2(*Nag*), 0(*Nas*), (6 or 7(*Nap*), λ*)* or (7 or 8(*Nap*), −1, if λ is given as *ap*), τ, (*Np*, (log *a*, *Btot*, *Ctot*, *Ltot*, *V*, I_o, I_1, (λ, if given as *ap*))$_{Np}$)$_{Ns}$

Day order follows.

Typ 7 or 8

It can be used for two-phase potentiometric titration from a three- or four-components system. *Typ* 7 is used for the three-components system $H_pB_qC_r$

and *Typ* = 8 for the four-components system $H_pB_qC_rL_t$. In the potentiometric titration either acid solution or alkaline solution is added to the two-phase system, and the value of – log [H^+] in the aqueous phase is measured.

Data: 14, text, 9, *Typ* (7 or 8), 6, *Ns*, 0(*Nag*), 0 (*Nas*), 5 or 6 (*Nap*), (*Np*, (log*a*, *Btot*, *Ctot*, *Ltot*, *V*, I_{exp})$_{Np}$)$_{Ns}$.

Day order follows.

Adapted from [58,59].

Note: + means that the value is not given in the input but calculated by the program.

TABLE 3.5
Chemical Model Search in Nd (III)—di-(2-Ethylhexyl) Phosphoric Acid System: Extraction to Hexane

Model Number	Species	σ(log *D*)	*U*, Sum of Squares of Residuals	log β
1	$NdA_3 \cdot 3HA$	0.301	5.256	17.33 ± 0.16
2	$NdA_3 \cdot 2HA$	0.189	2.060	13.24 ± 0.09
3	NdA_3	0.095	0.526	5.05 ± 0.04
4	$NdA_3 \cdot 3HA$	0.062	0.216	16.57
	NdA_3			15.99 ± 0.07
5	$NdA_3 \cdot 3HA$	0.062	0.216	15.99 ± 0.07
	NdA_3			5.01 ± 0.08
	$NdA_2 \cdot NO_3$			
6	$NdA_3 \cdot 2HA$	0.061	0.208	12.69 ± 0.015
	NdA_3			4.88 ± 0.07
	$NdA_2 \cdot NO_3$			Max 12.24
7	$NdA_3 \cdot 3HA$	0.063	0.223	16.55 ± 0.09
	$NdA_3 \cdot 2HA$			Max 12.24
	NdA_3			4.92 ± 0.06
8	$NdA_3 \cdot 3HA$	0.063	0.223	16.58 ± 0.17
	NdA_3			4.88 ± 0.07
	$Nd\,NO_3A_2$			
	$NdA_3\,(HA)_2$			
	$Nd\,NO_3\,A_2 \cdot 3HA$			
9	$NdA_3 \cdot 3HA$	0.063	0.216	16.57 ± 0.16
	NdA_3			4.94 ± 0.05
	$NdA_3 \cdot NO_3$			Max 3.50
	$NdA \cdot (NO_3)_2$			

Source: Data from [60].

water sorption equilibria on ion exchangers in connection with the program WSLET [69,70]. Some applications of this method in equilibria studies and problems involved are discussed in the monograph [2] and elsewhere [1].

Some results of the use of the ESI approach for liquid–liquid equilibrium data evaluation are presented to demonstrate this possibility for chemical model search. All calculations were performed with EXLET 92 program [71], accomplished with complete statistical analysis of residuals and several advanced minimizing routines in addition to pit mapping of LETAGROP.

TABLE 3.6
Experimental Data and the Form of Input for LETAGROP-DISTR Program

14,

Extraction of Nd(III) with D2EHPA into hexane

9,4,6,9,0,0,6,

5,–2.00,3.86E–5,1.0E–3,0.1,1.0,0.162,–2.03,3.86E–5,1.0E–3,0.1,1.0,0.216,

–2.08,3.86E–5,1.0E–3, 0.1,1.0,0.252, –2.17,3.86E–5,1.0E–3,0.1,1.0,0.606,

–2.22,3.86E–5,1.0E–3,0.1,1.0,0.763,

7,–2.28,3.86E–5,1.0E–3,0.1,1.0,1.209,–2.31,3.86E–5,1.0E–3,0.1,1.0,2.090,

–2.41,3.86E–5,1.0E–3,0.1,1.0,3.075,–2.50,3.86E–5,1.0E–3,0.1,1.0,5.691,

–2.55,3.86E–5,1.0E–3,0.1,1.0,6.941,–2.56,3.86E–5,1.0E–3,0.1,1.0,7.375,

–2.61,3.86E–5,1.0E–3,0.1,1.0,10.956,

3,–2.56,3.86E–5,1.0E–3,0.1,1.0,8.745,–2.49,3.86E–5,1.0E–3,0.1,1.0,4.552,

–2.44,3.86E–5,1.0E–3,0.1,1.0,4.454,

7,–1.60,4.11E–5,1.0E–3,0.1,1.0,0.012,–1.97,4.11E–5,1.0E–3,0.1,1.0,0.127,

–2.06,4.11E–5,1.0E–3,0.1,1.0,0.259,–2.18,4.11E–5,1.0E–3,0.1,1.0,0.516,

–2.20,4.11E–5,1.0E–3,0.1,1.0,0.665,–2.22,4.11E–5,1.0E–3,0.1,1.0,0.706,

–2.25,4.11E–5,1.0E–3,0.1,1.0,0.914,

6,–2.28,4.11E–5,1.0E–3,0.1,1.0,1.207,–2.32,4.11E–5,1.0E–3,0.1,1.0,1.609,

–2.36,4.11E–5,1.0E–3,0.1,1.0,2.178,–2.41,4.11E–5,1.0E–3,0.1,1.0,2.986,

–2.45,4.11E–5,1.0E–3,0.1,1.0,3.814,–2.50,4.11E–5,1.0E–3,0.1,1.0,5.220,

13,–2.40,4.86E–5,0.5E–3,0.1,1.0,0.455,–2.46,4.86E–5,0.5E–3,0.1,1.0,0.698,

–2.48,4.86E–5,0.5E–3,0.1,1.0,0.780,–2.55,4.86E–5,0.5E–3,0.1,1.0,1.391,

–2.62,4.86E–5,0.5E–3,0.1,1.0,1.852,–2.64,4.86E–5,0.5E–3,0.1,1.0,2.538,

–2.67,4.86E–5,0.5E–3,0.1,1.0,2.966,–2.74,4.86E–5,0.5E–3,0.1,1.0,4.364,

–2.75,4.86E–5,0.5E–3,0.1,1.0,4.767,–2.76,4.86E–5,0.5E–3,0.1,1.0,4.839,

–2.82,4.86E–5,0.5E–3,0.1,1.0,7.625,–2.91,4.86E–5,0.5E–3,0.1,1.0,12.490,

–2.92,4.86E–5,0.5E–3,0.1,1.0,13.333,

6,–2.41,5.45E–5,0.5E–3,0.1,1.0,0.561,–2.45,5.45E–5,0.5E–3,0.1,1.0,0.671,

–2.54,5.45E–5,0.5E–3,0.1,1.0,1.954,–2.62,5.45E–5,0.5E–3,0.1,1.0,2.027,

–2.73,5.45E–5,0.5E–3,0.1,1.0,3.346,–2.80,5.45E–5,0.5E–3,0.1,1.0,6.471,

3,–2.69,5.45E–5,0.5E–3,0.1,1.0,3.300,–2.70,5.45E–5,0.5E–3,0.1,1.0,3.663,

–2.81,5.45E–5,0.5E–3,0.1,1.0,5.973,

9,–2.34,5.12E–5,0.5E–3,0.1,1.0,0.390,–2.38,5.12E–5,0.5E–3,0.1,1.0,0.512,

–2.40,5.12E–5,0.5E–3,0.1,1.0,0.588,–2.44,5.12E–5,0.5E–3,0.1,1.0,0.731,

–2.50,5.12E–5,0.5E–3,0.1,1.0,1.191,–2.60,5.12E–5,0.5E–3,0.1,1.0,2.061,

–2.67,5.12E–5,0.5E–3,0.1,1.0,3.052,–2.68,5.12E–5,0.5E–3,0.1,1.0,3.029,

–2.80,5.12E–5,0.5E–3,0.1,1.0,7.247,

7,9,9,6,1.0,0,0,1,0,0,0,1.0,0,0,0,0,1,0,1,1.0,0,0,0,0,1,0,

6.31,–4,0,0,1,0,0,2.04,–5,–1,0,1,0,0,3.16,4,0,0,2,0,1,

TABLE 3.6 (Continued)
Experimental Data and the Form of Input for LETAGROP-DISTR Program

1.70,0,0,1,0,1,0,3.76,16,–3,1,6,0,1,8.62,4,–3,1,3,0,1,
0,0,0,
8,3,0.2,–6,1E–6,–6,1E–6,–6,1E–6,
3,1,9,0.1,5,5,5,5,5,5,5,13,15,2,–1,

Source: Data from [60].

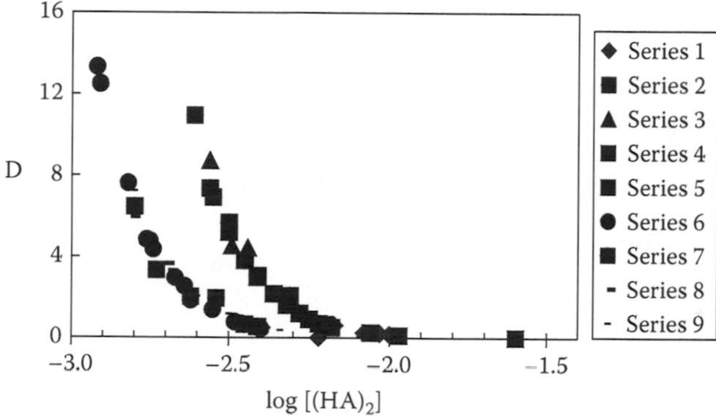

FIGURE 3.4 Distribution of Nd (III) concerning extraction with di-(2-ethylhexyl) phosphoric acid into hexane.

3.4.1.1 Theory

In a two-phase system with up to, for example, three components—H, M, L—assuming $H_pM_qL_r$ species (charges omitted) with formation and extraction constants β_{pqr}, formed either in aqueous or in the organic phase, the extraction equilibria are usually followed measuring the distribution ratio D

$$D = \frac{q\Sigma[\mathrm{H}_p\mathrm{M}_q\mathrm{L}_T]_{(\mathrm{org})}}{[M]+q\Sigma[\mathrm{H}_p\mathrm{M}_q\mathrm{L}_T]_{(\mathrm{aq})}} \tag{3.27}$$

The determination of the chemical model from extraction equilibrium data means to determine (1) the number of species, (2) stoichiometry, and (3) formation and extraction constants. Mostly, the trial-and-error method is used for the purpose of trying various combinations of p, q, and r indices, whereas the minimum for U, being the sum of squares of residuals

$$U = \Sigma\,(D_{\mathrm{exp}} - D_{\mathrm{cal}})^2 = \mathrm{f}\,(\beta_{\mathrm{pqr}};\,(p,\,q,\,r)_{\mathrm{i}}) \tag{3.28}$$

depends also on the stoichiometry. One can therefore search for the best model changing not only β's continuously but also changing indices simultaneously. The

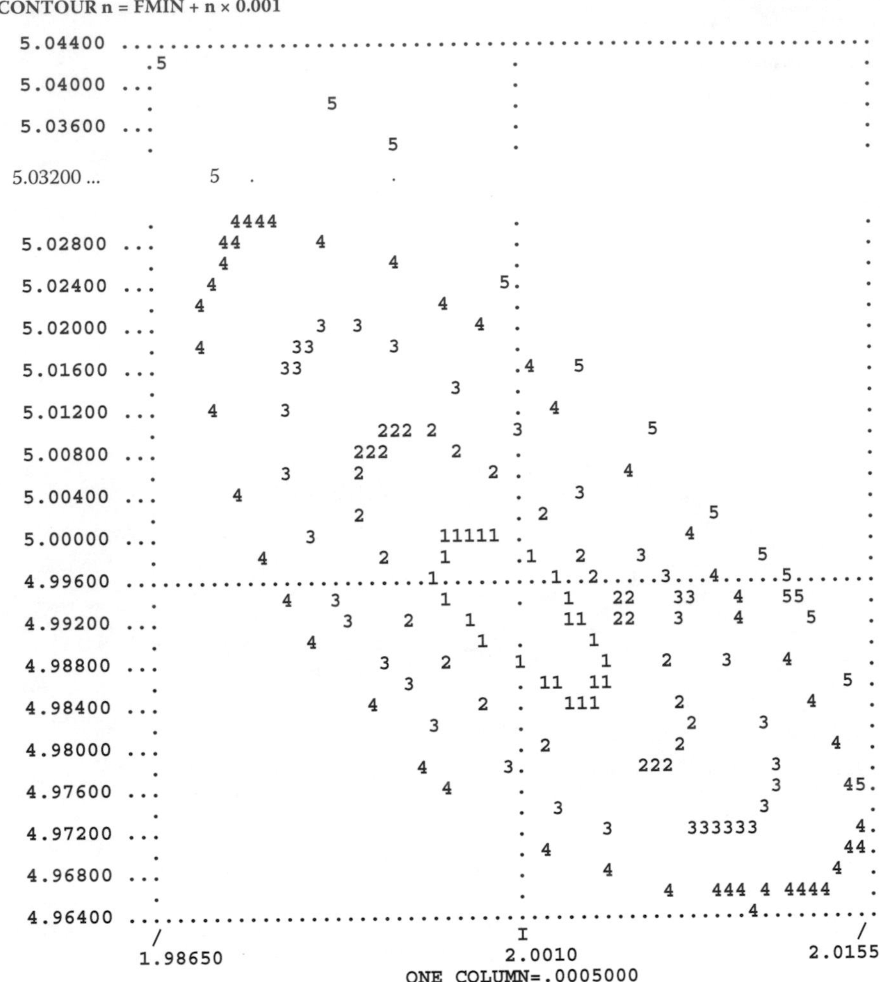

FIGURE 3.5 Contours $U=f(p,q)$ around minimum for case of lead(II) extraction with TLA into toluene at ionic strength 1.0. Extraction constants were those as found in literature [72] and were kept constant; species formed $Pb_pCl_q(TLA.HCl)_r$.

minimum—if it exists at all—is to be obtained for values (p,q,r), being now the real numbers, which should be close to integer values.

An example of the shape of the minimum when changing (p,q) values around minimum for data concerning extraction of lead(II) with TLA is shown in Figure 3.5. We have to stress that, naturally, only integer values have physical meaning, and afterward the real, noninteger, values of indices obtained in the minimum should be rounded to integer ones. Some examples are demonstrated following.

3.4.1.1.1 *Example 1: Recalculation of Data for Lead(II) Extraction with Trilaurylammonium (TLA) Chloride*

Masana and Valiente [72] studied extraction of lead (II) with TLA in different chloride media (I = 1, 2, and 3 mol l^{-1}) and found in the organic phase species $Pb_pCl_q(TLA.HCl)_r$ with *p*, *q*, *r*-values (1,2,2) and (1,2,5). Except the formation of this species, trimerization of amine salt in toluene under the formation $(TLA.HCL)_3$ takes place with aggregation constant K = 501. Also, $PbCl_n^{(2-n)}$ species formed in the aqueous phase with log β_n values 0.73, 0.99, 0.84, and 0.10 (corrected by applying the specific interaction theory) were taken into account. The contours map around the minimum for the sum of squares of deviations $U = \Sigma$ (log D_{exp} – log D_{cal}) as a function of the composition $U = f(r_1, r_2)$ was calculated by means of EXLET 92 program, whereas all the other parameters were kept constant at levels suggested in the original work [72]. The minimum is really obtained for values close to $r_1 = 2$ and $r_2 = 5$ (cf. Figure 3.5). This figure illustrates also the principal of the method; that is, starting with even a wrong estimate for stoichiometry, changing stoichiometric indices as adjustable parameters, correct values of indices near to integers are obtained in one run without the necessity to examine separate models. However, the results of a detailed computation with the EXLET 92 program [71] given in Table 3.7 show that this model is only apparently good. In Run 2 the same model as found before [72] was used, and the comparison to Run 1 shows that almost the same values of constants are obtained. Even if a rather good fit is found in both cases, in Run 3 the stoichiometric indices of TLA in extracted lead species were calculated simultaneously with corresponding extraction constants. The resulting values of indices are however not near integer; the first index of a dimmer changed to 1.31, and the second one moved from 5 to 5.87 (i.e., near to 6).

This is an indication that model (1,2,2) + (1,2,5) is not a complete one. It might mean that lower and higher values are the average of two different indices. A better fit is obtained assuming (1,2,1) + (1,2,5) (Run 4) or (1,2,2) + (1,2,6) (Run 5) species; the best one is obtained for the model containing all (1,2,1), (1,2,2), (1,2,5), and (1,2,6) species (Run 6). If now some indices are calculated, the values obtained are again near to integer, which proves the correctness of the model.

This new model has also a logical stoichiometry of a successive ligand addition, whereas (1,2,5) would result from the addition of TLAHCL trimmer to the former (1,2,2). Rearranging two trimmers of the extractant may form the species (1,2,6).

3.4.1.1.2 *Example 2: Extraction of Hydrochloric Acid by tri-n-Hexylamine (THA) to n-Octane*

Aguilar and Valiente [73] studied extraction of hydrochloric acid by THA and in this very detailed study the applied graphical methods [9] and MESAK evaluation of average *p*,*q* values and LETAGROP-DISTR program with species selector STYRE; all together, 19 models were examined. The best one found was including species $(HCl)_p(THA)_q$ with (*p*,*q*) values (1,1) and (3,3).

The data were simulated here for various instrumental errors s_{inst} (corresponding to the error in *Z*) in the range 0.001 to 0.25 and two concentrations of THA equal to 0.0097 and 0.1445 M in the pH range 2.5 to 4.5 (step 0.25). The data used for s_{inst} = 0.01 are presented in Table 3.8. For different initial composition of species, the

TABLE 3.7
Results of Computation of Extraction of Lead (II) with TLA into Toluene at Ionic Strength 1.0, Species Formed Pb_p Cl_q $(TLA.HCl)_r$

			Fitness Test							
Run	Species p,q,r	log k ± σ (log k)	U	σ (log D)	Chi2	Skewness	Curt.	Hamilton R_f	Ref.
1	1,2,2	4.62 ± 0.07	0.019	0.041					[72]
	1,2,5	8.51 ± 0.18							
2	1,2,2	4.63 ± 0.023	0.0164	0.037	2.67	−0.015	2.005	7.48%	This work
	1,2,5	8.52 ± 0.069							
3	1,2,2	0.73 ± 0.01	0.0050	0.020	4.00	−0.765	3.31	3.49%	This work
	1,2,5	13.40 ± 0.03	(r_1 = 1.31 ± 0.00, r_2 = 5.87 ± 0.00)						
4	1,2,1	2.92 ± 0.02	0.0058	0.022	5.33	−0.416	2.050	3.76%	This work
	1,2,5	8.80 ± 0.02							
5	1,2,1	2.96 ± 0.02	0.0076	0.025	5.33	0.2814	2.559	4.30%	This work
	1,2,6	10.15 ± 0.02							
6	1,2,2	4.00 ± 0.05	0.0051	0.021	4.00	−0.576	2.83	3.53%	This work
	1,2,5	8.19 ± 0.07							
	1,2,1	2.84 ± 0.01							
	1,2,6	9.96 ± 0.02							

Source: Data from Masana and Valiente [72].

stoichiometry and extraction constants were calculated with the EXLET 92 program. The results are given in Table 3.9, and it can be followed that even starting from different approximations for the composition (values from 0 to 4) in one computer run the correct stoichiometry can be obtained. In various runs the effect of the experimental uncertainty s_{inst}, which corresponds to standard deviation, $\sigma(Z)$, was followed. The calculated stoichiometry is near to correct values even for rather high random error like ± 0.1 assumed in Run 6, the last run.

Proposed method to calculate simultaneously extraction equilibrium constants and stoichiometry of species can be used to speed up the search of the chemical model. However, the method should be used with care. The number of parameters to be determined is considerably higher, and thus precautions should be taken.

It is advisable to divide the species into a group of (1) certain species, with known stoichiometry; and (2) unknown species, with uncertain or really unknown stoichiometry. In the event that the number of species in the chemical model is correct, then calculated values of indices are coming out near integer. On the other hand, if not near integer values of indices are obtained it indicates that the model is not correct.

TABLE 3.8
Simulated (Z, –log h) Data for Extraction of Hydrochloric Acid with THA into n-Octane

	c_{THA} = 0.0097 M			c_{THA} = 0.1445 M		
–log h	Z(Accurate)	Error*	Z(Loaded)	Z(Accurate)	Error*	Z(Loaded)
1.50	0.954683	–0.009245	.945438	.991834	–0.010417	0.981417
1.75	0.919915	–0.000098	.919817	.985408	–0.001605	0.983803
2.00	0.861210	0.021240	.882450	.974151	0.008234	0.982385
2.25	0.765102	0.000784	.765886	.954346	0.009674	0.964019
2.50	0.617817	–0.008169	.609648	.919777	0.008625	0.928402
2.75	0.423703	0.003395	.427098	.860702	0.009703	0.870405
3.00	0.232603	0.003001	.235604	.762443	0.007558	0.770001
3.25	0.110943	0.005408	.116350	.609036	0.006533	0.615569
3.50	0.054817	–0.005515	.049301	.399740	0.010548	0.410289
3.75	0.029199	–0.007127	.022072	.186594	–0.000731	0.185863
4.00	0.016163	0.000559	.016723	.059969	–0.002978	0.056991
4.25	0.009069	0.011835	.020904	.018018	0.005527	0.023544
4.50	0.005105	0.008999	.014104	.006758	0.000462	0.007220
4.75	0.002874	0.013832	.016706	.003172	–0.030596	–0.027423

* s_{inst} = 0.01

Source: Data according to Aguilar and Valiente [73].

The method used in the EXLET 92 program yields the standard deviations of the stoichiometric indices and thus estimates the uncertainty with which the composition can be calculated from a given set of extraction equilibrium data.

A new scheme for the chemical model determination in equilibria studies might include except statistical tests (1) the application of factor analysis to estimate directly the number of species in solution and (2) the method of direct computation of species stoichiometry via simultaneous calculation of stability constants and stoichiometric indices—the ESI approach. The ESI method might also be used as a new diagnostic tool when searching the best chemical model. The proposed novel scheme of CMD based on PCA and ESI is schematically given in Figure 3.6.

3.4.2 Hard Models versus Soft Modeling in Extraction

3.4.2.1 Soft Modeling

Recently, so-called soft modeling has become more and more important in chemistry, especially related to the partial least squares (PLS) and artificial neural network (ANN) methods. The PLS method has been used to evaluate chemical equilibria in potentiometry by Perůtka et al. [74]. The other, more advantageous possibility is ANNs.

An ANN is an attempt to apply knowledge about functioning of the human brain in computer science. From our somewhat limited knowledge we try to create a simple

TABLE 3.9
The Effect of Experimental Error on Standard Deviation of Calculated Stoichiometry Data: Hydrochloric Acid-tri-n-Hexylamine (THA) Extraction to n-Octane

		Composition Computed					
Run	$s_{inst}(Z)$	(p,q)		log $ß_{pq}$	σ(Z)	U × 100	Reference
1		1,1		2.21 ± 0.20			[73]
	3,3			11.93 ± 0.08			
2	0.001	1.000 ± 0.	1.000 ± 0.0	2.21 ± 0.			This work
		3.00 ± 0.	3.000 ± 0.	11.93 ± 0.	0.00096	0.00241	
3	0.005	0.996 ± 0.006	1.003 ± 0.005	2.198 ± 0.018			This work
		3.003 ± 0.003	2.997 ± 0.003	11.945 ± 0.009	0.0049	0.006189	
5	0.01	1	1	2.19 ± 0.04			This work
		3.003 ± 0.006	2.997 ± 0.006	11.958 ± 0.019	0.00983	0.0025105	
4	0.01	1.098 ± 0.012	1.101 ± 0.010	2.719 ± 0.037			This work
		3.104 ± 0.007	3.096 ± 0.007	12.394 ± 0.021	0.00976	0.00247	
5	0.01	0.793 ± 0.013	0.795 ± 0.011	1.189 ± 0.039			This work
		3.006 ± 0.006	2.998 ± 0.006	11.987 ± 0.018	0.00972	0.00246	
5	0.05	1.084 ± 0.067	1.116 ± 0.066	2.621 ± 0.103			This work
		3.122 ± 0.029	3.080 ± 0.029	12.520 ± 0.002	0.0471	0.0622	
6	0.1	1.113 ± 0.150	1.086 ± 0.100	2.525 ± 0.571			This work
		3.132 ± 0.056	3.070 ± 0.057	12.630 ± 0.162	0.0984	0.2519	

Notes: $s_{inst}(Z)$, standard deviation of random error superimposed to exact simulated data; σ(*Z*), calculated standard deviation of *Z*, the average number of extracted acid per 1 mol of the amine(THA); *U*, the sum of squares of residuals $\Sigma (Z_{exp} - Z_{calc})^2$.

Source: Data from [73].

artificial brain. The progress of neural net research has been inspired by the fact that the brain outperforms any computer in practically all information processing problems, maybe except for pure calculations. Usually, ANN computing is defined as "the study of networks of adaptable nodes, which through a process of learning from task examples, store experimental knowledge and make it available for use" [75].

ANN has recently become the focus of interest in chemistry [75]. ANN, as well as PLS, also represent soft modeling without the necessity to know or to determine chemical physicochemical parameters or to establish a mathematical model. Recently, ANN has successfully been applied in capillary zone electrophoresis [76–78], ion chromatography [79], electrokinetic micellar chromatography [80], and optimization in high performance liquid chromatography (HPLC) [81], for example. We have also used ANN in equilibria, and they can also be used in extraction data analysis.

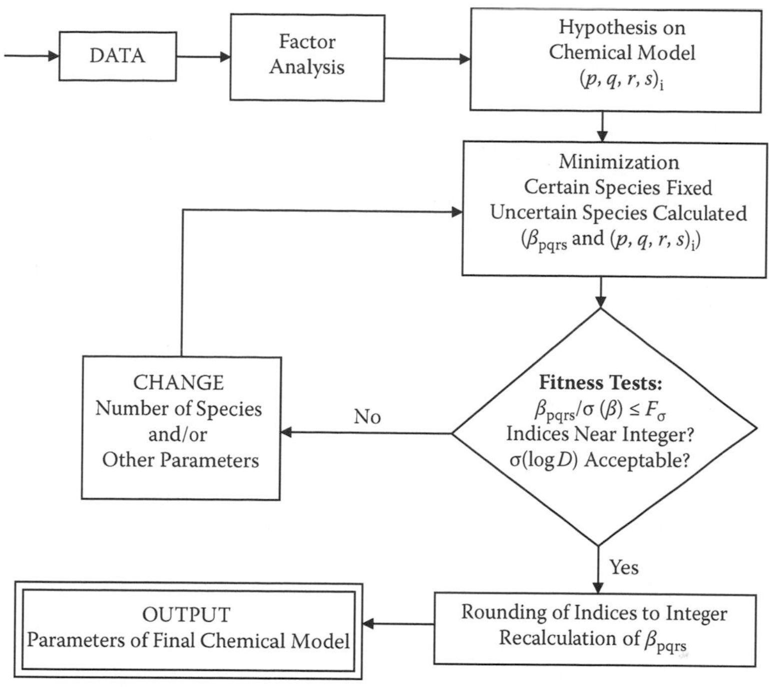

FIGURE 3.6 Novel scheme of chemical model search based on ESI approach.

3.4.2.2 Artificial Neural Networks in Chemical Equilibria and Extraction

ANNs represent one of the most powerful model free methods. It has been shown recently that the stability constants can be calculated even without solving mass balance equations applying a multivariate calibration approach and the ANN method. The procedure is in some way analogous to Sillén's so-called projection maps, described in the 1960s [82,83]. The projection maps method has even been computerized, but the evaluation is done graphically [84].

The use of the ANN approach was demonstrated for the first time on polarographic data [85,86]. It was first applied to the evaluation of equilibria (using ANN and experimental design methods) as a tool in electrochemical data evaluation for fully inert metal complexes [85] and later on for fully dynamic (labile) metal complexes [86]. The general application in chemical equilibria, for the evaluation of potentiometric or nuclear magnetic resonance (NMR) data, for example, was shown recently [87], where it was stressed that the method is general and can also be applied in extraction.

The method is based on a family of computer-generated curves for known values of stability and equilibrium constants. The parameters for the construction of curves are selected according to a suitable experimental design (ED). The family of curves is used in the training stage with appropriate ANN architecture. The second stage of the computation is then the prediction of the parameters (e.g., stability constants, E^0 of the electrode) from experimental data, which is almost instantaneous.

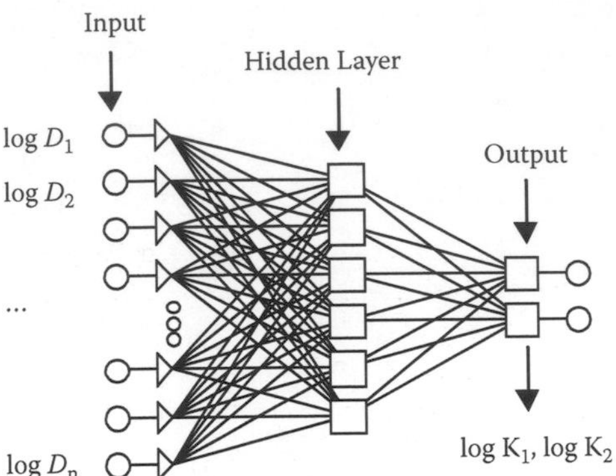

FIGURE 3.7 Scheme of artificial neural network architecture used to evaluate extraction equilibria.

It is demonstrated that the combination of a multivariate calibration done according to ED, and consequently ANN, modeling can be used to estimate equilibria constants from any kind of protonation or metal-ligand equilibrium data like potentiometry, polarography, spectrophotometry, NMR, and extraction [87]. The scheme of the method is given in Figure 3.7.

The method was tested on evenly or randomly distributed experimental error-free and error-corrupted data, and the results show that even rather higher experimental errors do not influence significantly the prediction power and correctness of ANN results. The ED–ANN approach can provide accurate prediction of the stability constants, whereas the computation is very robust.

A comparison with a hard-model evaluation based on nonlinear regression techniques shows excellent agreement. The proposed ANN method is of a general nature and, in principal, can be adopted to any instrumental technique used in equilibria studies including extraction.

3.5 SOME CONCLUSIONS CONCERNING COMPUTATION OF EXTRACTION EQUILIBRIA

General least squares programs like LETAGROP-DISTR and EXLET can be used for trial-and-error search of the chemical model in extraction, and the search can be quite efficient. Complete statistical analysis of residuals might be of some help. However, the experimental errors form the barrier that is not possible to break, and higher level of experimental errors might cause that some models are not distinguishable—that is, it is possible that several models should be considered as an equivalent. The highest data precision is therefore recommended.

Factor analysis, like in spectrophotometry, can help here to get estimate of the number of species involved in extraction without any a priori suggestions. The ESI method leads to multiple minima problem, but it can speed up the search of chemical model or to prove if the model found is correct. Soft modeling methods like PLS or ANN enable fast determination of stability constants without solving mass balance equations. These two approaches of data evaluation seem to be the perspective for computation of all types of equilibrium data and for the extraction data evaluation, as well.

REFERENCES

1. J. Havel, in Computer Applications in Chemistry VIII. Problem of Chemical Model Determination in Equilibrium Study, *Scripta Fac. Sci. Univ. Purk. Brun.* 17, 305–320 (1987).
2. M. Meloun, J. Havel, and E. Högfeldt, *Computation of Solution Equilibria: A Guide to Methods in Potentiometry, Extraction and Spectrophotometry*, Ellis Horwood, Chichester, 1987.
3. F. J. C. Rossotti and H. Rossotti, *The Determination of Stability Constants*, McGraw Hill, New York, 1961.
4. L. Sommer, V. Kubáň, and J. Havel, *Spectrophotometric Studies of the Complexation in Solution,* Folia Fac. Sci. Nat. Univ. Purk. Brun., Tom XI, Chemia, Op. 1, Brno, 1970.
5. F. Gaizer, *Coord. Chem. Rev.* 27, 195 (1979).
6. P. Gans, *Coord. Chem. Rev.* 19, 99 (1976).
7. F. R. Hartley, C. Burgess, and R. M. Alcock, *Solution Equilibria*, Ellis Horwood, Chichester, 1980.
8. W. R. Smith and R. W. Missen, *Chemical Reaction Equilibrium Analysis, Theory and Algorithms*, John Wiley & Sons, New York, 1982.
9. D. J. Leggett (ed.), Computational Methods for the Determination of Stability Constants, Plenum Press, New York, 1985.
10. M. Meloun and J. Havel, *Computation of Solution Equilibria 1: Spectrophotometry*, Folia Fac. Sci. Nat. Univ. Purk. Brun., Tom XXV, Chemia 17, Op. 7, Brno, 1984.
11. M. Meloun and J. Havel, in *Computation of Solution Equilibria 2: Potentiometry*, Folia Fac. Sci. Nat. Univ. Purk. Brun., Tom XXVI, Chemia 19, Op. 9, Brno, 1985.
12. L. G. Sillén and B. Warnqvist, *Arkiv Kemi* 31, 341 (1974).
13. A. Sabatini, A. Vacca, and P. Gans, *Talanta* 21, 53 (1974).
14. W. C. Hamilton, *Statistics in Physical Sciences*, Ronald Press, New York, 1964.
15. J. Havel, in *Coordination Chemistry in Solution*, ed. E. Högfeldt, Kungl. Tek Hög. Hand., Trans Royal Inst. Technol., Nr. 277, 371 (1972).
16. L. P. Varga, W. D. Wakley, L. S. Nicolson, M. L. Madden, and J. Patterson, *Anal. Chem.* 37, 1003 (1965).
17. P. Lubal, J. Perůtka, and J. Havel, *Chem. Papers* 54, no. 5, 289–295 (2000).
18. S. Ainsworth, *J. Phys. Chem.* 65, 1968 (1961).
19. C. Weber, *Nature* 190, 27 (1961).
20. S. Ainsworth, *J. Phys. Chem.* 67, 1613 (1963).
21. J. L. Simmonds, *Opt. Soc. Amer.* 53, 968 (1963).
22. R. M. Wallace, *J. Phys. Chem.* 64, 599 (1964).
23. R. M. Wallace and S. M. Katz, *J. Phys. Chem.* 68, 3890 (1964).
24. D. Katakis, *Anal. Chem.* 37, 876 (1965).
25. G. Wernimont, *Anal. Chem.* 39, 554 (1965).
26. L. P. Varga and F. C. Veatch, *Anal. Chem.* 39, 1101 (1967).
27. J. S. Coleman, L. P. Varga, and S. H. Martin, *Inorg. Chem.* 9, 1015 (1970).
28. J. J. Kankare, *Anal. Chem.* 43, 1322 (1970).
29. Z. Z. Hugus and A. A. El-Awady, *J. Phys. Chem.* 75, 2964 (1971).
30. J. Havel and M. Vrchlabský, Euroanalysis V, Cracow, Poland, August 26–31, 1984, Paper 18-20, 465.

31. J. Havel, M. Vrchlabský, E. Högfeldt, and M. Muhammed, *Chemica Scripta* 1989, 29, 133–137.
32. E. Högfeldt, in *Treatise on Analytical Chemistry*, ed. I. M. Kolthoff and P. J. Elving, Part I, vol. 2, p. 63, Wiley, New York, 1979.
33. D. H. Liem, *Acta Chem. Scand.* 25, 1521 (1971).
34. J. Havel and M. Meloun, in *Computational Methods for the Determination of Formation Constants*, ed. D. J. Leggett, p. 269, Plenum Press, New York, 1985.
35. L. G. Sillén and B. Warnqvist, *Arkiv Kemi* 31, 341 (1969).
36. J. Havel, E. Högfeldt, and B. Warnqvist, *Chemica Scripta* 29, 249 (1989).
37. E. Högfeldt, *Acta Chem. Scand.* 6, 610 (1952).
38. J. Havel and M. Meloun, *Talanta* 32, 171 (1985).
39. E. Högfeldt and F. Fredlund, Trans. Royal Inst. Technol., No. 226 (1964).
40. L. Kuča, E. Högfeldt, and L. G. Sillén, in *Solvent Extraction Chemistry*, ed. D. Dyrssen, J.-O. Liljenzin, and I. Rydberg, p. 454, North-Holland, Amsterdam, 1967.
41. L. Kuča and E. Högfeldt, *Acta Chem. Scand.* 22, 183 (1968).
42. J. M. Madariaga, M. Muhammed, and E. Högfeldt, *Solvent Extraction and Ion Exchange* 4, 1 (1986).
43. D. H. Liem, in *Solvent Extraction Chemistry*, eds. D. Dyrssen, J.-O. Liljenzin, and J. Rydberg, p. 264, North-Holland, Amsterdam, 1967.
44. D. H. Liem, *Acta Chem. Scand.* 22, 773 (1968).
45. G. Scibona, S.Basol, P. R. Danesi, and F. Orlandini, *J. Inorg. Nucl. Chem.* 28, 1441 (1966).
46. J. M. Madariaga and H. Aurrekoetxea, *Chem. Scripta* 22, 90 (1983).
47. E. Högfeldt, in *Developments in Solvent Extraction*, ed. S. Allegret, p. 75, Ellis Horwood, Chichester, 1988.
48. J. Paatero, *Acta Academiae Aboensis* 34, 1 (1974).
49. B. Warnqvist, *Chem. Scripta* 1, 49 (1971).
50. J. Havel, N. Miralles, A. Sastre, and M. Aguilar, *Computer Chem.,* 16, no. 4, 319–323 (1992).
51. M. Martínez, N. Miralles, A. Sastre, M. Lubalová, and J. Havel, *Chem. Papers* 50, no. 3, 109–114 (1996).
52. J. Havel and M. Meloun, *Talanta* 33, 435 (1986). J. Havel and M. Meloun, *Talanta* 33, 525 (1986).
53. F. James and M. Roos, *Comp. Phys. Commun.* 10, 343 (1975).
54. J. Havel, CPLET program manual, Department of Analytical Chemistry, Masaryk university, unpublished results (The manual can be obtained on request).
55. Y. Marcus and A. S. Kertes, Ion Exchange and Solvent Extraction of Metal Complexes, Wiley, New York, 1969.
56. M. Muhammed and R. Arnek, *Chem. Scripta* 8, 187 (1975).
57. M. Aguilar and E. Högfeldt, *Chem. Scripta* 2, 149 (1972).
58. D. H. Liem, *Acta Chem. Scand.* 25, no. 5, 1521–1534 (1971).
59. D.H. Liem and R. Ekelund, *Acta Chem. Scand.* A 33, 481–483 (1979).
60. M. Sanchez, M. Hidalgo, V. Salvado, and M. Valiente, *Solvent Extn. Ion Exchange* 17, no. 3, 455–474 (1999).
61. M. Aguilar, Graphical Treatment of Liquid–Liquid Equilibrium Data, in *Developments In Solvent Extraction*, ed. S. Alegret, Ellis Horwood, Chichester 1988.
62. L. G. Sillén, *Acta Chem. Scand.* 15, 1981 (1961).
63. M. Aguilar and D. H. Liem, *Acta Chem. Scand.* A30, 313 (1976).
64. J. Havel and M. Meloun, *Talanta* 33, 435 (1986).
65. J. Havel and M. Meloun, *Talanta* 33, 525 (1986).
66. M. Bartušek, J. Havel, and D. Matula, *Coll. Czechoslov. Chem. Commun.* 51, 2702 (1986).
67. M. Meloun, M. Javůrek, and J. Havel, *Talanta* 33, 513 (1986).
68. J. Havel and J. L. González, *React. Kinet. Catal. Lett.* 39, no.1, 141–146 (1989).

69. J. Havel, E. Högfeldt, in *Proc. Conf. Chemometrics II,* ed. J. Havel and M. Holík, Brno, September 3–6, 1990, p. 72.
70. J. Havel and E. Högfeldt, *Talanta* 39, 517–522 (1992).
71. J. Havel, M. Aquilar, and M. Valiente, unpublished results.
72. A. Masana and M. Valiente, *Solvent Extn. Ion Exchange* 5, no. 4, 667–685 (1987).
73. M. Aguilar and M. Valiente, J. *Inorg. Nucl. Chem.* 42, no. 3, 405–410 (1980).
74. J. Perůtka, J. Havel, L. Galindo, F. G. Montelongo, and J. J. Arias Leon, *Chem. Papers* 50, no. 4, 162 (1996).
75. J. Gasteiger and J. Zupan, Neural Networks in Chemistry, Angewandte Chemie 32, 503–527 (1993).
77. J. Havel, E. M. Peña-Méndez, A. Rojas-Hernández, J.-P. Doucet, and A. Panaye, *J. Chromatogr.* A 793, 317 (1998).
78. M. Farková, E. M. Peña-Méndez, and J. Havel, *J. Chromatogr.* A 848, 365 (1999).
79. V. Dohnal, M. Farková, and J. Havel, Chirality 11, 616 (1999).
80. J. Havel, J. E. Madden, and P. R. Haddad, *Chromatographia* 49, 481 (1999).
81. J. Havel, M. Breadmore, M. Macka, and P. R. Haddad, *J. Chromatogr.* A 850, 345 (1999).
82. L. G. Sillén, *Acta Chem. Scand.* A 10, 186 (1956).
83. G. Biedermann and L. G. Sillén, *Acta Chem. Scand.* 10, 203 (1956).
84. D. Ferri, O. Wahlberg, and E. Hogfeldt, *Acta Chem. Scand.* 27, 3591–3610 (1973).
85. I. Cukrowski and J. Havel, Electroanalysis 12, no. 18,1481–1492 (2000).
86. I. Cukrowski, M. Farková, and J. Havel, Electroanalysis 13, no. 4 (2001).
87. J. Havel, M. Farková, and P. Lubal, *Polyhedron* 21, 1375–1384 (2002).

4 Hollow Fiber Membrane-Based Separation Technology

Performance and Design Perspectives

Anil Kumar Pabby and Ana María Sastre

CONTENTS

4.1 INTRODUCTION

Membrane separation processes are used in a wide variety of industrial and medical applications for separation of ions, macromolecules, colloids, and cells. The most important advantages of membrane processes are their unique separation capabilities and ability to adapt in diversified conditions including easy scale-up. In some fields membranes are already proven technology and are incorporated in various production lines or purification processes [1–4]. Membrane technology dealing with various applications has generated businesses totaling more than one billion U.S. dollars annually [5]. Systems based on hollow fiber membrane contactors can make chemical plants more compact, more energy efficient, cleaner, and safer by providing a lower equipment size-to-production capacity ratio, by reducing energy requirements, by improving efficiency, and by lessening waste generation, with the correct choice of membrane material. The loss of organic from the membrane (by entrainment) can be reduced to levels lower than those expected in other type of contactors. In addition, traditional stripping, scrubbing, absorption, and liquid–liquid extraction processes can be carried out in this new configuration.

Hollow fiber membrane modules promise more rapid mass transfer than is commonly possible in conventional equipment. For example, mass transferred per equipment volume is about thirty times faster for gas absorption in hollow fibers than in packed towers [6,7]. Liquid extraction is 600 times faster in fibers than in mixer settlers [8–12]. This fast mass transfer in hollow fibers is due to their large surface area per volume, which is typically 100 times bigger than in conventional equipment. In addition, membrane contactors fit the process intensification approach for liquid–liquid extractions, scrubbing, or stripping by permitting independent variation of flow rates without problems of flooding, eliminating postprocess separations. Contactors show potential for membrane crystallizers using membrane distillation to produce crystals from supersaturated solutions, concentrating the solution and removing solvent in the vapor phase. Membrane crystallizers fit the process intensification scheme with a larger mass transfer area enclosed in a smaller volume than conventional crystallizers.

In recent times, knowledge of membrane materials and manufacturing methods has advanced dramatically. Among several types of modules available, hollow fiber and spiral wound types have proven to be the most economical to construct. Some membrane process applications require robust modules to handle high pressures, high temperatures, or severe chemical conditions. More compatibility-based membranes are manufactured for specific applications, and better membrane materials are being formulated to accomplish a wide variety of separations. In addition, there is an incentive to improve the mass transfer efficiency of current membrane module designs that can accommodate increasing membrane performance capabilities. In the early days, hollow fiber (HF) membrane technologies were used with HF contactors consisting of thick HF membranes. These membranes were the source of resistance with the HF contractor and were attributed to lower permeabilitity values. Nowadays, membranes with much lower mass transfer resistance are being developed with better material. This is done to make them able to withstand stringent chemical conditions during membrane processing. In addition, manufacturers have also focused attention on the other mass transfer resistances encountered in parallel flow contactors, particularly shell-side boundary layer and module design consideration. Although we have decided to present performance evaluation and design aspects in this chapter, important applications of this technology taking place in the food industry, beverage industry, gases-separation, organic removal, and hydrometallurgy and others are described in the next section to provide readers the current scenario and status of the technology. This will help us to know why we should need improvement in design, which is again related to throughput and performance.

Extensive studies on hollow fiber membrane based separation technology (HFMST) carried out for efficient removal of toxic heavy metals like Cr(VI), Cd, Zn, Ni, separation and concentration of gold from alkaline hydrometallurgical solution [13–26] were described in [2]. Further, recovery of valuable solutes from aqueous phases—for example, citric acid, carboxylic acid, amino acids, and $_L$-phenylalanine [27,28]—such as removal of phenol from industrial waste water [29] are well demonstrated by using this technology and are described in detail in a later section of this chapter.

Among interesting recent applications are the removal of traces of oxygen (at level of 10 ppb) from water for ultrapure water preparation for the electronic industry [30], the removal of CO_2 from fermentation broth, and the supply of CO_2 as gas-to-liquid phases (e.g., carbonation of soft drinks) [31] have gained paramount importance for large-scale commercialization. Additional examples include the removal of alcohol from wine and beer, the concentration of juice via osmotic or membrane distillation [31], the nitrogenation of beer [32], and the degassing of organic solutions and water ozonation [33]. Since 1993, a bubble-free membrane-based carbonation line has processed about 112 gal/min of beverage by membrane contactors having a total interfacial area of 193 m^2 (Pepsi Bottling Plant, West Virginia) [30].

In a very interesting study of osmotic distillation dealing with juice concentration, Hogan et al. [34] reported a total process cost of osmotic distillation concentration of the order of \$1.00/L of concentrate. From 1 L of fresh juice, it is possible to achieve about 200 mL of 70° Brix concentrate. The value of the concentrate is between \$2.50 and \$7.50/L. For larger-scale and semicommercial studies in osmotic distillation-based fruit juice concentration in a pilot system in Australia [31] containing 224" × 28" (surface area 425 m^2) Liqui-Cel modules (Vineland and Hoechst Celanese) were studied in detail. Commercialization prospects for osmotic distillation-based fruit juice concentration look good; module cleaning after each batch concentration is a concern, which is separately studied by several groups to minimize the membrane fouling to get consistent performance of the HF contactor. From these data, the economical advantages of the integrated membrane process seem evident.

As far as the protein extraction is concerned, this field has received increased attention in recent years. Dahurun and Cussler [35] studied protein extraction in membrane contactors under various experimental conditions. Solutions of cytochrome-c, myoglobin, α-chymotrypsin, catalase, and urease in phosphate buffer were extracted using an immiscible aqueous phase, polyethylene glycol (PEG). Also, membrane-based adsorption processes have been commercialized for the separation and recovery of proteins as a replacement for a packed column of adsorbent beads. Often such processes are called membrane chromatography [36a] or adsorptive membrane chromatography [36b]. Dealing with pharmaceutical applications, Prasad and Sirkar [12] and Basu and Sirkar [37] carried out the purification of mevinolinic acid [37] using an actual commercial stream provided by Merck. Similarly, simultaneous extraction of 4-methylthiazole (MT) and 4-Cyaothiazole (CNT) were performed from an actual process stream supplied by Merck using hollow fiber membrane contactor (12). Using hollow fiber contactors, several industrially important organic acids produced by fermentation have been subjects of membrane contactor studies, including acetic [35,38–43], lactic [44a], succinic [41–43,45], and citric acid [46].

In an important environmental application, pilot-scale evaluation of microporous membrane-based solvent extraction of a wide range of organic contaminants from industrial waste water at two industrial plant sites in the Netherlands was successfully carried out by hollow fiber contactors (Liqui-Cel modules, area 2-3 m^2, flow rate 75 l/h). Process economics appeared quite competitive with other conventional technologies [47]. In a very recent study, p-nitrophenol was extracted from 1-octanol into an aqueous buffered solution using membrane-supported extraction in hollow fiber liquid contactors containing hydrophobic, microporous propylene fibers [48]. Likewise, Aziz et al. [49] studied the biodegradation of trichlroethylene in a microporous polypropylene membrane contactor. They passed contaminated water (containing up to 709 μg/l trichloroethylene) through the tube side of the fibers while circulating *Methylosinus trichosporium* OB3b, a methanotrophic bacterium, through the shell side. Some 78.3 to 99.9% of the trichloroethylene was removed from the fibers at the residence time of 3 to 15 minutes. The authors concluded that the hollow fiber membrane bioreactor is a promising technology for degradation of chlorinated solvents.

Membrane processes for gaseous mixture separation is today technically well consolidated and apt to substitute for traditional techniques [50]. The gas separation business was evaluated in 1996 [51] at $85 million in the United States with growth of about 8% per year. For important applications dealing with gas–liquid contacting mode, blood oxygenation studies were initiated in 1975 by Qi and Cussler using gas-filled hydrophobic hollow fibers [6,52]; and were also pursued by several other researchers in this field. Gas absorption and gas stripping [4] have been successfully commercialized on a large scale using hydrophobic fibers and gas filled pores [53]. Stripping of CO_2 to extend anion bed life, adding CO_2 to carbonate beverages [31], and deoxygenation of water to prevent corrosion [54] are being implemented in commercial hydrophobic microporous membrane-based contacting devices (Celgard Celanese, Charlotte, North Carolina) having continuous flow rate of several thousand liters per minute.

This chapter provides a state-of-the-art review of HFMST including a general review of hollow fiber membrane contactors, operating principles, design consideration, commercial availability of hollow fiber membrane, and module for scale-up and large-scale studies. Application of HFMST in pharmaceutical, biotechnological, gas absorption and stripping, wastewater treatment, and few latest studies of metal ion extraction are described in detail.

4.2 HOLLOW FIBER MEMBRANE-BASED TECHNOLOGY: IMPORTANT COMPONENTS AND DESIGN ASPECTS

4.2.1 Different Types of Contacting Devices and Commercial Availability

The mass transfer operations can be conducted using a number of different configurations, including spiral wound, rotating, annular, plates, pleated sheets, and hollow fiber contactors. Among these contacting devices, hollow fiber contactors have received the maximum attention.

The hollow fiber modules are available from a variety of sources, although some are designed for pressure-driven filtration processes rather than concentration-driven mass transfer. The most well-known modules designed for concentration-driven mass transfer are the Liquid-Cel® Extra flow module (4 in. × 28 in. and 10 in. × 28 in.) offered by CELGARD LLC (Charlotte, North Carolina), shown in Figures 4.1A and 4.1B, respectively. The smallest Liquid-Cel modules are 2.5 inches in diameter and contain 1.4 m^2 of contact area (approximately 10,000 lumens), whereas the largest are 10 inches in diameter and offer 130 m^2 of contact area by virtue of 225,000 fibers. A large module can handle liquid flow rates of several thousand liters per minute [53,55,56]. Modules with epoxy potting and polypropylene casings can withstand most chemicals except chlorinated solvents, ketones, and dimethylformamide (DMF). Two types of microporous hydrophilic hollow fiber are available from AKZO (Asheville, North Carolina): (1) Cuprophan® hollow fiber devices consisting of these hydrophilic fibers and (2) nylon hollow fibers. Hollow fiber devices of these fibers have been prepared and studied by Prasad and Sirkar [43] and Basu et al. [14]. Commercially available, single-ceramic (alumina) (ALCOA, Warrendale, Pennsylvania) membrane tubes with an asymmetric structure and micropouous glass tubes (ASAHI, New York) have also been used for nondispersive extraction [57].

For gas–liquid mass transfer applications, Membrane Corporation (Minneapolis, Minnesota) offers modules that are designed for this specific purpose. These modules fit within standard PVC pipes and contain multiple fiber bundles, each containing around 500 fibers. The fibers are potted into polyurethane at one end only and are individually sealed at the other end so that there is no exhaust stream. That is, all enter gas exits by diffusing across the membrane into the surrounding water, leading to 100% gas transfer efficiency. The packing density is only 10%, and fibers are composed of a 1 μm layer of polyurethane sandwiched between two layers of microporous polyethylene; inside and out diameters are 220 and 270 μm, respectively [55].

For ozonation of semiconductor cleaning water application, W. L. Gore & Associates (Elkton, Maryland) markets a module designed for bubble-free gas–liquid mass transfer for production of ultrapure water for the semiconductor industry. Their DISSO³LVE™ module features expanded polytetraflouroethylene (PTFE) fibers that are compatible with highly corrosive chemicals such as ozone. Among various modules available, the one suitable for mainly semiconductor applications is 10 cm in diameter and 80 cm in length and contains about 100 fibers housed in a PVDF shell. Each fiber has an inside diameter of 1.7 mm, a wall thickness of 0.5 mm, and a pore size of 0.003 μm. The fibers are arranged as a helix—a geometry that

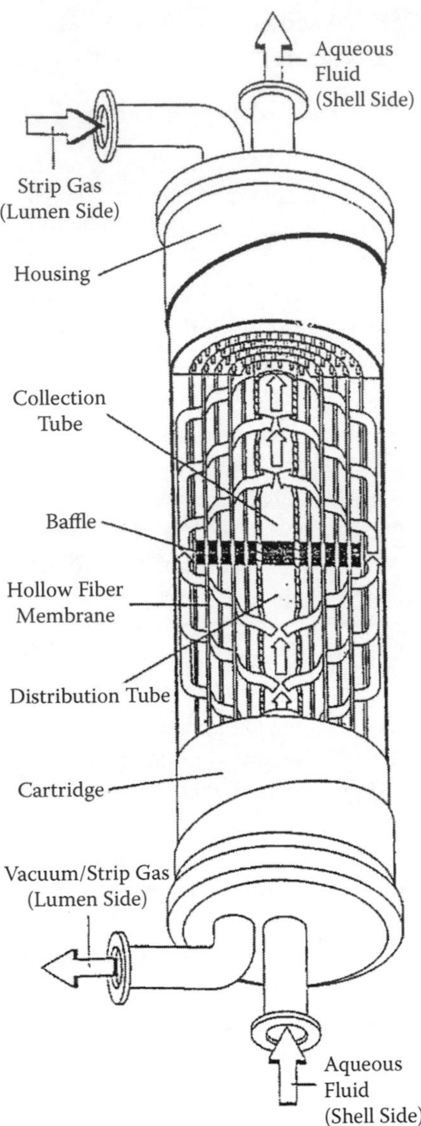

FIGURE 4.1(a) Liquid-Cel extra-flow 4 × 28 membrane contactor. (From Liquid-Cel Contactors, Separations Products Division, Hoechst Celanese Corporation, 1996. With permission.)

offers higher shell-side mass transfer coefficients than one with fibers parallel to the shell. Furthermore, the nature of expanded PTFE allows each fiber to serve as a point of use particle filter, an important advantage in semiconductor manufacturing. Gas and liquid flow rates of 3 and 10 –20 l/min are typical, with the gas on the tube side [55]. Likewise, Pall Corporations (East Hill, New York) offers the Separel TM EFM-530 module for use in ultrapure water applications. The product is made from nonporous polyolefin fibers that are woven into a fabric and wrapped around

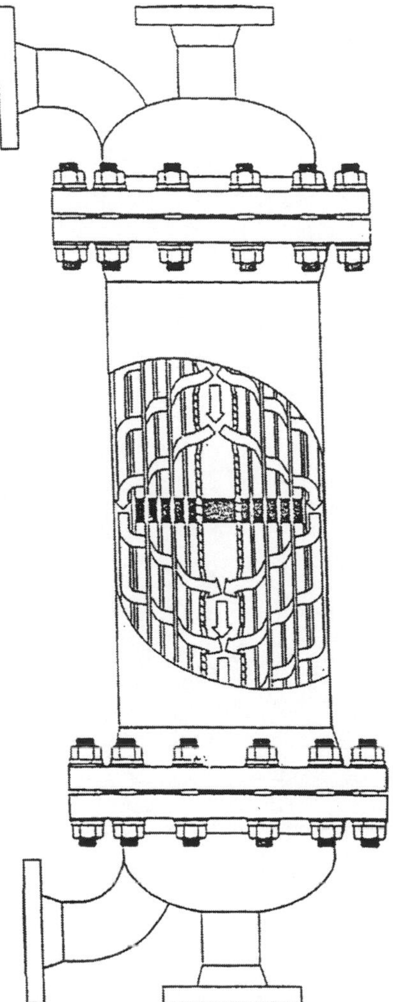

FIGURE 4.1(b) Liquid-Cel extra-flow 10 × 28 membrane contactor. (From Liquid-Cel Contactors, Separations Products Division, Hoechst Celanese Corporation, 1996. With permission.)

a central core. Water introduced through central core flows normal to fibers, which are either maintained under vacuum, flushed with nitrogen sweep gas, or both. The shell is made from a clean type of PVC known as Esloclean™, but PVDF will also be available in the near future. The nonporous membrane sets this product apart from others discussed in this section, which all use microporous membranes. Nonporous membranes give higher selectivity of oxygen and other dissolved gases over water [55]. Commercially available hollow fiber contactors, their source, and characteristics of modules are presented in Table 4.1. Similarly, hollow fiber membranes and their source and characteristics are summarized in Table 4.2.

TABLE 4.1
Characteristics of Commercially Available Hollow Fiber Modules

Fiber Diameter (μm)			Shell Dimensions (cm)				
Inner Diameter	Outer Diameter	Number of Lumina	Diameter	Length	Area per Unit Volume (cm^{-1})	Hollow Fiber Membrane Material	Source/ Supplier
405	464	900	1.9	15.8	46.3	polypropylene	Liqui-Cel (Hoechst)
244	298	7500	4.7	24.1	40.4	polypropylene	Liqui-Cel
405	464	3200	4.7	24.1	26.8	polypropylene	Liqui-Cel
244	298	7500	4.7	24.1	40.4	polypropylene	Liqui-Cel
405	464	3200	4.7	54.6	26.8	polypropylene	Liqui-Cel
240	300	2100	2.5	20.0	A = 0.23 m^2	polypropylene	Liqui-Cel
240	300	10000	8	28	29.3A = 1.4 m^2	polypropylene	Liqui-Cel
240	300	32500	10	71	36.4A = 19.3 m^2	polypropylene	Liqui-Cel
240	300	—	26.4	71	A = 193 m^2	polypropylene	Liqui-Cel
250–3000	—	—	—	17.8–182.9 cm	A = 0.019–69.7 m^2	polysulfone polyacryl-onitrile	Koch Membrane Systems (Wilmington, DE)
200–5500	—	—	—	25–304.9 cm	0.02–25 m^2	Microdyn Technologies (Wuppertal, Germany)	Polypropylene sulfonated polyether sulfone, polyethylene, regenerated cellulose
500–1100	—	—	—	—	0.03–5	Millepore (New Bedford, MA)	polysulfone with polypropylene fiber wrap

Source: From A. Kumar and A. M. Sastre, in *Ion Exchange and Solvent Extraction,* eds. Y. Marcus and A. K. Sengupta, Marcel Dekker, New York, 331–469, 2001. (With permission.)

TABLE 4.2
Characteristics of Commercially Available Hollow Fiber Membranes

Membrane	Pore Size (μm)	Fiber Porosity (%)	Wall Thickness (μm)	Material	Source
Celgard X-10	0.03ID = 100	20	25	polypropylene	Hoechst Celanese
Celgard X-20	0.03ID = 240 or 400	40	25	polypropylene	Hoechst Celanese
Celgard X-30	0.03ID = 240	30	30	polypropylene	Hoechst Celanese
Accurel PP 50/280	0.2ID = 600 ± 90	—	50 ± 10	polypropylene	Enka, Germany
Accurel PP Q3/2	0.2ID = 280	—	200 ± 45	polypropylene	Enka, Germany
Accurel PPS6/2	0.2	60–65	450 ± 70	polypropylene	Enka, Germany
Accurel PPV8/2HF	— ID = 5500 ± 300	—	1550 ± 150	polypropylene	Enka, Germany

Note: ID, inner diameter.

Source: From A. Kumar and A. M. Sastre, in *Ion Exchange and Solvent Extraction,* eds. Y. Marcus and A. K. Sengupta, Marcel Dekker, New York, 331–469, 2001. (With permission.)

4.2.2 Alternate Hollow Fiber Geometries, Hollow Fiber Fabric, Design Consideration, Optimum Fiber Diameter, Flow Direction, and Performance Evaluation

Crowder et al. [58] presented a comprehensive study to focus a new type of transverse flow: a hollow fiber module, spiral-wound prototype module. Authors have claimed that the transverse flow arrangement is likely to be the most attractive because high mass transfer coefficients can be achieved at lower Reynolds numbers, which should correspond to lower feed-pump energy costs. Hoechst-Celanese has commercialized a radial flow hollow fiber device [59] that is similar in concept to the module shown in Figure 4.2a. In this case, however, Celgard® microporous polypropylene fibers are knitted in a fabric using smaller multifilament polyester cross-strands, and the fabric is wound around the central tube. The small, solid cross-strands hold the hollow fibers apart and in place. These modules are suitable for a variety of applications, including blood oxygenation and liquid–liquid extraction fibers. A similar module made with silicone rubber fibers might prove to be a step forward in perevaportion efficiency. Other transverse flow hollow fiber modules have been patented or developed for commercial use. One of the oldest ideas was patented by Strand in 1967 [60a]. As shown in Figure 4.2b, Strand used a fabric in which both wrap and weft were hollow fiber membranes. Liquid flowed through the mat, which was held in a square holder, and onto the next mat. The permeate flowed inside the fibers to exit channels at the periphery. The packing density (membrane area per unit module volume) of these plate-and-frame units would be rather low compared with more typical hollow fiber and even spiral-wound designs. A similar approach is described in a patent by Nichols [60b] in 1989. Nichols's module was made of a series of circular

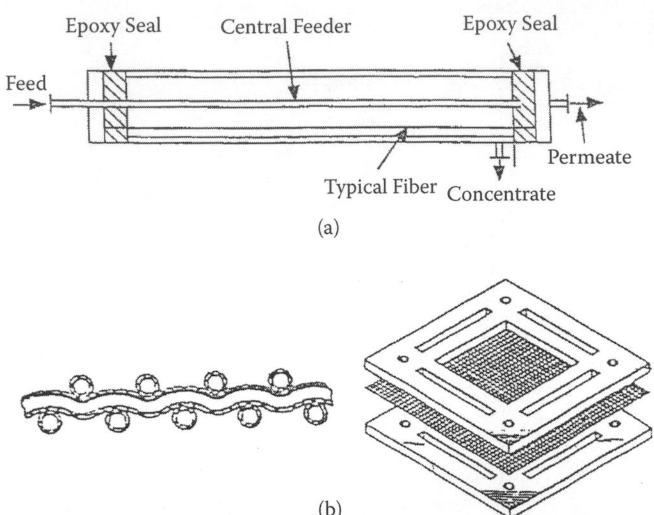

FIGURE 4.2(a)(b) (a) Schematic of a radial flow, hollow fiber membrane module. (b) Essential portion of Strands's patented module. (From M. L. Crowder and C. H. Gooding, *J. Membr. Sci.*, 137, 17, 1997. With permission.)

wafers with each wafer consisting of fibers mounted chord-wise and parallel to one another. Nichols mentions a woven cross-fiber as one means for keeping the hollow fibers parallel. Zenon Environmental of Canada has developed modules similar to the Strand patent for perevaporation [61]. These consist of many square plates, each of which contains hollow fibers mounted parallel to one another. The fibers are transverse to the liquid flow, and alternate plates are rotated 90° so that the fibers in adjoining plates are perpendicular to one another. Preliminary results reported by Zenon are shown in the transverse mass transfer correlation in Figure 4.2c. Zenon has a perevaporation system available for commercial use, but details of the system cost and mass transfer performance under field conditions have yet to be reported. For the small cell tests, 8 × 10 cm fabric sections were woven in the author's laboratory using a handloom with monofilament nylon as the warp (cross-strand) and poly(dimethylsiloxane) (PDMS) hollow fibers as the weft. Seven different fabrics were woven and tested by authors. Table 4.3 summarizes important characteristics of the fabrics. Fabrics 1 through 6 were plain weaves; Fabric 7 was twill. Compared to Fabric 1, Fabrics 2 and 3 tested the effect of reducing the space between the PDMS hollow fibers. Fabric 4 was similar to Fabric 3 in spacing of the hollow fibers but contained only half as many solid cross-strands. Fabric 5 incorporated cross-strands of alternating size, and Fabric 6 used the larger cross-strands throughout. Figure 4.2d is an scanning electron microscope (SEM) photomicrograph of Fabric 6 (see Table 4.3 for details of Fabric 6). A rigorous comparison of alternative fabric designs and of the fabric concept to conventional modules will have to take into account all capital and operating costs for a specific application, but a preliminary evaluation can be made by considering two important characteristics: the mass transfer coefficient and

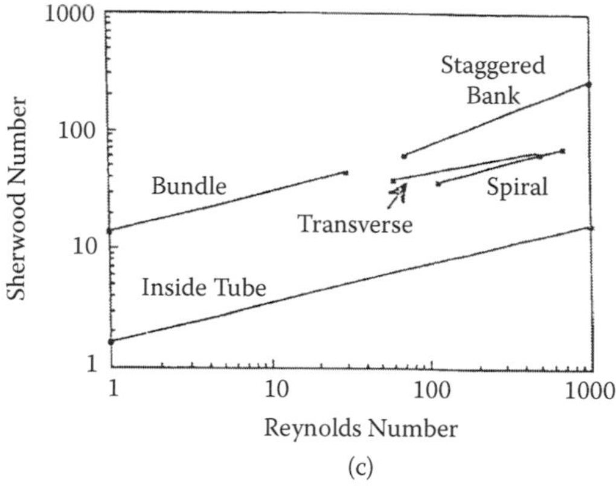

FIGURE 4.2(c) Mass transfer correlations for five module configurations. (From M. L. Crowder and C. H. Gooding, *J. Membr. Sci.*, 137, 17, 1997. With permission.)

(d)

FIGURE 4.2(d) SEM micrograph of Fabric 6 (see Table 4.3 for more details)—top view. (From M. L. Crowder and C. H. Gooding, *J. Membr. Sci.*, 137, 17, 1997. With permission.)

TABLE 4.3
Detailed Fabric Descriptions

Fabric	Fiber Spacing (fibers/cm)	Cross-Strand Diameter (mm)	Cross-Strand Spacing (strands/cm)	Effective Surface (cm2)	Channel Height	Void Fraction
1	3.6[a]	0.33	4.3	18.5	1.80	0.91
2	6.2[a]	0.33	4.2	25.0	1.80	0.86
3	16.5[a]	0.33	4.5	67.9	1.35	0.57
4	15.7	0.33	2.6[a]	67.9	1.35	0.61
5	15.5	0.33, 0.43[a]	2.7	61.1	1.65	0.67
6	16.0	0.43[a]	2.5	60.8	1.67	0.69
7[b]	19.9[a]	0.33	2.6	81.0	1.65	0.59

[a] Silent fabric feature
[b] Fabric 7 was a twill weave. All others were plain weave.
Source: From M. L. Crowder and C. H. Gooding, *J. Membr. Sci.*, 137, 17, 1997. (With permission.)

the pressure drop [62]. Higher mass transfer coefficients reduce capital costs, and lower pressure drops reduce operating costs. A module efficiency defined as $k/(\Delta P/L)$ was calculated from the regression equations for each fabric at a single superficial velocity of 0.16 m/s. The membrane area per module was calculated for a nominal 4-inch module, actually 1 m long by 10 cm in diameter. The results are shown in Table 4.4. This analysis indicates that Fabric 6 is the best design of the fabrics tested in this work. At reasonable pressure drops, mass transfer coefficients for this fabric exceeded 0.01 cm/s, which is 20% higher than the maximum value reported by Hickey and Gooding for spiral-wound spacers evaluated at comparable—and even higher—velocities [63]. The fabric concept competes well with an optimized spiral-wound module in terms of mass transfer coefficient and friction loss, and it

TABLE 4.4
Performance and Packing Density Compositions of Tested Fabrics and Spiral-Wound Modules

Fabric	Descriptions	Module Efficiency [k(cm/s)/ΔP/L (Pa/m] × 10^8	Area/Module (m^2)
1	Well spaced	7	3
2	Spaced	9	5
3	Close packed	1	18
4	Packed	5	17
5	Big and small nylon	4	14
6	Tall channel (packed)	20	14
7	Twill (close packed) Hickey's optimum spiral-wound spacer [11]	120	194

Source: From M. L. Crowder and C. H. Gooding, *J. Membr. Sci.*, 137, 17, 1997. (With permission.)

provides more than three times the membrane area per module. It should be noted that several researchers have successfully used silicone rubber membranes for the perevaportive removal of tricholoro ethane (TCE) from water [63–66]. More care in material selection should eliminate the fouling problem observed in the prototype tests. In fact, some successful experiments involving TCE removal and additional deoxygenation experiments were conducted using Fabric 4 in the small test cell. The test-cell apparatus contained no problem jelly, no metal feed lid, no rubber o-rings, and a chemically resistant pump impeller. Results for these evaluations indicated that k values for both O_2 and TCE increased with velocity as expected. Though the k values for TCE removal were lower than those for O_2 removal, they were not even as low as would be expected considering the difference in diffusivity between the two compounds. The k values for O_2 were lower than those in previous evaluations with the same fabric by about 20%, probably due to the slightly larger channel height used for the latter experiments. From these experiments, authors have claimed that TCE can be removed as efficiently as O_2 using woven fabrics made of silicone rubber hollow fibers.

As described in Section 4.2.1, the hollow fiber contactor Celgard-LLE has been designed to improve the performance of hollow fiber contactors (extra flow contactors), which contain a shell-side baffle, a feature offering two advantages. First, the baffle improves efficiency by minimizing shell-side bypassing; second, it provides a component of velocity normal to the membrane surface, which results in a higher mass transfer coefficient than that achieved with a strictly parallel flow. With regard to the alternate membrane geometries, researchers have continuously performed studies in this direction to check the influence of membrane geometry on mass transfer coefficient and overall performance. For example, Wickramasinghe et al. [67] stripped oxygen from water into nitrogen using four module configurations, each with water flowing outside and across the microporous polypropylene hollow fibers. The four geometries were a cylindrical tube bundle, a helically wound bundle, a rectangular bed of fibers, and crimped flat membrane. The schematic drawing of the modules used in this study are shown in Figure 4.3. As clear from Table 4.5, for fast flow inside the fiber, the author's results are consistent both with the theory of Lévêque and earlier experiments by authors and others. Authors found that this correlation was so well established that the observed consistency seems to be more a justification of the author's experimental procedure than a new verification of his established results. The authors compared performance of the various module geometries studied in terms of oxygen removal efficiency on the basis of equal flow per unit membrane area. The best results were obtained with the rectangular bundle, which stripped 98% of the incoming oxygen. The cylindrical bundle, helical bundle, and crimped flat plate removed 82%, 86%, and 72%, respectively. All of these cross-flow designs were substantially more efficient than a parallel flow cylindrical module, which stripped 7% of the oxygen in the incoming water. Findings were similar when module performance was compared on the basis of equal flow per unit volume. To compare module operated at equal flow per membrane area, authors made a mass balance on the module [68] to find the fraction removed θ.

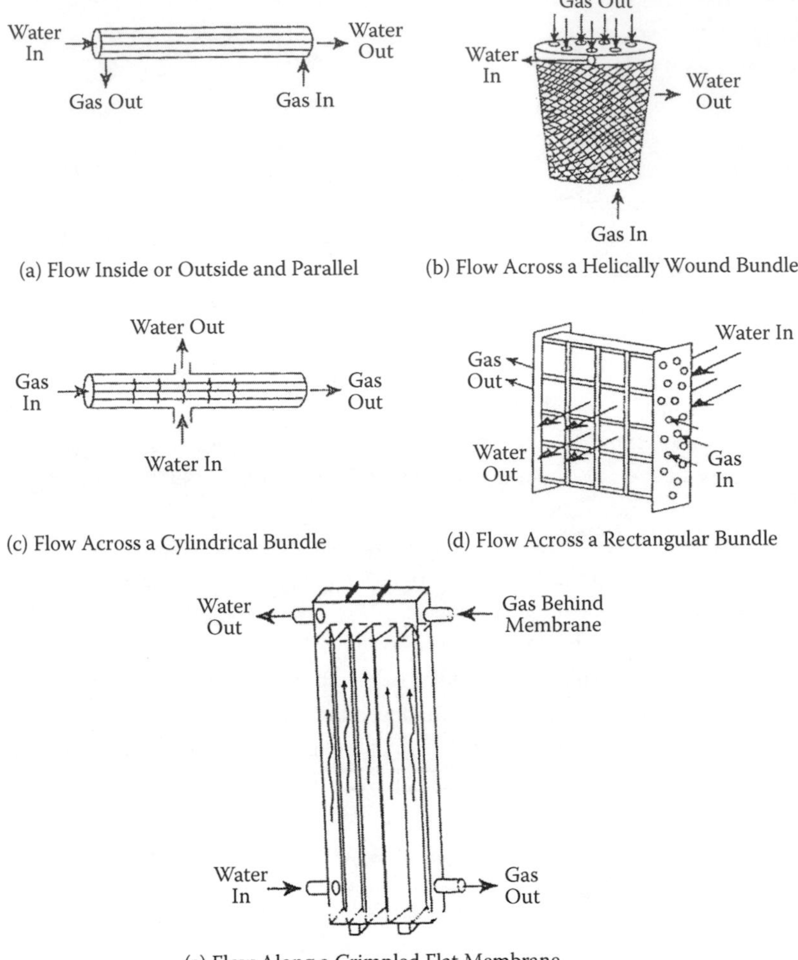

FIGURE 4.3 Schematic drawings of the module used. Three modules have the form in (a); one has the form in each of (b), (c), and (d); and three have the form in (e) (for more details see text). (From S. R. Wickramasinghe, M. J. Semmens, and E. L. Cussler, *J. Membr. Sci.*, 69, 235, 1992. With permission.)

$$\theta = 1 - \frac{\langle c \rangle}{c_o} = 1 - e^{\langle k \rangle A/Q} \tag{4.1}$$

where A is the total area in the membrane module and $\langle c \rangle$, c_0, Q, and $\langle k \rangle$ are average concentration, inlet concentration, average water flow, and average mass transfer coefficient, respectively. These results are the general case of the mass balance given by

TABLE 4.5

Mass Transfer Correlations for Hollow Fiber Modules of Varying Geometry

Flow Geometry	Flow Range	Experimental Result[a]	Inferred Correlation	Literature Correlation
Flow inside fibers	Gr > 4	$\langle k \rangle = 4.3 10^{-5} (\upsilon/l)^{1/3}$	$Sh = 1.62\ Gr^{1/3}$	$Sh = 1.62\ Gr^{1/3}$
	Gr < 4	$\langle k \rangle = 1.5 10^{-4} (\upsilon/l)$	$Sh = Sh_0\left[1 - \left\{\frac{18 Sh_0}{Gr} + 7\right\}\varepsilon_0^2 + \ldots\right]$	—
Flow outside and parallel to fibers[b]	Gr < 60	$\langle k \rangle = 2.5 10^{-5}\ \upsilon$	$Sh = 0.019\ Gr^{1.0}$	$Sh = 1.25\left(\frac{de^2\upsilon}{\nu L}\right)^{0.93}\left(\frac{\nu}{D}\right)^{1/3}$
Flow outside and across fibers	Re > 2.5	$\langle k \rangle = 8.1\ 10^{-4}\ \upsilon^{0.8}$	$Sh = 0.15\ Re^{0.8}\ Sc^{0.33}$	$Sh = 0.39\ Re^{0.59}\ Sc^{0.33}$
	Re < 2.5	$\langle k \rangle = 2.0\ 10^{-3}\ \upsilon^{0}$	$Sh = 0.12_2\ Re^{1.0}\ Sc^{0.33}$	
Flow along a crimped flat membrane[c]	Gr > 11	$\langle k \rangle = 0.0025\ (Q/A)^{0.35}$	$Sh = 6.0\ Gr^{0.35}$	$Sh = 6.4\ Gr^{1/3}$
	Gr > 11	$\langle k \rangle = 3.0\ (Q/A)$	$Sh = 1.25\ Gr^{1.0}$	

Notes: Dimensionless groups are defined for the hollow fiber modules as follows: Sherwood numbers, $Sh = \langle k \rangle d/D$; Graetz number $Gr = d^2\upsilon/Dl$; Reynolds number $Re = d\upsilon/\nu$, Schmidt number $Sc = \nu/D$. Here, d is the fiber diameter. First row: "Flow in side fibers," theory and experimental do not agree. Second row: "Flow outside and parallel to fibers," the correlation obtained here, which agrees with the earlier result, can be written either versus Graetz number or versus Reynolds and Schmidt numbers. Third row: "Flow outside and across fibers," the values are less than those for well-spaced fibers but increase more with increasing velocity. Fourth row: "Flow along a crimped flat membrane," the results at high flow agree closely with the literature correlation, but those at low flow do not.

[a] Units are k:m/s, υ:m/s, l:m, Q: m^3/s, A:m^2.

[b] The characteristic length for this geometry is the equivalent diameter d_e, equal to four times the cross-section for flow divided by the wetted perimeter.

[c] The characteristic length for this geometry is the crimp length b; the Graetz number is defined as b^2Q/DV.

Source: From S. R. Wickramasinghe, M. J. Semmens, and E. L. Cussier, *J. Membr. Sci.*, 69, 235, 1992. (With permission.)

TABLE 4.6
Relative Performance for Different Geometries with Equal Flow per Membrane Area

Flow Geometry	Module Type	Membrane Area (m²)	Bed Length (cm)	Water Flow (cm³/s)	k(10⁻³ cm/s)	Percent Recovered
Inside fibers	Shell and tube (Figure 4.3a)	1.0	18	52	4.0	55
	Shell and tube (Figure 4.3a)	2.3	25	116	4.0	55
	Regular bundle (Figure 4.3d)	3.2	14	160	4.3	57
Outside fibers	Cylindrical bundle (Figure 4.3c)	1.8	10	90	8.5	82
	Helical bundle (Figure 4.3b)	2.0	12	100	9.8	86
	Shell and tube (Figure 4.3a)	2.9	25	146	0.37	7
	Rectangle bundle (Figure 4.3d)	3.7	14	185	19	98
Parallel to flat membrane	Crimped membrane (Figure 4.3e)	0.4	25	20	6.3	72
	Crimped membrane (Figure 4.3e)	3.0	25	150	6.3	72

Notes: The flow per area of 0.005 cm/s is typical of that used in absorption or extraction. All physical properties assume oxygen dissolved in water being transferred across a microporous membrane into rapidly flowing, water-saturated nitrogen.

Source: From S. R. Wickramasinghe, M. J. Semmens, and E. L. Cussier, *J. Membr. Sci.*, 69, 235, 1992. (With permission.)

$$\left(\langle k \rangle = \frac{RQ}{2V} \ln \frac{C_0}{\langle c \rangle}\right)$$

for flow inside the hollow fibers, where the membrane area for fiber volume is $(2/R)$. R and V denote the average fiber radius and average volume of the lumen, respectively. Enhancement in maximizing mass transfer will result in more fraction transferred. Because the flow per area Q/A is constant this means that one wants to maximize the mass transfer coefficient. The results in Table 4.6 suggest that better performance will come from modules operated with flow across deep beds of hollow fibers. Such beds will have a higher pressure drop and hence a higher pumping cost. Based on other work [69], the pumping cost will become important for beds with fiber diameters around 200 μm and a membrane cost of $10 /m²-yr. Results summarized in Table 4.7 have established that cross-flow modules are most effective, followed first by crimped membranes and then by shell-and-tube modules with flow inside the fibers. Shell-and-tube modules with flow outside and parallel to the fibers are least

TABLE 4.7
Relative Performance for Different Geometries with Equal Flow per Membrane Volume

Flow Geometry	Module Type	Module Volume (cm^3)	Module Length (cm)	Water Flow (cm^3/s)	$\langle k \rangle$ a (10^{-3} cm/s)	Percent Removed
Inside fibers	Shell and tube (Figure 4.3a)	62	18	62	0.71	51
	Shell and tube (Figure 4.3a)	140	25	140	0.71	51
	Regular bundle (Figure 4.3d)	180	14	180	0.80	55
Outside fibers	Cylindrical bundle (Figure 4.3c)	150	10	150	1.53	78
	Helical bundle (Figure 4.3b)	150	12	150	1.81	83
	Shell and tube (Figure 4.3d)	165	14	165	4.0	98
	Rectangle bundle (Figure 4.3a)	240	25	240	0.086	7
Parallel to flat membrane	Crimped membrane (Figure 4.3e)	70	25	70	0.57	44
	Crimped membrane (Figure 4.3e)	450	25	450	0.63	47

Notes: The flow per volume of 1.0 s^{-1} is typical of that in membrane oxygenators. All physical properties assume oxygen dissolved in water being transferred across a microporous membrane into rapidly flowing, water-saturated nitrogen.

Source: From S. R. Wickramasinghe, M. J. Semmens, and E. L. Cussier, *J. Membr. Sci.*, 69, 235, 1992. (With permission.)

effective under these conditions. The most effective modules among the cross-flow devices are those with the greatest membrane area per volume a. A large value of a increases the fraction removed; it also implicitly increases the velocity past the fibers and hence the mass transfer coefficient. The authors have taken caution in concluding that some blood oxygenators are better than others solely on the basis of the results in Table 4.7. One has to keep in mind that these results are for oxygen being removed from water, not for oxygen diffusing into blood. It is clear that choice of a blood oxygenator also depends on factors like clinical convenience and blood damage, factors that are not investigated by the authors. Blood damage in particular may be increased by factors like high shear, factors that also increase mass transfer rates. Also, the module performance is independent of membrane properties. For blood rather than water, module performance is more complicated. The partition coefficient H now drops, and the mass transfer coefficient $\langle k \rangle$ can be accelerated by the oxygen-haemoglobin reaction. Authors have suggested that better membrane modules for chemical processing should try to include both local flow across the fibers and countercurrent flow in the module. In a similar approach for gas absorption applications, Jansen et al. [70] proposed a rectangular design with gas flow outside and normal to the fibers. According to the authors, such a design offers a number of advantages, including high mass transfer, low pressure drop, well-defined flow conditions on both sides of the membrane, gas flow unobstructed by fiber potting, and ability to stack modules. A preliminary optimization study suggested that mass transfer was more efficient with an in-line rather than a suggested fiber arrangement.

The best results were obtained when the transversal pitch was twice the outside diameter and the longitudinal pitch divided by the outside diameter was between 1.1 and 3.0—with this optimal geometry for *Re* (Reynolds Number) and *Sc* (Schmidt Number) each between 1 and 1000, with *Re* based on the external fiber diameter. In another study, Yang [71] used hydrophobic hollow fibers for the removal of oxygen or CO_2 from water into nitrogen. The authors obtained the following correlations for closely and loosely packed fibers, respectively:

$$Sh = 1.38\ \mathrm{Re}^{0.34}\ Sc^{0.33}, \tag{4.2}$$

$$Sh = 0.90\ \mathrm{Re}^{0.40}\ Sc^{0.33}, \tag{4.3}$$

As mentioned earlier, baffled contactors were developed to overcome shortcomings of parallel flow such as shell-side bypassing. However, Seibert and Fair [45] found that bypassing does in fact occur with these modules in some cases. These researchers proposed a stage efficiency model that was an excellent predictor of mass transfer coefficients for a shell-side controlled system (octanol/hexanol/water) and a mixed resistance system (butanl/hexano/water) at high shell-side flow rates but a poor predictor at low flow rates. They suggested poor shell-side distribution at low flow rates as a possible explanation for the discrepancy between expected and observed results. According to the authors, such problems are more severe in larger modules; little improvement is obtained by decreasing the fiber density because fibers tend to part and provide channels for bypassing.

Some of the observation and important studies already reported in this section predicted that uneven flow distribution led to a reduction in the mass transfer coefficient. Ho and Sirkar [4] and Cussler [72] pointed out that the inevitable nonuniform fiber spacing in commercial modules results in nonuniform flow and in turn lower mass transfer coefficients than those obtained with hand-built laboratory units, where precise spacing is more easily achieved. Cussler [72] suggested the use of fibers woven into a fabric to obtain more uniform spacing and presented data showing that mass transfer coefficients obtained using this approach were nearly as high as those with hand-built contactors. In another study, Wickramasighe et al. [73] stripped oxygen from water using microporous polypropylene hollow fibers arranged in one of the four configurations; in all cases the gas flowed to the tube side. The membrane modules used in this work were constructed by the authors and not manufactured commercially. The first configuration (Figure 4.4a) was an annular bed of hollow fiber wound helically around a central core; liquid entering the central core was forced radially outward by a plug so that the flow was perpendicular to the fibers. The second configuration was similar to the first except that the shell contained plugs and o-rings to provide multiple shell-side passes. The third configuration (Figure 4.4b) was similar to the second except that it was made with a knitted hollow fiber fabric rather than with individual fibers. Finally, in the fourth configuration (Figure 4.4c), hollow fiber fabric was mounted diagonally in an open-ended rectangular box, and the liquid entered through a tubular manifold. The whole box was towed through the liquid, forcing the liquid to flow both through and around the box. The authors tested

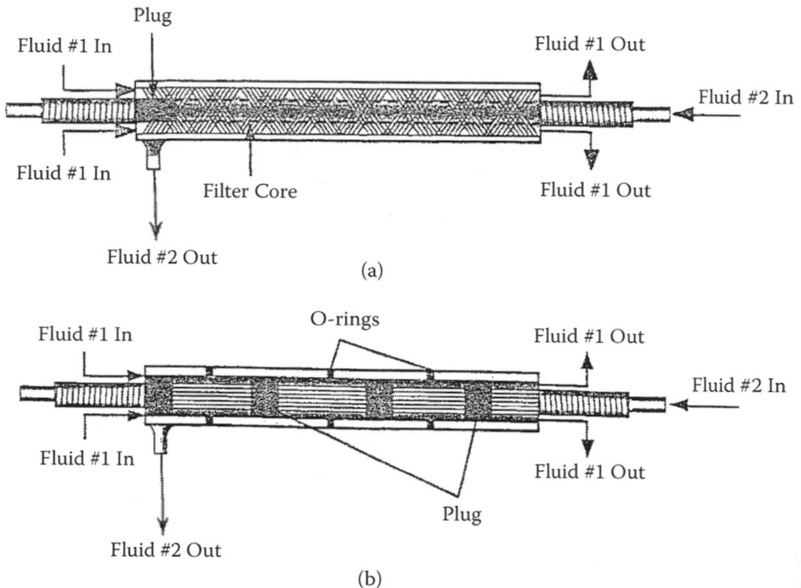

FIGURE 4.4(a)(b) (a) A module containing a hollow fiber wound helically around a central core. Liquid entering the core is forced radially outward by the plug so that the flow is perpendicular to the fibers. (b) A module containing woven hollow fiber fabric wound helically around a central core. The plugs and O-rings provide multiple small passes. (From S. R. Wickramasinghe, M. J. Semmens, and E. L. Cussler, *J. Membr. Sci.*, 84, 1, 1993. With permission.)

the hypothesis that hollow fiber module performance requires better spacing of the hollow fibers. To test this hypothesis, mass transfer coefficients in four different modules were compared. The mass transfer coefficients for axially wound modules are shown as function of velocity in Figure 4.4d. The coefficients are reported as Sherwood number (kd/D), and velocities are given as Reynolds numbers ($d\upsilon/\nu$). The data from two modules are indistinguishable. Authors have claimed that the mass transfer coefficient is not significantly changed by the baffles. Also, commercially

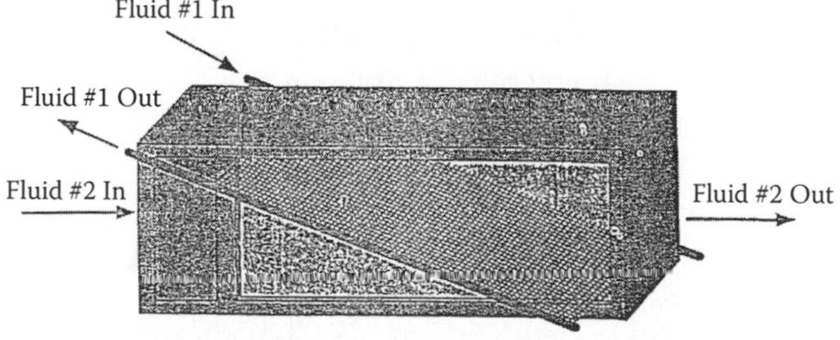

FIGURE 4.4(c) A module containing a woven hollow fiber fabric mounted diagonally in an open-ended box. (From S. R. Wickramasinghe, M. J. Semmens, and E. L. Cussler, *J. Membr. Sci.*, 84, 1, 1993. With permission.)

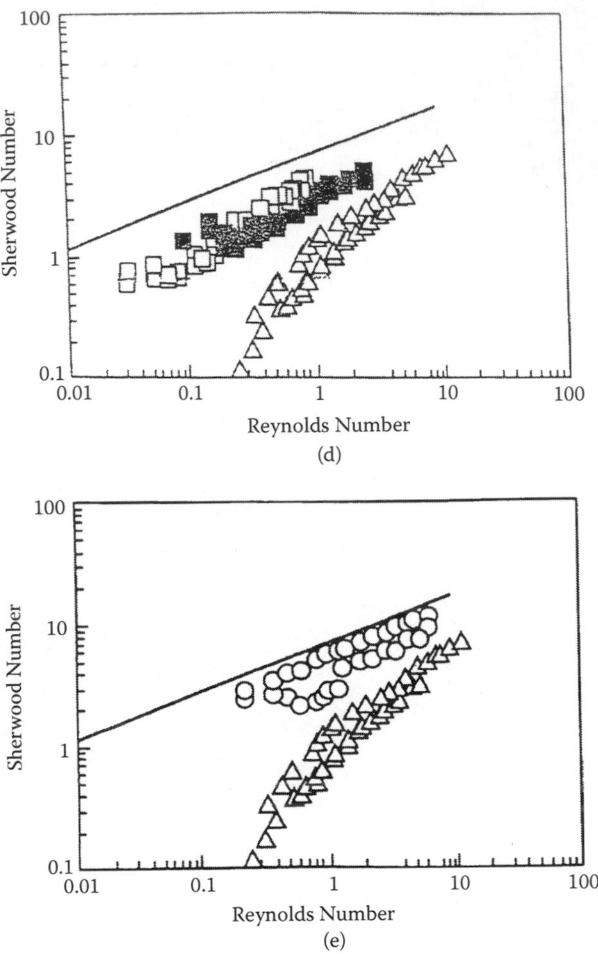

FIGURE 4.4(d)(e) (d) Mass transfer coefficients for axially wound modules. The unbaffled modules (open squares) are similar to those for baffled modules (filled squares). These coefficients are greater than those for commercial modules (triangles) but less than those of modules built with one fiber at time (the solid line). (e) Mass transfer coefficients for hollow fiber fabric. These coefficients, shown as open circles, are above those of commercial modules, shown as triangles, and slightly below those of modules built one fiber at a time, shown as a solid line. (From S. R. Wickramasinghe, M. J. Semmens, and E. L. Cussler, *J. Membr. Sci.*, 84, 1, 1993. With permission.)

assembling modules find difficulty in maintaining uniform fiber spacing as Reynolds numbers obtained with modules having one fiber were slightly different than commercially available modules. Figure 4.4e depicts mass transfer coefficients for a hollow fiber fabric module. At low flows, the data in Figure 4.4e mean that a fabric-based module oxygenator can achieve the same oxygen transfer as a commercial unit that has ten times more active membrane. These data present a strong support for building modules with fabric. The fabric based mass transfer coefficients fall slightly

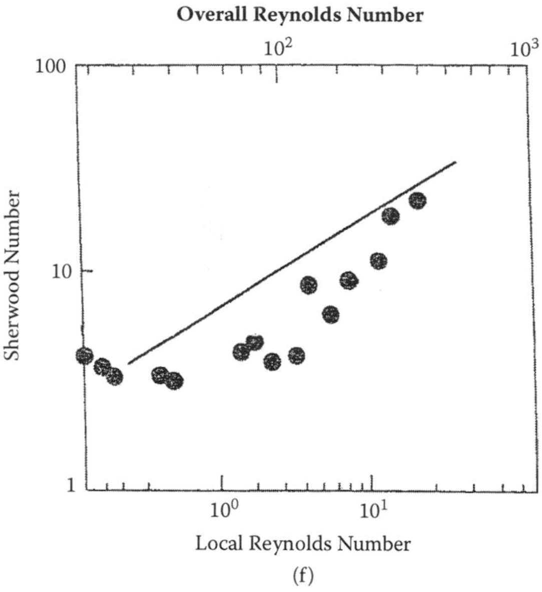

FIGURE 4.4(f) Mass transfer coefficients for a vane module. The coefficients for vane module can be correlated in terms of overall Reynolds number based on the incident velocity υ_0 or in terms of an estimated local velocity in the fiber bed υ_1'. The latter choice gives rough agreement with the correlation for modules built one fiber at a time (the solid line). (From S. R. Wickramasinghe, M. J. Semmens, and E. L. Cussler, *J. Membr. Sci.,* 84, 1, 1993. With permission.)

below those for modules built with one fiber at a time and slightly above those of the axially wound modules. Higher and higher mass transfer coefficient were achieved by using models that carefully space the hollow fibers. Last, mass transfer coefficients for the vane module are plotted in Figure 4.4f. The mass transfer coefficients obtained for the hard-built modules are summarized by the solid lines in Figure 4.4f. This comparison is much less exact than those in Figure 4.4d and 4.4e because of the significant approximation in using related equations to calculate the velocity and hence the exact Reynolds number around the fiber. The mass transfer correlations inferred from the various modules are summarized in Table 4.8. The first column in this table describes the module; the second gives the actual variation measured, and third gives the range of flows used in this measurement. The fourth column gives the correlation inferred, where authors have paralleled other studies of mass transfer in assuming a cube root dependence on the Schmidt number. Similarly, Wang and Cussler [74] also compared gas stripping with hollow fiber contactors containing a woven fabric of microporous polypropylene fibers to modules built from individual fibers. The three module configurations they used are shown in Figures 4.5a, 4.5b, and 4.5c. The first, Figure 4.5a, was conventional parallel fibers; liquid and gas were placed on the tube and shell side, respectively. The second and the third configurations (Figures 4.5b and 4.5c) used the woven fabric instead of individual fibers, with shell-side fluid flow primarily across rather than parallel to the fibers. The module depicted in Figure 5b was rectangular with two baffles; liquid and gas flowed

TABLE 4.8
Correlations of Mass Transfer Coefficients in High-Performance Modules

Module Geometry	Measured Correlations[a]	Flow Range[b]	Inferred Correlations[b,c]	Remarks
Module built one fiber at a time[d]	—	$10^{-2} < \mathrm{Re} < 10$	$Sh = 0.90\ \mathrm{Re}^{0.40}\ Sc^{0.33}$	This result is very close to the heat transfer correlation for a single tube.
Axially wound (Figure 4a)	$k = 4.5\ 1.0^{-3}\ \upsilon^{0.53}$	$0.03 < \mathrm{Re} < 3$	$Sh = 0.49\ \mathrm{Re}^{0.53}\ Sc^{0.33}$	These tubes were assembled by hand, slowly winding six fibers at a time.
Fabric (Figure 4b)	$k = 7.0\ 1.0^{-3}\ \upsilon^{0.49}$	$0.1 < \mathrm{Re} < 10$	$Sh = 0.82\ \mathrm{Re}^{0.49}\ Sc^{0.33}$	These modules are most easily built.
Vane (Figure 4c)	$k = 4.4\ 1.0^{-3}\ \upsilon_1'^{0.46}$	$0.1 < \mathrm{Re} < 10$[d]	$Sh = 0.8\ \mathrm{Re}^{0.46}\ Sc^{0.33}$	The velocity used in the Reynolds number is that estimated from equation presented in footnotes.[e]

[a] Units are cm/s for both k and υ.

[b] The Reynolds number shown is based on an outer fiber diameter of 0.03 cm, a kinematic viscosity for water of 0.01 cm2/s, and superficial velocity across the fibers.

[c] The Schmidt number for oxygen transfer in water is taken as 480, implying a diffusion coefficient of 2.1 10^{-5} cm^2/s.

[d] These values are based on the local Reynolds number ($d\upsilon_1'/\nu$).

[e]

$$\upsilon_1 = \left(\frac{S_1'}{S_1}\right)\upsilon_1$$

$$= \left[\frac{fd^2}{300\nu l}\frac{\varepsilon^3}{(1-\varepsilon)^2}\frac{S_0}{(S_0 - S_1)^2}\frac{S_1'}{S_1}\right]\upsilon_0^2$$

where υ_1, υ_1', υ_0, S_1, S_1', S_0, d, f,l and ν are denoted by velocity within-vane module, local velocity across vane module, flume velocity, flume cross-section, fabric cross-section, flume cross-section, friction factor outside the vane module, outer fiber diameter, length of fiber bed in direction flow, and kinematic viscosity.

Source: From S. R. Wickramasinghe, M. J. Semmens, and E. L. Cussler, *J. Membr. Sci.,* 84, 1, 1993. (With permission.)

through the shell and fibers, respectively. The third configuration (Figure 4.5c) was a fully baffled cylindrical module; again liquid and gas were placed on the shell and tube sides.

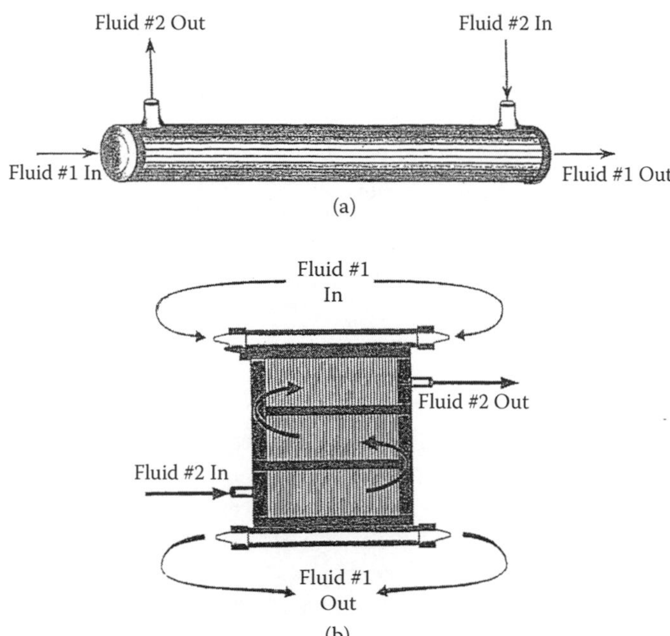

FIGURE 4.5(a)(b) (a) A parallel flow hollow fiber module. (b) A rectangular module containing two baffles. (From K. L. Wang and E. L. Cussler, *J. Membr. Sci.,* 85, 265, 1993. With permission.)

In mass transfer with membrane contactors the design objective depends on the application. In addition, for industrial applications the objective is clearly to minimize the cost per amount of mass transferred, which does not necessarily coincide with the maximum solute transferred per unit volume [69]. In a key design specification, module length is an important aspect because it is directly related to the number of transfer units *(L =* HTU × *NTU).* However, the higher efficiency offered by longer modules comes at the cost of increased pressure drop. If the module is too long, the pressure of the nonwetting fluid may fall below that of wetting fluid, allowing bulk flow of the wetting into the nonwetting fluid. There may be some recourse in increasing the inlet pressure, but pressure is limited by breakthrough pressure. Moreover, other important features must be considered to obtain the required capacity and number of transfer units while meeting these pressure constraints: tube diameter,

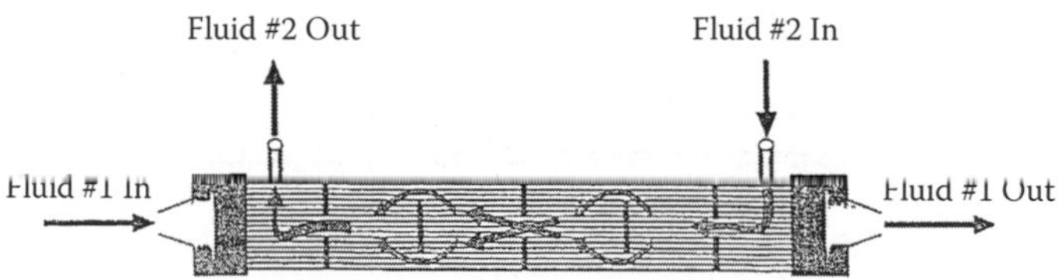

FIGURE 4.5(c) A fully baffled cylindrical module. (From K. L. Wang and E. L. Cussler, *J. Membr. Sci.,* 85, 265, 1993. With permission.)

wall thickness, porosity, tortuocity, packing factor, flow rates, inlet concentration, distribution coefficient, and fluid physical properties. Prasad and Sirkar [75] suggested a design procedure that considers these factors and give us the required number of fibers, module diameter, and length for a given design problem. For cases where single module designs are not practical, Prasad and Sirkar [4] suggested the use of series and parallel cascades. Reed et al. [76] gave several design examples.

With regard to fiber diameter, it is important to mention that this parameter plays an important role in cost estimation. For example, Wickramasighe et al. [69] looked at the cost of stripping gas from water into nitrogen using a hollow fiber contactor with the water flowing through the tubes. The cost was taken as the sum of the membrane and pumping costs; the amount of mass transferred was calculated using shell- and tube-side mass transfer obtained by standard equations. On examination of results, in nearly all cases, there was an optimum fiber diameter that corresponded to a minimum cost. That is, pumping cost decreased with increasing fiber diameter, but membrane cost increased due to the increase in membrane area. This means that pumping costs dominated at small diameters whereas membrane costs dominated at large ones, hence existence of an optimum. The cost varied as d^{-4} at low diameters and linearly with d at large diameters. Furthermore, the optimum fiber diameters increased with increasing length; this was explained by the increase in pumping cost resulting from the increase in length, which in turn required larger diameter before the membrane cost began to dominate. It is interesting that in most cases the optimum diameter was around a few hundred microns, which coincides with the diameters of commercially available fibers.

To examine flow direction in HF contactors, Wang and Cussler [74] explained that conventional parallel flow modules offer true countercurrent flow and are preferred when the membrane- or the tube-side boundary-layer resistance controls. However, with these modules mass transfer coefficients can be reduced, or flows can become uneven if the shell-side resistance is significant; in this case a cross-flow module is preferred. Flow normal rather than parallel to the fibers leads to the higher mass transfer coefficient, but the price is loss of efficiency compared to countercurrent designs. Some efficiency is regained by the use of baffled modules, which provide elements of both countercurrent and cross-flow; efficiency increases as the number of baffles increases, but pressure drop increases as well, and the modules become more difficult to build. Also, the same authors looked at the effect of the number of baffles on mass transfer performance. They measured the fraction of toluene stripped from water into nitrogen using a rectangular module (Figure 4.5b) equipped with one, two, or four baffles, operated in either cocurrrent or countercurrent mode with the liquid on the tube side. Toluene removal was higher for countercurrent flow as expected. Interestingly, oscillation in the fraction of toluene removed versus number of baffles was observed for cocurrent but not countercurrent flow. The authors explained that in cocurrent flow, exit liquid contacts less saturated gas if there is an even number of baffles and more saturated gas if there is an odd number. Consequently, more toluene is removed with an even number and less with an odd number—hence the oscillation. No oscillations were observed with countercurrent flow since exit liquid always contacts the entering gas. In a similar approach, the effect of baffling on oxygen stripping from water into nitrogen using cylindrical modules was assessed with a predictive model. The highest oxygen removal was achieved using countercurrent flow with five baffles, although performance with two baffles was almost as good. Cross-flow with no baffles was superior to cocurrent flow with two.

In a recent publication by Gawronski et al. [77], the effect of liquid flow rate, membrane type, and packing density of modules on the extraction kinetics have been studied by using several commercial modules for the extraction of ethanol from aqueous solution into *n*-octanol. In a case of laminar flow through the fiber lumen, the mass transfer rate can be expressed by the Graetz solution of mass transfer problem for developed velocity profile and developing concentration distribution [78]. The different correlations (inside fiber) are described in [77]. Similarly, more comprehensive coverage to this subject was given by Gableman et al. [55] and Pabby et al. [2]. More attention needs to be given for the other configurations dealing with mass transfer on the shell side. Gawronski et al. [77] discussed mass transfer on the shell side more elaborately with the support of experimental data. The random distribution of the fibers makes it very difficult to develop a mathematical model for the external mass transfer in a module. The existing theoretical descriptions of mass transfer on the shell side are based on the simplified models that assume uniform configuration of fibers. Nevertheless, some conclusions can be drawn from them [79,80].

- The shell-side Sherwood number strongly depends on the fiber density. There is an optimal value of diameter to pitch ratio. In loosely packed modules the distance that the solute molecules must travel is too long. On the other hand, in tightly packed modules the disadvantageous fiber-to-fiber interaction takes place.Even in uniformly packed modules wide residence time distribution of fluid elements does exist.
- The fiber Sherwood number is independent of the module packing density.
- Mass transfer is influenced by the inlet ports design and module geometry.

The literature correlations are given in [2,55,77]. Authors have carried out few experiments to test important factors associated with mass transfer correlations. For this purpose, one chemical system was selected dealing with extraction of ethanol into *n*-octanol from aqueous solution, simulating a fermentation broth in hollow fiber contactor. This extraction was studied in several hollow fiber modules with different surface areas, packing densities, and membrane materials. The experiments were carried out with the commercial modules produced by Chemical Fibers Institute, Poland (Module 3), DIDECO (s.p.a., Italy, Module 1), and EURO-SEP Ltd., Poland (Modules 2, 4, and 6). The modules did not include any baffles. To facilitate the uniform flow distribution the inlet sections of the modules had a bigger diameter than the rest of the module. Additionally, a specially profiled ring was placed near the inlet ports to reduce the flow maldistribution. The detailed description of the membrane modules employed in experiments is given in Table 4.9. The experimental set-up is shown in Figure 4.6a. The aqueous ethanol solution at concentration of 80 to 110 g/dm^3 was fed outside of the fibers. The n-octanol extractant was introduced to the hollow fibers at a given flow rate in a countercurrent and once-through mode. Figure 4.6b shows a comparison of the mass transfer coefficient Kw in the modules

TABLE 4.9
Data of the Studied Hollow Fiber Modules

Module/Fabric Number[a]	1	2	3	4	5	6
Material	Cellulose acetate 2 1/2 CA	Cellulose acetate 2 1/2 CA	Polysulfone	Polypropylene PILX	Polypropylene PILX	Polypropylene K600
Pore size	1000 kDa	1000 kDa	40 kDa	0.2 μm	0.2 μm	0.2 μm
Fiber inside diameter, d_i (mm)	0.35	0.35	1.00	0.33	0.33	0.60
Wall thickness, t (mm)	0.15	0.15	0.16	0.15	0.15	0.20
Fiber length, L (mm)	200	200	500	200	200	200
Module diameter, D (mm)	44	40	44	40	40	40
Number of fibers, N	3600	2730	500	1411	1814	770
Specific surface area, a (m^2/m^3)	2603	2389	1003	1164	1497	1155
Packing density, Φ	0.79	0.72	0.45	0.35	0.45	0.48

[a] More details of Module/Fabric are given in Table 4.4.

Source: From R. Gawronski and B. Wrzesinska, *J. Membr. Sci.*, 168, 213, 2000. (With permission.)

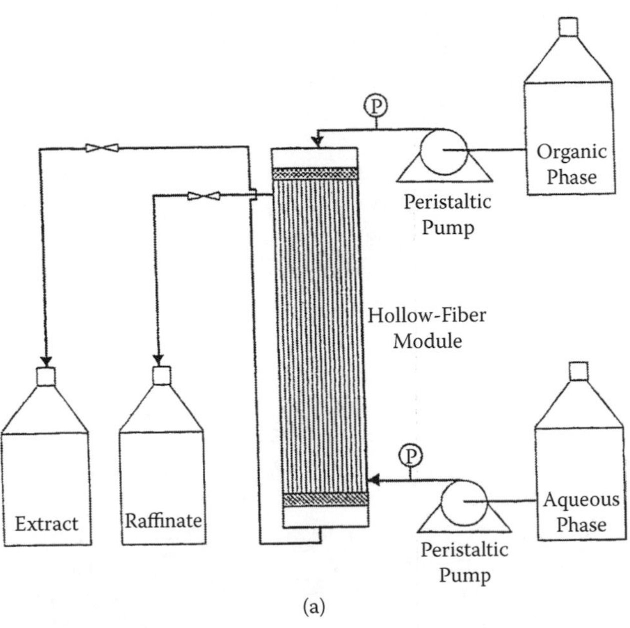

FIGURE 4.6(a) Experimental set-up for ethanol extraction using hollow fiber contactors. (From R. Gawronski and B. Wrzesinska, *J. Membr. Sci.*, 168, 213, 2000. With permission.)

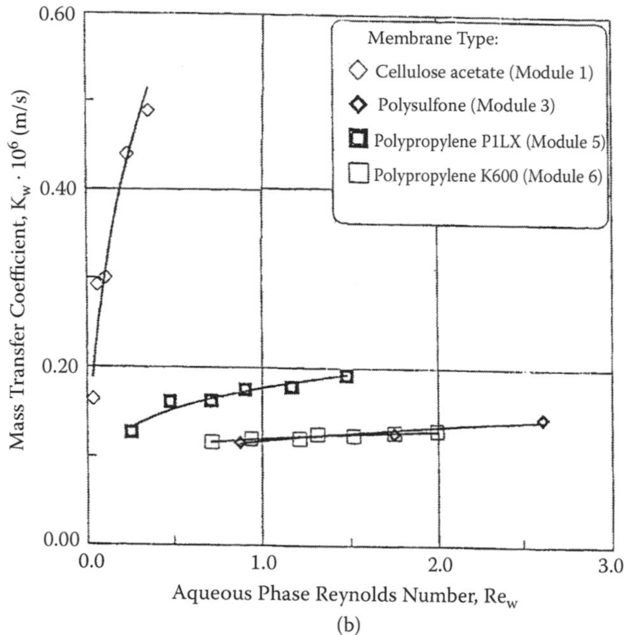

FIGURE 4.6(b) Performance of various membranes, $\upsilon_o = 3.6\ 10^{-3}$ m/s. (From R. Gawronski and B. Wrzesinska, *J. Membr. Sci.,* 168, 213, 2000. With permission.)

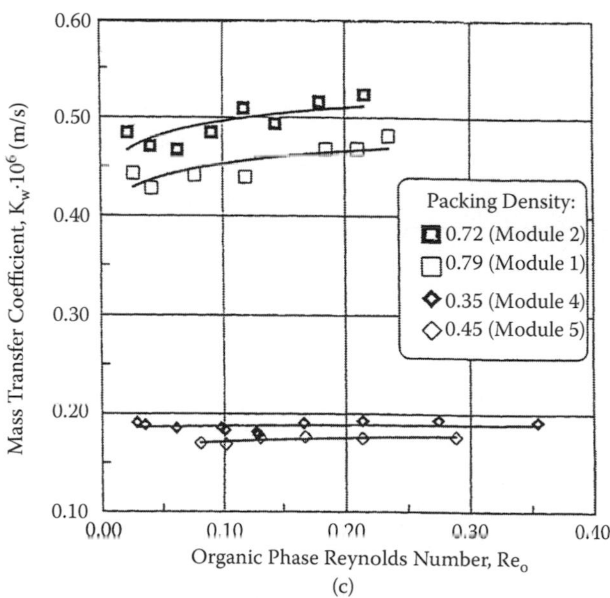

FIGURE 4.6(c) Effect of packing density on mass transfer rate, $\upsilon_w = 1.8\ 10^{-3}$ m/s. (From R. Gawronski and B. Wrzesinska, *J. Membr. Sci.,* 168, 213, 2000. With permission.)

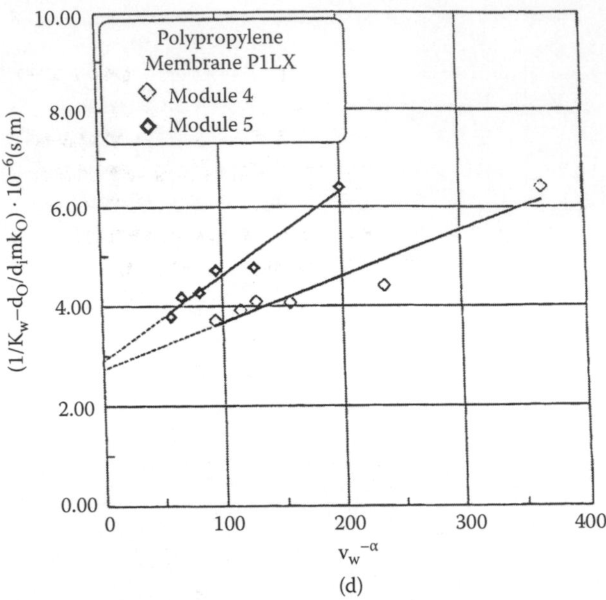

FIGURE 4.6(d) Wilson plot for Modules 4 and 5 (see Table 4.9 for more details) with polypropylene membrane (PILX). (From R. Gawronski and B. Wrzesinska, *J. Membr. Sci.*, 168, 213, 2000. With permission.)

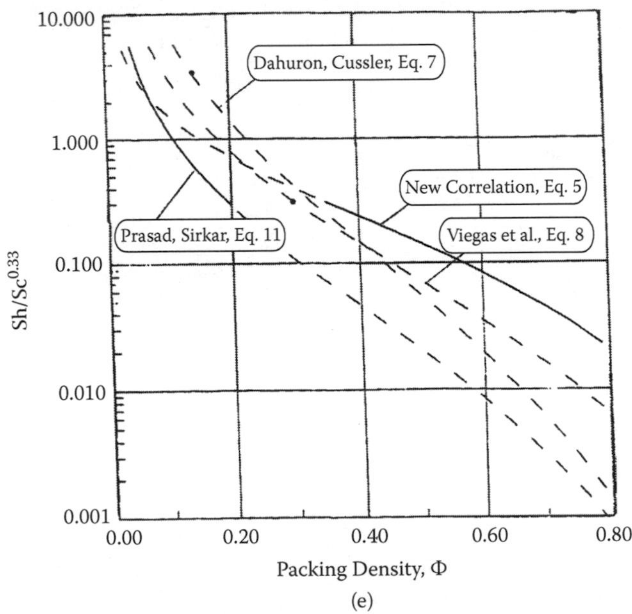

FIGURE 4.6(e) Comparisons of shell-side mass transfer correlations; solid line: verified range of packing densities, dashed line: extrapolated results. (From R. Gawronski and B. Wrzesinska, *J. Membr. Sci.*, 168, 213, 2000. With permission.)

made of different hollow fiber membranes. Generally, it was observed that for the hydrophilic, cellulose acetate membrane the *Kw* are twenty-four times higher than those for the hydrophobic membranes. Since the pores of hydrophilic membranes are filled with the aqueous solution, in which the diffusion of ethanol is faster than in the organic phase, the mass transfer coefficient in the hydrophilic module is also higher. The effect of packing density on the performance of the membrane extraction unit is shown in Figure 4.6c. The data indicate that the increase of the packing density results in decrease in the mass transfer rate. The small fiber-to-fiber distance in the tightly packed modules creates interactions between the neighboring fibers. Additionally, irregular distribution of the fibers generates a nonuniform liquid flow outside the fibers. In the section where the membranes almost touch each other the fluid flow is restricted and the active mass transfer area decreases. On the contrary, channeling occurs in the zone of loosely packed fibers. The presence of shell-side bypassing was verified visually by the authors using a dye tracer technique. To estimate the mass transfer coefficients in liquid filling the membrane pores, *km,* the Wilson plot method [81] was applied (Figure 4.6d). The sum of the mass transfer resistances in the membrane and on the shell side was plotted against the (υ_w^{-a}). The values of the exponent (-α) were determined for each module to obtain a linear function. The mass transfer coefficient in membrane km was calculated from the intersection of the straight line with the vertical axis. The intercept corresponds to an infinite aqueous phase rate and thus the negligible resistance of aqueous phase. The membrane resistances and *km* coefficients of these experiments indicated that the hydrophilic membrane made of cellulose acetate was the best one. For practical applications the following empirical correlation was used in which the Sherwood number is a function of the Reynolds and Schmidt numbers:

$$Sh = A\ \mathrm{Re}^{\alpha}\ Sc^{\beta} \tag{4.4}$$

where A, α, and β are the constants determined experimentally. The authors have realized that the values of the coefficient *A* and the exponent α obtained for different modules could be described by the linear functions of packing density Φ. As a result the following dimensionless equation was developed using the data for all investigated modules:

$$Sh = 0.09\ (1 - \Phi)\ \mathrm{Re}^{(0.08-1.16\Phi)}\ Sc^{0.33} \tag{4.5}$$

The Sherwood and Reynolds numbers were calculated using the hydraulic diameter and the average linear velocity of liquid outside the fibers. The comparison of Equation (4.5) with the other shell-side correlations (Table 4.10) developed for extraction systems is shown in Figure 4.6e. It was observed by the authors that for tightly packed modules the mass transfer rate is much higher than predicted by known equations. In the range of low packing densities the results calculated on the basis of Equation (4.5) are in good agreement with the literature data.

Design aspects described by Crowder et al. [58] have already been described in this chapter. Several correlations were compared by the author for already existing geometeries, and among them five correlations obtained from prior work are sum-

TABLE 4.10
Literature Shell-Side Mass Transfer Correlations

Authors [Reference]	Remarks	Conditions	Correlation	Equation Number
Prasad and Sirkar [43]	Results for solvent extraction A = 5.85 hydrophilic membrane A = 6.1 hydrophobic membrane	$\Phi < 0.20$ Re < 500 300 < Sh < 1000	$Sh = A(1-\Phi)\left(\frac{d_h}{L}\right)\mathrm{Re}^{0.6}\,Sc^{0.33}$	(4.6)
Dahuron and Cussler [35]	Results for solvent extraction, for superficial velocity	$\Phi = 0.15$	$Sh = 8.8\left(\frac{d_h}{L}\right)\mathrm{Re}\,Sc^{0.33}$	(4.7)
Viegas et al. [44b]	Results for solvent extraction	$\Phi = 0.30$	$Sh = 8.71\left(\frac{d_h}{L}\right)\mathrm{Re}^{0.74}\,Sc^{1/3}$	(4.8)

Source: From R. Gawronski and B. Wrzesinska, *J. Membr. Sci.*, 168, 213, 2000. (With permission.)

marized in Figure 4.2c. To plot each line, the Schmidt number was assumed to be 1,000, a reasonable value for dilute, low molecular weight organics in water. The .correlation labeled *inside tube* and *bundle* were obtained by Yang and Cussler [71] from experiments on gas desorption from absorption into water, using microporous polypropylene fibers. The inside tube side line is for laminar feed flow inside hollow fibers. The equation $Sh = 1.64$ (Re $Sc\ d/L$) is nearly identical to the classic Lévêque correlation [83]. An inside diameter to length ratio of 10^{-3} was assumed for this illustration, which shows that laminar flow in a tube yield relatively low Sherwood numbers and thus low mass transfer coefficients. If the bundle correlation is extrapolated, a mass transfer coefficient of 10^{-2} cm/s would be achieved at a Reynolds number of about 50, assuming the organic diffusivity of 10^{-5} cm^2/s and an outside diameter of 0.005 cm. For this case, the transverse flow velocity required is only 10 cm/s, which should result in a relatively low feed-side pressure drop. The line labeled *transverse* is from data reported by Lipski et al. [84]. Their work involved the perevaporation of dilute aqueous solutions of toluene through an array of silicone rubber-coated hollow fiber mounted transverse to the liquid flow. The results were similar to the bundle and staggered bank correlations but were somewhat lower. The final line in Figure 4.2c, labeled *spiral*, is for a flat channel containing a spiral-wound feed spacer and was obtained by Hickey and Gooding [63,84]. This line represents the best spacer or turbulence promoter that Hickey and Gooding have tested in their work, based on high mass transfer coefficients and low pressure drop. Similarly, hollow fibers with inside diameters of 0.2 cm and modestly turbulent flow (Re = 4000) can be achieved at a reasonable velocity (< 2 m/s) and pressure drop. With the Schmidt number and diffusivity as before, the mass transfer coefficient should be

10^{-2} cm/s. Unfortunately, fibers with such a large inside diameter yield relatively low surface-area-to-volume ratios.

Kubaczka et al. [85] studied the effect of all the individual mass transfer mechanism on the overall mass fluxes during membrane extraction. In multicomponent systems each of the constituent mechanisms is defined by an appropriate matrix of mass transfer coefficients. The method developed to assess the relative importance of the individual transport mechanisms is validated using the multicomponent extraction studied experimentally by Kubaczka et al. as a test case. The authors established that if the resistances identified by the norm as negligible are omitted, the overall accuracy of the simulations remains unaffected. Even when the membrane resistance can be neglected (as was the case in our study), the membrane determines the mass fluxes in the stationary systems of coordinates by defining the coefficients of the bootstrap (transformation matrix). As the analysis reveals, the omission of this effect leads to a visible drop in the accuracy of calculations. In membrane extraction, where the diffusivities depend strongly on concentrations along the diffusion path, the omission of this dependence also generates considerable errors. As the authors claimed, the method for calculating mass transport resistances based on the evaluation of matrix norms has been found quite useful in the extraction process investigated. Also, as the problem of assessing transport resistances in multicomponent systems is common to all mass transport processes, it is highly probable that the method will prove helpful in dealing with other processes of this type.

4.2.3 Mass Transfer Modeling in Different Hollow Fiber Operating Modes

Models are presented covering three configurations of hollow fiber membranes as given in the following sections.

4.2.3.1 Hollow Fiber Impregnation Mode or Hollow Fiber Supported Liquid Membrane

For the recycling mode, both feed solution and stripping solution are recycled, as shown in Figure 4.7. To model the recycling mode, in 1984 Danesi [86] proposed a simple model with a constant permeation coefficient. The difficulties involved in describing a nonstudy state process with variation of the concentration in the axial and radial directions making use of the continuity equation lead to the use of the macroscopic mass balance of the permeating solute in a certain volume of fiber in a given time interval [87,88]. The model for the transport of solute in a hollow fiber supported liquid membrane system operating in a recycling mode consists of four equations describing (1) the change of solute concentration in the feed and stripping streams when circulating through the membrane module and (2) the change of solute concentration in the feed and stripping tanks, where the aqueous solutions are continuously recirculated based on the complete mixing hypothesis.

Assuming linear concentration gradients and the lack of back-mixing, these equations are formulated as follows.

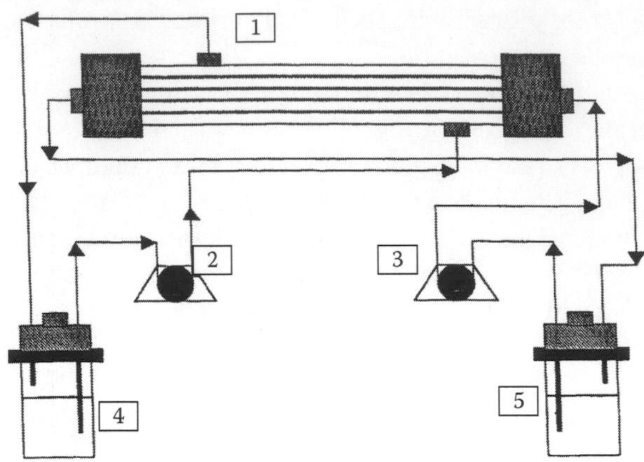

FIGURE 4.7 Schematic view of hollow fiber supported liquid membrane run in recycling mode. (1) hollow fiber contactor, (2,3) feed and strip pump, (4,5) feed and strip reservoir, respectively. (From A. Kumar and A. M. Sastre, *Ind. Eng. Chem. Res.*, 39, 146, 2000. With permission.)

4.2.3.1.1 For the Feed Solution

Module mass balance.

$$\frac{\partial C_f^m}{\partial t} = -\upsilon_f \frac{\partial C_f^m}{\partial z} - \left(\frac{A}{V_m}\right)_{in} P(C_f^m - C_s^m) \tag{4.9}$$

Tank mass balance

$$\frac{dC_f^T}{dt} = \frac{Q_f}{V_f}(C_{f,z=L}^m - C_{f,z=0}^m) \tag{4.10}$$

4.2.3.1.2 For the Stripping Solution

Module mass balance.

$$\frac{\partial C_s^m}{\partial t} = -\upsilon_s \frac{\partial C_s^m}{\partial z} + \left(\frac{A}{V_m}\right)_{out} P(C_f^m - C_s^m) \tag{4.11}$$

Tank mass balance:

$$\frac{dC_s^t}{dt} = \frac{Q_s}{V_s}\left(C_{s,z=L}^m - C_{s,z=0}^m\right) \tag{4.12}$$

where P is the overall permeability coefficient (cm/s), C is the solute concentration (g/cm^3), L is the fiber length (cm), Q is the flow rate (cm^3/s), υ is the linear velocity (cm/s), and V is the tank volume (cm^3). The subscripts f and s refer to the feed and stripping solutions, respectively. The superscripts m and T refer to the membrane module and phase tank, respectively.

A/V_m is the ratio of the area of the volume of mass transfer of the fiber:

(1) For the feed phase circulating through the inside of the fiber:

$$(A/V_m)_{in} = \frac{2\pi n_f r_i L}{\pi n_f r_i^2 L} = \frac{2}{r_i} \tag{4.13}$$

4.2.3.1.3 For the Stripping Phase Circulating along the outside of the Fiber

$$(A/V_m)_{out} = \frac{2\pi n_f r_0 L}{\pi(R_c^2 - n_f r_o^2)L} = \frac{2r_o n_f}{R_c^2 - n_f r_o^2} \tag{4.14}$$

where n_f is the number of fibers contained in the membrane module, R_c is the inner radius of the module cell, and r_i and r_o are the inner and outer radii of the hollow fiber, respectively.

The integration of the system of differential Equations (4.9–4.12) for concurrent flow can be obtained by numerical methods. When a stripping solution is used as the stripping agent, an instantaneous reaction is assumed to occur on the outside of the fiber, leading to $C_s^m = 0$ and $C_s^T = 0$. In this case, the solution to Equations (4.9–4.12) is simplified to

$$V_f \ln\left(\frac{C_{f,t=0}}{C_f}\right) = Q_f \left\{1 - \exp\left(\frac{2PL}{\upsilon_f r_i}\right)\right\} t \tag{4.15}$$

Experimental results can thus be fitted to a first-order kinetic law

$$V_f \ln\left(\frac{C_{A,t=0}^f}{C_A}\right) = St \tag{4.16}$$

where S is the factor dependent on the geometry of the fibers and the module, the linear velocity of the fluids, and the overall permeability of the system. The overall permeability coefficient can easily be obtained from the experimental value of the slope S as follows:

$$P = \frac{\upsilon_f r_i}{2L}\left[\ln\left(1 - \frac{S}{Q_f}\right)\right] \tag{4.17}$$

for a system run in a recycling mode.

The design of the hollow fiber supported liquid membrane modules for the separation concentration of solute using overall permeability coefficient P centers on three mass transfer resistances. One of them occurs in the liquid flowing through the hollow fiber lumen. The second corresponds to the solute-complex diffusion across the liquid membrane immobilized on the porous wall of the fiber. The third resistance is due to the aqueous interface created on the outside of the fiber.

The reciprocal of the overall permeability coefficient is given by

$$\frac{1}{P} = \frac{1}{k_i} + \frac{r_i}{r_{lm}} \frac{1}{P_m} + \frac{r_i}{r_o} \frac{1}{k_o} \tag{4.18}$$

where r_{lm} is the hollow fiber log mean radius and k_i and k_o are the interfacial coefficients corresponding to the inner and outer aqueous boundary layers. P_m is the membrane permeability.

4.2.3.2 Membrane Diffusion

The effective diffusion coefficients (D_{eff}) of solute complexes through the organic membrane phase were determined through the model. An effective diffusion coefficient (D_{eff}) for the solute in the immobilized organic liquid membrane can be defined as follows:

$$D_{eff} = k_m t_m \tau \tag{4.19}$$

4.2.3.3 Hollow Fiber Membrane-Based Liquid–Liquid Extraction (Nondispersive Membrane Extraction) and Model Equation

The following assumptions were made for developing the model:

- The system is at a steady state.
- Equilibrium exists at the fluid–fluid interface.
- Pore size and wetting characteristics are uniform throughout the membrane.
- The curvature of the fluid–fluid interface does not significantly affect the rate of mass transfer, the equilibrium solute distribution, or the interfacial area.
- No bulk flow correction is necessary—that is, mass transfer is described adequately by simple film-type mass transfer coefficients.
- No solute transport occurs through the nonporous parts of the membrane.
- The two fluids are virtually insoluble in each other.
- The equilibrium solute distribution is constant over the concentration range of interest.

As derived by D´Elia et al. [28], the key equation for the calculation of K_E or K_S for cocurrent flow is

$$\left[\frac{\dfrac{1}{Q_f} + \dfrac{1}{Q_{e/s}H}}{\dfrac{1}{V_f} + \dfrac{1}{V_{e/s}H}}\right] \ln\left[\frac{C^0_{e/s} / H - C^0_f)}{(C^0_{e/s} - C^0_f) + (V_f / HV_{e/s})(C^0_f - C_f)}\right] \tag{4.20}$$

For countercurrent contact, the following equation was used to calculate K_E or K_S [2,11,24]:

$$\ln\left[\frac{\left(C^0_{e/s}/H - C^0_f\right)}{\left(C^0_{e/s}/H - C^0_f\right) + \left(V_f/HV_{e/s}\right)\left(C^0_f - C_f\right)}\right]$$

$$= t\,\frac{\left[1-\exp\left(-4K_E V_m/d\left(1/Q_f - 1/Q_{e/s}H\right)\right)\right]\left[1/V_f + 1/V_{e/s}H\right]}{\dfrac{1}{Q_f} - \dfrac{1}{Q_{e/s}}\exp\left[-\dfrac{4K_E V_m}{d}\left(\dfrac{1}{Q_f} - \dfrac{1}{Q_{e/s}H}\right)\right]} \tag{4.21}$$

where Q_f and $Q_{e/s}$ are the feed and extract/strip flows; V_f and $V_{e/s}$ are the feed and extract/strip volumes; C_f^0 and $C^0_{e/s}$ are the concentrations of the solute in the feed and in the extract/strip solutions at time zero; C_f is the concentration of the solute in the feed at time t; V_m is the volume of all the hollow fibers; and d is the diameter of one fiber.

The system consists of an aqueous phase containing solute flowing in the tube side of microporous hollow fiber membranes, the pores of which are filled with the organic extractant, which flows cocurrently or countercurrently in the shell side (Figure 4.8). The reaction takes place at the inside wall of membrane where the phase interface is located. The various steps in the extraction process are assumed to be as follows [89–91]:

Step 1: The gold cyanide in the aqueous phase (tube side) diffuses from the bulk to the aqueous–organic interface (the inside wall of fiber) through the boundary layer.

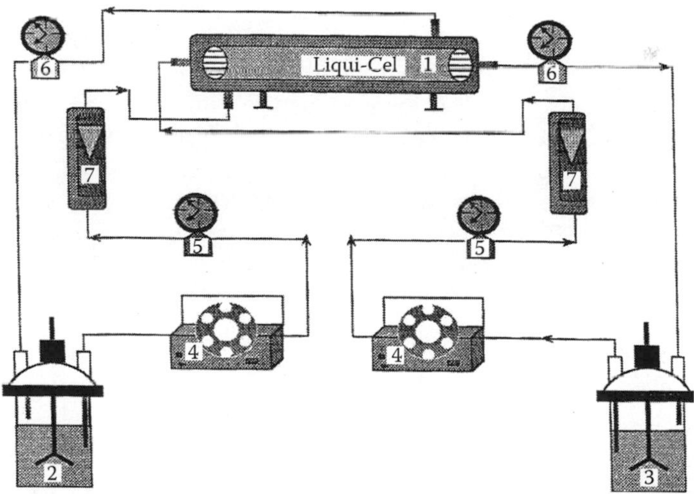

FIGURE 4.8 A schematic view of the membrane-based extraction process. (1) hollow fiber contactor, (2,3) organic extractor and feed, (4) feed and organic pump, (5,6) inlet and outlet pressure gauge, respectively, for organic and feed, (7) flow meters for feed and organic. (From A. Kumar, R. Haddad, G. Benzal, and A. M. Sastre, *Ind. Eng. Chem. Res.*, 41, 613, 2002. With permission.)

Step 2: At the aqueous–organic interface, desired solute to be separated or removed in the organic phase in the membrane pore to form the metal extractant complex.
Step 3: Solute complex diffuses from the aqueous organic interface to the outside wall of fiber through the organic filled membrane pore; free LIX79 diffuses in the opposite direction from the shell side into the pore.
Step 4: Solute complex diffuses from the outside fiber wall to the organic-phase bulk (the shell side), flowing cocurrently or countercurrently to the aqueous phase.

Therefore, the expression for overall mass transfer coefficient, K_E, can be written as [68]

$$\frac{1}{K_E} = \frac{1}{k_i} + \frac{1}{k_f} + \left(\frac{r_i}{r_{lm}}\right)\frac{1}{Hk_m} + \left(\frac{r_i}{r_o}\right)\frac{1}{k_s H} \tag{4.22}$$

and k_i is the effective rate of interfacial reaction at the surface and k_f, k_m, and k_s are the mass transfer coefficients in the aqueous feed, membrane, and organic solvent, respectively. The term r_{lm} denotes logarithmic mean radius. The first term on the right-hand side of Equation (4.22) represents the resistance due to interfacial reaction; the second term indicates the mass transfer resistance in the aqueous phase; the third term is the membrane resistance; and the fourth term is the resistance of the organic extraction solvent. The partition coefficient, *H*, appears in the membrane resistance, because in our experiments the organic solvent wets the membrane but water does not.

The overall mass transfer coefficient can be calculated from the individual transfer coefficients k_i, k_f, k_m, and k_s. The tube- and shell-side mass transfer coefficient are known to depend on the flow conditions in the fiber lumen and shell fluid, respectively, and correlations are available in the literature expressing these dependencies.

As described in previous work, correlations for mass transfer in hollow fiber or small tubes were established for the tube side and the shell side by Prasad et al. [12] and Dahuron et al. [35], respectively. For the tube-side mass transfer coefficient, the following correlation was given:

$$N_{Sh} = \frac{k_f d_i}{D_t} = 1.64\left(\frac{D_t}{d_i}\right)\left(\frac{d_i^2 \upsilon_t}{LD_t}\right)^{1/3} \tag{4.23}$$

In general, Equation (4.23) predicts mass transfer coefficients with reasonable accuracy for $N_{Gz} > 4$ but overestimates them for $N_{Gz} < 4$ [82]. Under the present experimental conditions, N_{Gz} ranges from 5.5 to 8.2. Similarly, for the shell-side mass transfer coefficient, the following correlation was given [14,16,43]:

$$N_{Sh} = \frac{k_s D_h}{D_s} = \beta\left[D_h(1-\phi)/L\right] N_{Re}^{0.6} N_{Sc}^{0.33} \tag{4.24}$$

where the Reynolds number,

$$N_{\mathrm{Re}} = \frac{\upsilon_s D_h}{\eta_s},$$

and the Schmidt number,

$$N_{Sc} = \frac{\eta_s}{D_s},$$

and β is 5.85 for hydrophobic membranes and $0 < N_{Re} < 500$ and $0.04 < \phi < 0.4$. In the present system, both the conditions are met as N_{Re} ranges between 3.0 and 11.0 and $\phi = 0.35$ and where D_h is the hydraulic diameter, D_t is the diffusion coefficient of solute in tube side, d_i is the inner fiber diameter, L is the fiber length, D_s is the diffusion coefficient of solute on the shell side, υ_t and υ_s are the velocity of liquid inside the fiber and shell side, respectively, and ϕ is the packing fraction. The membrane mass transfer coefficient can be determined from the following expression or by using Equation 4.19 [39,40]:

$$k_m = \frac{D_m \varepsilon}{t_m \tau} \tag{4.25}$$

For the membrane and solvent here, t_m is the membrane thickness (30 μm), τ is the tortuosity (3; value obtained from Celgard GmbH), and D_m is the diffusion coefficient of the solute complex in membrane. The membrane tortuosity was also determined by the Wakao-Smith relation, expressed as the inverse of the membranc porosity, which almost matches the value suggested by Celgard GmbH [92].

One example is discussed for comparing the effectiveness of the reciprocating palate extraction column with hollow fiber membrane modules for extracting phenol from solution using MIBK. The data for the example are drawn from B.W. Reed et al. [76]. These authors tested a 3.8 m × 5.2 cm diameter column to treat a water (pilot scale) containing 440 ppm of phenol to compare this data with hollow fiber contactors. The newer membrane contactors using a more efficient, cross-flow design are now available, which offer a larger capacity than previous contactors. The same contactors having the following characteristics are employed in these experiments:

- Overall dimensions: 2-in. diameter × 24-in. length
- Number of fibers: 9,000
- Internal fiber diameter: 240 μm
- Fiber wall thickness: 30 μm
- Membrane porosity: 30%
- Effective membrane pore size: 0.05 μm
- Fiber material polypropylene, membrane tortuosity: 2.6
- Effective fiber length: 54.6 cm.

The conditions tested and relevant data are as follows:

- Influent phenol concentration: 440 mg/L
- Effluent phenol concentration: 1 mg/L
- The partition coefficient K for MIBK: 40
- Water flow rate: 24 cm^3/s
- MIBK flow rate: 27.9 cm^3/s
- Temperature: 105°C
- Diffusivity of phenol in water: 10^{-5} cm^2/s
- Diffusivity of phenol in MIBK: 3 10^{-5} cm^2/s.

The authors found NTU to be 6.61 and the HTU to be 55.4 cm from the experiments performed on the reciprocating plate extraction column. How many modules are required if modules as described earlier (2-in. diameter × 24-in. length) are used in this application?

The water velocity υ within the fibers may be calculated as

$$\upsilon = 24\ \text{cm}^3\ \text{s}^{-1} \cdot 4/\pi \cdot (0.024\ \text{cm})^2 \cdot 7500$$

$$7.1\ \text{cm s}^{-1}$$

We may also calculate a superficial water velocity, υ_o, for the module and the solvent velocity, υ_s.

$$\upsilon_o = 24\ \text{cm}^3\ \text{s}^{-1} \cdot 4/\pi \cdot (4\ \text{cm})^2$$

$$= 1.9\ \text{cm s}^{-1}$$

$$\upsilon_s = 27.9\ \text{cm}^3\ \text{s}^{-1} \cdot 4/\pi \cdot (4\ \text{cm})^2 \cdot 0.43$$

$$\upsilon_s = 5.2\ \text{cm s}^{-1}$$

The mass transfer coefficients are calculated as

$$k_{feed} = 1.64\left(\frac{D_t}{d_i}\right)\left(\frac{d_i^2\upsilon_t}{LD_t}\right)^{1/3}$$

$$= 1.64\left[\frac{1\cdot 10^{-5}\,cm^2/s}{0.024}\right]\left[\frac{(0.024)^2(7.1\,cm/s)}{(1\cdot 10^{-5}\,cm^2/s)(54.6\,cm)}\right]^{1/3} \tag{4.26}$$

$$= 1.3\,10^{-3}\,\text{cm/s}$$

$$k_{membrane} = D\varepsilon K/l\tau = \left(\frac{3\cdot 10^{-5}\,cm^2/s\cdot 0.3\cdot 40}{0.003\,cm\cdot 2.6}\right)$$

$$= 4.6\,10^{-2}\,\text{cm/s}$$

$k_{receiving}$ is estimated using the equation of Prasad and Sirkar [43]:

$$k_{receiving} = 5.85\left(\frac{HD}{d_e}\right)\left[\frac{d_e(1-\phi}{d_e}\right]\left[\frac{\rho\upsilon_s d_e}{\mu}\right]^{0.6}\left[\frac{\nu}{D}\right]^{0.3}$$

$$= 5.85\cdot\left(\frac{40.3\times10^{-5}\,cm^2/s}{0.53}\right)\left(\frac{0.53cm(1-0.43)}{54.6cm}\right) \tag{4.27}$$

$$\left(\frac{5.2cm/s\cdot0.53cm}{0.75\,cm^2/s}\right)^{0.6}\times\left(\frac{0.075\,cm^2/s}{3\times10^{-5}\,cm^2/s}\right)$$

$$= 3.33\cdot10^{-3}\text{cm/s}$$

Summing the resistances,

$$\frac{1}{k} = \frac{1}{1.3\times10^{-3}} + \frac{1}{3.33\times10^{-3}} + \frac{1}{4.6\times10^{-2}}$$

$$k = 9.4\times10^{-4}\text{cm/s}$$

Therefore, substituting in,

$$c/c_o = e^{-kaL/\upsilon}$$

$$c/c_o = 0.023 = e^{\left(\frac{-9.4\cdot10^{-4}\,cm/s\times166cm^{-1}\times L}{7\,cm/s}\right)}$$

$$\text{L} = 272\,\text{cm}$$

and five modules are needed in series having each of 54.6 cm in length.

The HTU is then calculated for membrane module: 272/6.61 = 41.3 cm. A comparison of contactors with reciprocal plate contactor are presented in Table 4.11.

Further, Qin et al. [93] suggested that modules in series are superior to a long single module or modules in parallel for a given separation, since it allows an improved lumen mass transfer rate. Moreover, the authors provided more accurate design equations than those reported in the literature. In membrane extraction, supported liquid membrane, supported gas membrane, perevporation, gas (vapor) permeation, pertraction, and permaborption processes, the mass transfer often occurs with a country external resistance and constant shell concentration or with a constant external resistance and a varied shell concentration, resulting from a cocurrent or countercurrent operational mode. Introducing dimensionless parameter, such as Sh_w (Sherwood number based on the external mass transfer coefficients K), m′ (the separating factor), and Z′ (dimensionless length of the module), and using the correlation of lumen mass transfer coefficient as described in a previous study [94], a set of algebraic expressions has been derived so as to design a single module or modules

TABLE 4.11
Comparison of Contactors

	A(cm^{-1})	(HTU) (cm)	Volume Required (cm^3)
Membrane contactors	167	41.3	3430
Reciprocal plate contactor	2–4?	55.4	7777
Membrane/column ratio	(80/1	0.75/1	0.44/1

Source: From B. W. Reed, M. J. Semmens, and E. L. Cussler, in *Membrane Separation Technology, Principles and Applications,* eds. R. D. Noble and S. A. Stern, Elsevier, Amsterdam, 1995, p. 474. (With permission.)

in series and to predict their operating performance. To calculate the number of hollow fiber contactors for the study dealing with phenol removal was reported earlier in this Section 4.2.3.3 [76]. Now based on the modified approach, recalculation of number of modules was performed in following manner. The parameters given by them are already listed in that example; a few additional parameters are Z = 0.546 m, k_m/m = 4.6 10^{-4} m/s, k_s/m = 3.33 10^{-5} m/s, $d_m = (d_o\text{-}d_i)/\ln(d_o/d_i)$ = 2.69 10^{-4} m. Similarly, few other parameters are $K = (\mathrm{m}d_i/d_m k_m + \mathrm{m}d_i/d_o k_s)^{-1}$ = 3.85 10^{-5} m/s, $Sh_w = d_i K/D$ = 6.78, $Z' = ZD/d_i^2u$ = 0.1469, $m' = mQ_l/Q_s$ = 0.021 ≈ 0, $Sh_{avg,Z}$ = 4.24 (for calculation details see [93]), $Sh_{avg,Z,o} = (1/Sh_w + 1/Sh_{avg,Z})^{-1}$ = 2.61, when $C_{s,in}$ = 0.0, and m′ = 0.0. The final equation will be as follows:, $C_{out}/C_{in} = \exp(-4NZ'Sh_{avg,Z,o}$, thus N can be calculated to be 3.97, so 4 modules are needed in the series compared with N = 4.98 where 5 modules are needed as concluded in an earlier article [76] by the simplification that that Lévêque's equation is valid even when Sh_w has limited value and Z′ is large.

4.2.3.4 Hollow Fiber Contained Liquid Membrane (HFCLM)

In the facilitated transport of desired solute from an aqueous feed solution through a hydrophobic HFCLM permeator containing organic extractant on the shell side of the HFCLM permeator, water or aqueous stripping solution was used as stripping or receiving phase. Both mobile phases are in cocurrent flow through the lumen of the two separate fiber sets. The general assumptions are similar to those described in Sengupta et al. [95].

The local value of the total flux of species A, N_A^T, across an organic film of thickness δ that contains the water-insoluble species A and C is obtained by an analysis similar to those by Olander [96], Ward [97], and Cussler [98]. They studied analytically a system involving mass transfer and equilibrium reaction. If D_{AO} and D_{CO} are assumed to be equal, the total local flux of A though a film of thickness δ is

$$N_A^T = \frac{D_{Co}}{\delta}\left(C_{Co}^F - C_{Co}^S\right) + \frac{D_{Ao}}{\delta}\left(C_{Ao}^F - C_{Ao}^S\right) \tag{4.28}$$

All important notations are presented in list of symbols provided at the end of the chapter.

For facilitated transport through an HFCLM, if one assumes negligible aqueous phase boundary layer resistances, the estimated value of the organic-phase-based overall transfer coefficients K for each species C and A can be obtained from organic species transport [95,99] as

$$\frac{1}{K_{0,C}} = \frac{d_i}{d_{lm}} \frac{2}{D_{Co}\varepsilon_s / \left[\tau_s (d_o - d_i)/2\right]} + \frac{d_i}{d_o} \frac{1}{D_{Co}/\delta_m} = \frac{t_{eff}}{D_{Co}} \tag{4.29}$$

and

$$\frac{1}{K_{0,A}} = \frac{d_i}{d_{lm}} \frac{2}{D_{Ao}\varepsilon_s / \left[\tau_{ss} (d_o - d_i)/2\right]} + \frac{d_i}{d_o} \frac{1}{D_{Ao}/\delta_m} = \frac{t_{eff}}{D_{Ao}} \tag{4.30}$$

Some diffusion in the organic liquid in hollow fiber pores and in the organic contained liquid membrane of effective of effective thickness t_w are in the series, it is obvious that (D_{Co}/t_{eff}) is $K_{o,C}$ and (D_{Ao}/t_{eff}) is $K_{o,A}$. Consequently, in the expression for $N_A{}^T$ in Equation (4.28), (D_{Co}/δ) may be replaced by either (D_{Co}/t_{eff}) or $K_{o,C}$, and (D_{Ao}/δ) by (D_{Ao}/t_{eff}) or $K_{o,A}$. It is to be noted that it is illustrative to represent $K_{o,C}$, for example, by (D_{Co}/t_{eff}) instead of the complete expression in Equation (4.29). The aforementioned developments assume that (1) there are no interfacial resistances and (2) r_s is same for both species since the pore size are orders of magnitude larger.

4.3 CONCLUDING REMARKS AND FUTURE DIRECTIONS

In the past few years, additional innovations in hollow fiber module design and performance related factors have been explored in view of adopting HFMST for commercial applications. As already evidenced in this presentation, significant progress has been made in gas–liquid, liquid–liquid, and gas–gas applications covering design aspects and improvements recorded for better throughput and efficiency with these microporous hollow fiber contactors. Several membrane-based processes in the food industry, beverage industry, and large-scale gases separations are already commercialized [2,4,5,55,102]. Still, many academic and industrial projects in this area are in progress that in the near future will bring more commercial success in different areas as stated by Pabby and Sastre [2], Drioli [5], and Baker [102] in their latest published reviews. As stated in [102], most of the today's gas separation membranes are formed into hollow fiber modules, with perhaps fewer than 20% being formed into spiral-wound modules. The low cost of hollow fiber contactors is a major advantage of these types of contactors. In recent modifications and design-related improvements, a cross-flow configuration with baffles was replaced by the Lique-Cel contactor with an old existing hollow fiber contactor.

With the help of hollow fiber contactors, commercial success of membrane stripping, membrane-based solvent extraction, membrane gas adsorption, and other related processes could be possible. True membrane separation processes (e.g., reverse osmosis [RO], ultrafiltration [UF], nanofiltration [NF]) have, in general, practical limitations to achieving very high levels of purifications of fluid streams being processed. On the other hand, conventional equilibrium-based separation processes such as solvent extraction, absorption, stripping, or distillation suffer from no such limitations. Membrane-based equilibrium or contacting processes are as capable of purification as their nondispersive counterpart. Membrane -based equilibrium processes have many additional benefits including the fact that they are nondispersive.

Conventional contactors may be fabricated from a wide range of materials, including metals, polymers, ceramics, and glass, in which the chemical resistance is suitable for many operating environments. The chemical compatibility of the materials of construction must be carefully assessed in advance for each application.

It is then realistic to affirm that new wide perspectives considering design and performance of hollow fiber membrane contactors are equally important, and it is needless to mention that HFMST will contribute toward sustainable growth of membrane-based technology.

To facilitate further commercialization, challenges that need to be addressed in the future include the following:

1. More and more pilot plant scale and plant scale experience with hollow fiber contactors (duration two to three years); the frequency of replacing hollow fiber membrane during total period would be evaluated and analyzed.
2. In order to evaluate the performance of hollow fiber membrane technology, more case studies from chemical industries should be performed and critical analyses should be provided with regard to the inherent problems of conventional technologies.

LIST OF SYMBOLS

a Mass transfer area per unit permeator volume, cm^{-1}
A or A_m Total area in the membrane module (cm^2)
C Solute concentration (mol/cm^3)
$\langle c \rangle$ Average concentration (mol/cm^3) defined in Equation 4.1
c_0 Inlet concentration (mol/cm^3) defined in Equation 4.1
d_a Thickness of the aqueous feed boundary layer (cm).
d Diameter of one fiber (cm)
d_i and d_o Inner and outer fiber diameter, respectively
D_{eff} Effective membrane diffusion coefficient of the gold-containing species
D_{Co} and D_{Ao} Diffusion coefficient of species in organic and aqueous phase ($cm^2\ s^{-1}$)
Dr or H Partition coefficient of extraction/stripping
D_t Diffusion coefficient of solute in tube side ($cm^2\ s^{-1}$)
D_s Diffusion coefficient of solute in shell side
D_m Diffusion coefficient of metal complex in membrane
D_h Hydraulic diameter (cm)
d_i Inner fiber diameter
HTU Height of transfer units
H or D_r Partition coefficient of gold
$\langle c \rangle$ Average mass transfer coefficient (cm/s)
K_{oC} and K_{oA} Overall mass transfer coefficient for total citric acid transport based on organic phase, aqueous phase, cm/s; defined in Equations (4.29) and (4.30)
k_f Mass transfer coefficient of aqueous feed (cm/s)
k_i Rate of interfacial reaction defined in Equation 22
k_m Membrane mass transfer coefficient (cm/s)
k_s Organic mass transfer coefficient (cm/s)
k_{st} Mass transfer coefficient for stripping (cm/s)
k_i Aqueous mass transfer coefficient (cm/s); defined in Equation (4.18)
k_o Mass transfer coefficient in the strip (cm/s); defined in Equation (4.18)
K^E, K^S Overall mass transfer coefficient for extraction and stripping, respectively (cm/s)
L Fiber length (cm)
NTU Number of transfer units
N_{Re} or Re Reynolds number for feed

$$\left(N_{Re} = \frac{\upsilon\, d}{\eta} \right)$$

N_{Sc} or Sc Schmidt number

$$\left(N_{Sc} = \frac{\eta}{D}\right)$$

N_{Gz} or Gz Graetz number

$$\left(N_{Gz} = \frac{d^2 \upsilon}{DL}\right)$$

N_{Sh} or Sh Sherwood number for tube side

$$N_{Sh} = \frac{k_f d_i}{D_t}$$

and shell side

$$N_{Sh} = \frac{k_s D_h}{D_s}$$

respectively.

n_f Number of fibers
N_A^T The total flux of species A (mol/cm^2.s)
P Overall permeability coefficient (cm/s)
Q Flow rate (cm^3/s)
r Hollow fiber radius (cm)
r_i and r_o Inner and outer hollow fiber radius (cm)
t_m thickness of the fiber membrane (cm)
V_e Organic tank volume (cm^3)
V_f Feed tank volume (cm^3)
V_s Stripping tank volume (cm^3)
V_m Volume of hollow fibers (cm^3)

Subscripts

i For inner radii
e/s For extract/strip
f Feed
m Membrane
o For outer radii
s Strip side or shell side
t Tube side

Superscripts

0 Concentration at time zero
m and T Membrane module and phase tank

Greek Letters

τ Tortuosity of the membrane
ϕ Packing fraction of HF module
ε Porosity
η_t Viscosity of aqueous feed /stripping solution, cP or gm/cm s.
η_s Viscosity of organic solution, cP or (gm/cm s).
υ_t **and** υ_s Velocity of liquid in side fiber and shell side (cm/s)

ACKNOWLEDGMENTS

This work was supported by the Spanish Ministry of Science and Culture, CICYT (QUI 99-0749). Dr. Anil Kumar Pabby acknowledges financial support from the Comisión Interministerial de Ciencia y Tecnología, Spain for awarding a visiting scientist fellowship. Thanks are also due to Shri. V. P. Kansra, director, Nuclear Recycle Group, BARC, India, and Shri. A. Ramanujam, head, FRD, PREFRE, BARC, India, P. K. Dey, plant superintendent, PREFRE, BARC, and Shri. A. K. Venugopalan, PRFFRE, BARC, for their valuable suggestions during the preparation of this manuscript.

REFERENCES

1. M.-B. Hägg, Membranes in chemical processing: A review of applications and novel developments, *Separ. Purif. Method.*, 27, 51–168 (1998).
2. A. Kumar and A. M. Sastre, Developments in non-dispersive membrane extraction-separation processes, in A. K. Sengupta and Y. Marcus, eds., *Ion Exchange and Solvent Extraction*, New York: Marcel Dekker, 331–469 (2001).
3. A. M. Sastre, A. Kumar, J. P. Shukla, and R. K. Singh, Improved techniques in liquid membrane separations: An overview, *Separ. Purif. Method.*, 27, 213–298 (1998).
4. W. Ho and K. K. Sirkar, *Membrane Handbook*, New York: Van Nostrand Reinhold (1992).
5. E. Drioli and M. Romano, Progress and new perspectives on integrated membrane operations for sustainable industrial growth, *Ind. Eng. Chem. Res.*, 40, 1277–1300 (2001).
6. Q. Zhang and E. L. Cussler, Microporous hollow fibers for gas absorption. I. Mass transfer in the liquid, *J. Membrane Sci.*, 23, 321–332 (1985).
7. Q. Zhang and E. L. Cussler, Microporous hollow fibers for gas absorption. II. Mass transfer across the membrane, *J. Membrane Sci.*, 23, 333–345 (1985).
8. A. Kiani, R. R. Bhave, and K. K. Sirkar, Solvent extraction with immobilized interfaces in a microporous hydrophobic membrane, *J. Membrane Sci.*, 20, 125–145 (1984).
9. N. A. D'Elia, L. Dahuron, and E. L. Cussler, Liquid–liquid extractions with microporous hollow fibers, *J. Membrane Sci.*, 29, 309–319 (1986).
10. J. Gyves and E. R. San Miguel, Metal ion separations by supported liquid membranes, *Ind. Eng. Chem. Res.*, 38, 2182–2202 (1999).

11. L. Dahuron and E. L. Cussler, Protein extractions with hollow fibers, *AIChE J.*, 34, 130 (1988).
12. R. Prasad and K. K. Sirkar, Hollow fiber solvent extraction: Performances and design, *J. Membrane Sci.*, 50, 153–175 (1990).
13. D. O. Cooney and M. S. Poufos, Liquid–liquid extraction in a hollow-fiber device, *Chem. Eng. Commun.*, 61, 159–167 (1987).
14. R. Basu, P. Prasad, and K. K. Sirkar, Nondispersive membrane solvent back extraction of phenol, *AIChE J.*, 36, 450–460 (1990).
15. C. H. Yun, R. Prasad, and K. K. Sirkar, Membrane solvent extraction removal of priority organic pollutants from aqueous waste streams, *Ind. Eng. Chem. Res.*, 31, 1709 (1992).
16. C. H. Yun, R. Prasad, A. K. Guha, and K. K. Sirkar, Hollow fiber solvent extraction removal of toxic heavy metals from aqueous waste streams, *Ind. Eng. Chem. Res.*, 32, 1186 (1993).
17. A. Alonso, A. M. Urtiaga, and A. Irabien, Extraction of Cr(VI) with Aliquat 336 in hollow fiber contactors: Mass transfer analysis and modeling, *Chem. Eng. Sci.*, 49, 901 (1994).

18(a). A. I. Alonso, B. Galan, A. Irabien, and I. Ortiz, Separation of Cr(VI) with Allquat 336: Chemical equilibrium modeling, *Sep. Sci. Tech.*, 32, 1543 (1997).

18(b). A. I. Alonso, A. M. Urtiaga, S. Zamacona, A. Irabien, and I. Ortiz, Kinetic modelling of cadmium removal from phosphoric acid by non-dispersive solvent extraction, *J. Membrane Sci.*, 130, 193 (1997).

19. K. Yoshizuka, R. Yasukawa, M. Koba, and K. Inoe, Diffusion model accompanied with aqueous homogeneous reaction in hollow fiber membrane extractor, *J. Chem. Eng. Jpn.*, 28, 59–65 (1995).
20. I. Ortiz, B. Galan, and A. Irabien, Membrane mass transport coefficient for the recovery of Cr(VI) in hollow fiber extraction and back-extraction modules, *J. Membrane Sci.*, 118, 213 (1996).
21. I. Ortiz, A. I. Alonso, A. M. Urtiaga, M. Dermircioglu, N. Kocacik, and N. Kabay, An integrated process for the removal of Cd and U from wet phosphoric acid, *Ind. Eng. Chem. Res.*, 38, 2450 (1999).
22. G. R. M. Broembroek, G. J. Witcamp, and G. M. Van Rosmalene, Extraction of cadmium with trilaurylamine-kerosine through a fiat-sheet-supported liquid membrane, *J. Membrane Sci.*, 147, 195 (1998).
23. A. Kumar and A. M. Sastre, Hollow fiber supported liquid membrane for the separation/concentration of gold(I) from aqueous cyanide media: Modeling and mass transfer evaluation, *Ind. Eng. Chem. Res.*, 39, 146 (2000).
24. A. Kumar, R. Haddad, and A. M. Sastre, Integrated membrane process for gold recovery from hydrometallurgical solutions, *AIChE J.*, 47, 328 (2001).
25. G. W. Stevens and H. R. C. Pratt, Solvent extraction equipment design and operation: Future directions from an engineering perspective, *Solvent Extr. Ion Exch.*, 18, 1051 (2000).
26. R.-S. Juang and J.-D. Chen, Mass transfer modeling of citric and lactic acids in a microporous membrane extractor, *J. Membrane Sci.*, 164, 67 (2000).
27. R. Basu and K. K. Sirkar, Citric acid extraction with microporous hollow fibers, *Solvent Extr. Ion Exch.*, 10, 119 (1992).
28. H. Escalante, A. I. Alonso, I. Ortiz, and A. Irabien, Separation of L-phenylalanine by nondispersive extraction and backextraction: Equilibrium and kinetic parameters, *Sep. Sci. Tech.*, 33, 119 (1998).

29(a). A. M. Urtiaga, M. I. Ortiz, E. Salazar, and I. Irabien, Supported liquid membranes for the separation-concentration of phenol. 1. Viability and mass-transfer evaluation, *Ind. Eng. Chem. Res.*, 31, 877 (1992); Supported liquid membranes for the separation-concentration of phenol. 2. Mass-transfer evaluation according to fundamental equations, *Ind. Eng. Chem. Res.*, 31, 1745 (1992).

29(b). A. M. Urtiaga, Applicacion de las membranes liquias a soportadas a la recuperacion de fenol en modulos de fibras huecas, Ph.D. Dissertation, Universidad del Pais Vasco, Santander, Spain (1991).

30. A. Sengupta, P. A. Peterson, B. D. Miller, J. Schneider, and C. W. Fulk, Large-scale application of membrane contactors for gas transfer from or to ultrapure water, *Sep. Purif. Tech.*, 14, 189 (1998).

31. K. K. Sirkaar, Membrane separations: Newer concepts and applications for the food industry, in R. K. Singh and S. S. Rizvi, eds., *Bioseparation Processes in Food*, New York: Marcel Dekker (1995), 353–356.

32. K. Scott, *Handbook of Industrial Membranes*, Oxford: Elsevier (1995).

33. M. J. Wikol, M. Kobayashi, and S. J. Hardwick, International Conference on Complex Systems 14th International Symposium on Contamination Control, 14th Annual Technology meeting, Phoenix, AZ (April 26–May 1, 1998).

34. P. A. Hogan, R. P. Cannin, P. A. Peterson, R. Johnson, and A. S. Michaelis, A new option: Osmotic distillation, *Chem. Eng. Prog.*, 94, 49 (1998).

35. L. Dahuron and E. L. Cussler, Protein extractions with hollow fibers, *AIChE J.*, 34, 130 (1988).

36(a). J. Thömmes and M. R. Kula, Membrane chromatography—An integrative concept in the downstream processing of proteins, *Biotechnol. Prog.*, 11, 357 (1995).

36(b). D. K. Roper and E. N. Lightfoot, Separation of biomolecules using adsorptive membranes, *J. Chromatogr. A.*, 702, 3 (1995).

37. R. Prasad and K. K Sirkar, Hollow fiber solvent extraction of pharmaceutical products: A case study, *J. Membrane Sci.*, 47, 235 (1989).

38. N. A. D'Elia, L. Dahuron, and E. L. Cussler, Liquid–liquid extractions with microporous hollow fibers, *J. Membrane Sci.*, 29, 309 (1986).

39. A. Kiani, R. R. Bhave, and K. K. Sirkar, Solvent extraction with immobilized interfaces in a microporous hydrophobic membrane, *J. Membrane Sci.*, 20, 125 (1984).

40. R. Prasad, A. Kiani, R. R. Bhave, and K. K. Sirkar, Further studies on solvent extraction with immobilized interfaces in a microporous hydrophobic membrane, *J. Membrane Sci.*, 26, 79 (1986).

41. R. Prasad and K. K. Sirkar, Microporous membrane solvent extraction, *Sep. Sci. Tech.*, 22, 619 (1987).

42. R. Prasad and K. K. Sirkar, Solvent extraction with microporous hydrophilic and composite membranes, *AIChE J.*, 33, 1057 (1987).

43. R. Prasad and K. K. Sirkar, Dispersion-free solvent extraction with microporous hollow-fiber modules, *AIChE J.*, 34, 177 (1988).

44. L. M. Coelhoso, P. Silvestre, R. M. C. Viegas, and J. G. S. P. Crespo, Membrane-based solvent extraction and stripping of lactate in hollow-fibre contactors, *J. Membrane Sci.*, 134, 19 (1997).

45. A. F. Seibert and J. R. Fair, Scale-up of hollow fiber extractors, *Sep. Sci. Tech.*, 32, 573 (1997).

46. R. Basu and K. K. Sirkar, Hollow fiber contained liquid membrane separation of citric acid, *AIChE J.*, 37, 383 (1991).

47. L. Graetz, Uber die Wärmeleitungsfähigkeit von Flüssigkeiten, *Ann. Phys. Chem.*, 25, 337 (1885).

48. S. W. Peretti, C. J. Tomkins, J. L. Goodall, and A. S. Michaels, Extraction of 4-nitrophenol from 1-octanol into aqueous solution in a hollow fiber liquid contactor, *J. Membrane Sci.*, 195, 193 (2002).
49. C. E. Aziz, M. W. Fitch, L. K. Linguist, J. G. Pressman, and G. Georiou, Methanotrophic biodegradation of trichloroethylene in a hollow fiber membrane bioreactor, *Environ. Sci. Tech.*, 29, 2574 (1995).
50. W. J. Koros and R. T. Chern, Separation of gaseous mixtures using polymer membranes, in R. Rousseau, ed., *Handbook of Separation Processes Technology*, New York: John Wiley and Sons, 862–953 (1987).
51. P. Puri, Membrane gas separations, An opportunity for gas industry or just a niche market, Preprints of the International Conference on Membrane Science and Technology (ICMST:98), Beijing, China (June 9–13, 1998).
52. K. K. Sirkar, Other new membrane processes, in W. S. Ho and K. K. Sirkar, eds., *Membrane Handbook*, New York: Chapman & Hall (1992).
53. K. K. Sirkar, Membrane separation technologies: Current developments, *Chem. Eng. Commun.*, 157, 145 (1997).
54. A. Sengupta, R. A. Sodaro, and B. W. Reed, Oxygen removal from water using process-scale extra-flow membrane contactors and systems, Paper presented at the 7th annual meeting of NAMS, Portland, OR (May 23, 1995).
55. A. Gabelman and S.-T. Hwang, Hollow fiber membrane contactors, *J. Membrane Sci.*, 159, 61 (1999).
56. A. Sengupta, B. W. Reed, and F. Seibert, Liquid–liquid extraction studies on semi-commercial scale using recently commercialised large membrane contactors and systems, Paper presented at the AIChE annual meeting, San Francisco, CA (November 16, 1994).
57. R. Prasad, S. Khare, A. Sengupta, and K. K. Sirkar, Novel liquid-in-pore configurations in membrane solvent extraction, *AIChE J.*, 36, 1592 (1990).
58. M. L. Crowder and C. H. Gooding, Spiral wound, hollow fiber membrane modules: A new approach to higher mass transfer efficiency, *J. Membrane Sci.*, 137, 17 (1997).
59. A. W. Mancusi, J. C. Delozier, R. Prasad, C. J. Runkle, and H. F. Shuey, U.S. Patent 5,186,832, assigned to Hoechst Celanese Corp. (Feb. 16, 1993).
60. N. S. Strand, U.S. Patent 3,342,729, assigned to The Dow Chemical Company (September 19, 1967).
61. P. L. Côté, R. P. Mourion, and C. J. Lipski, U.S. Patent 5,104,535, assigned to Zenon Environmental, Inc. (April 14, 1992).
62. P. J. Hickey and C. H. Gooding, The economic optimization of spiral wound membrane modules for the pervaporative removal of VOCs from water, *J. Membrane Sci.*, 97, 53 (1994).
63. P. J. Hickey and C. H. Gooding, Mass transfer in spiral wound pervaporation modules, *J. Membrane Sci.*, 97, 59 (1994).
64. R. Psaume, P. Aptel, T. Aurelle, J. C. Mora, and J. L. Bersillon, Pervaporation: Importance of concentration polarization in the extraction of trace organics from water, *J. Membrane Sci.*, 36, 373 (1988).
65. C.-L. Zhu, C.-W. Yuang, J. R. Fried, and D. B. Greenberg, Pervaporation membranes—a novel separation technique for trace organics, *Environ. Prog.*, 2, 132–143 (1983).
66. H. H. Nijhuis, M. H. V. Mulder, and C. A. Smolders, Removal of trace organics from aqueous solutions. Effect of membrane thickness, *J. Membrane Sci.*, 61, 99 (1991).
67. S. R. Wickramasinghe, M. J. Semmens, and E. L. Cussler, Mass transfer in various hollow fiber geometries, *J. Membrane Sci.*, 69, 235 (1992).
68. E. L. Cussler, Diffusion: Mass transfer in fluid systems, 2nd ed., Cambridge, UK: Cambridge University Press (1997).

69. S. R. Wickramasinghe, M. J. Semmens, and E. L. Cussler, Better hollow fiber contactors, *J. Membrane Sci.*, 62, 371 (1991).
70. A. E. Johnsen, R. Klaassen, P. H. M. Feran, J. H. Hanemaaijer, and B. Ph. Ter Meulen, Membrane gas absorption processes in environmental applications, in J. G. Crespo and K. W. Böddeker, eds., *Membrane Process in Sepration and Purification*, Dordrecht: Kluwer Academic Publishers, 343–356 (1994).
71. M.-C. Yang and E. L. Cussler, Designing hollow-fiber contactors, *AIChE J.*, 32, 1910 (1986).
72. E. L. Cussler, Hollow fiber contactors, in J. G. Crespo and K. W. Boddeker, eds., *Membrane Processes in Separation and Purification*, Dordrecht: Kluwer Academic Publishers (1995).
73. S. R. Wickramasinghe, M. J. Semmens, and E. L. Cussler, Hollow fiber modules made with hollow fiber fabric, *J. Membrane Sci.*, 84, 1 (1993).
74. K. L. Wang and E. L. Cussler, Baffled membrane modules made with hollow fiber fabric, *J. Membrane Sci.*, 85, 265 (1993).
75. R. Prasad and K. K. Sirkar, Membrane based solvent extraction, in W. S. Ho and K. K. Sirkar, eds., *Membrane Handbook*, New York: Chapman & Hall, 727–763 (1992).
76. B. W. Reed, M. J. Semmens, and E. L. Cussler, Membrane contactors, in R. D. Noble and S. A. Stern, eds., *Membrane Separation Technology, Principles and Applications*, Elsevier (1994).
77. R. Gawronski and B. Wrzesinska, Kinetics of solvent extraction in hollow-fiber contactors, *J. Membrane Sci.*, 168, 213 (2000).
78. A. H. P. Skellnd, *Diffusional mass transfer*, New York: Wiley (1974).
79. I. Noda and C. C. Gryte, Mass transfer in regular arrays of hollow fibers in countercurrent dialysis, *AIChE J.*, 25, 113 (1979).
80. J. Lemanski and G. G. Lipscomb, Effect of shell-side flows on hollow-fiber membrane device performance, *AIChE J.*, 41, 2322 (1995).
81. E. E. Wilson, A basis for rational design of heat transfer apparatus, *Trans. ASME*, 37, 47 (1915).
82. X. Lévêque, Les lois de la transmission de chaleur par conduction, *Ann. Mines*, 13, 201–299 (1928).
83. C. Lipski, P. L. Côté, and H. F. Fleming, Transverse feed flow for hollow fibers significantly improves mass transfer at low energy consumption, in R. Bakish, ed., *Proceedings of 5th International Conference on Perevaporation Processes in the Chemical Industry*, Englewood, NJ: Bakish Materials Corp., 134 (1991).
84. P. J. Hickey and C. H. Gooding, Mass transfer in spiral wound pervaporation modules, *J. Membrane Sci.*, 92, 59 (1994); Erratum, *J. Membrane Sci.*, 98, 293 (1994).
85. A. Kubaczka and A. Burghardt, Effect of mass transport resistances in multicomponent membrane extraction on the overall mass fluxes, *Chem. Eng. Sci.*, 55, 2907 (2000).
86. P. R Danesi, A simplified model for the coupled transport of metal ions through hollow-fiber supported liquid membranes, *J. Membrane Sci.*, 20, 231–248 (1984).
87. A. M. Urtiaga, M. I. Ortiz, and J. A. Irabien, Mathematical modeling of phenol recovery with supported liquid membranes, Abstract of paper of the International Solvent Extraction Conference, Kyoto, Japan (July 16–21, 1990).
88. A. M Urtiaga and J. A. Irabien, Internal mass transfer in hollow fiber supported liquid membranes, *AIChE J.*, 39, 521 (1993).
89. P. R. Danesi, E. P. Horwitz, G. F. Vandegrift, and R. Chiarizia, Mass transfer rate through liquid membranes — interfacial chemical reactions and diffusion as simultaneous permeability controlling factors, *Sep. Sci. Tech.*, 16, 201 (1981).
90. M. Teramoto and H. Tanimoto, Mechanism of copper permeation through hollow fiber liquid membranes, *Sep. Sci. Technol.*, 18, 871 (1983).

91. A. B. Haan, P. V. Bartels, and J. Graauw, Extraction of metal ions from wastewater. Modelling of the mass transfer in a supported-liquid-membrane process, *J. Membrane Sci.*, 45, 281 (1989).
92. S. B. Iversion, V. K. Bhatia, K. Dam-Johansen, and G. Jonsson, Characterisation of microporous membranes for use in membrane contactors, in Proceedings of the 7th International Symposium on Synthetic Membranes in Science and Industry, Decchema e,v. Frankfurt am Meim: Alemanha, 22–23 (1994).
93. Y. Qin and J. M. S. Cabral, Theoretical analysis on the design of hollow fiber modules and modules cascades for the separation of diluted species, *J. Membrane Sci.*, 143, 197 (1998).
94. Y. J. Qin and J. M. S. Carbral, Lumen mass transfer in hollow-fiber membrane processes with constant external resistances, *AIChE J.*, 43, 1975 (1997).
95. A. Sengupta, R. Basu, and K. K. Sirkar, Separation of solutes from aqueous solutions by contained liquid membranes, *AIChE J.*, 34, 1698 (1988).
96. D. R. Olander, Simultaneous mass transfer and equilibrium chemical reaction, *AIChE J.*, 6, 233 (1960).
97. W. J. Ward, III, Analytical and experimental studies of facilitated transport, *AIChE J.*, 16, 405 (1970).
98. E. L. Cussler, Membrane which pump, *AIChE J.*, 17, 1300 (1971).
99. A. Sengupta, R. Basu, R. Prasad, and K. K. Sirkar, Separation of liquid solutions by contained liquid membranes, *Sep. Sci. Tech.*, 23, 1735 (1988).
100. Liquid-Ce® Cotactors, Information Brochure, Separations Products Division, Hoechst Celanese Corporation (now known as Celgard LLE) (1996).
101. A. Kumar, R. Haddad, G. Benzal, and A. M. Sastre, Dispersion-free solvent extraction and stripping of gold cyanide with LIX79 using hollow fiber contactors: Optimization and modeling, *Ind. Eng. Chem. Res.*, 41, 613 (2002).
102. R. W. Baker, Future directions of membrane gas separation technology, *Ind. Eng. Chem. Res.*, 41, 1395 (2002).

5 Solvent Extraction in the Hydrometallurgical Processing and Purification of Metals

Process Design and Selected Applications

Kathryn C. Sole

CONTENTS

5.1 INTRODUCTION

Sixty years ago, solvent extraction (SX) was employed mainly as an analytical tool for the separation and analysis of elements with very similar chemical properties. The discovery and isolation of the lanthanide and actinide elements provided impetus for the further development of these technologies, as the closely related properties of these *f*-group elements stretched the boundaries of knowledge in this young field (Nash 1993). During World War II, the Manhattan Project required significant quantities of high-purity isotopes of uranium and plutonium. The chemistry and engineering developed for these bulk separations were the first industrial applications of SX (Jenkins 1979; Seaborg 1963).

Today, SX is widely employed in a variety of industries for both the upgrading and purification of a range of elements and chemicals. The technology is used in applications as diverse as pharmaceuticals, agriculture, industrial chemicals, petrochemicals, the food industry, the purification of base metals, and the refining of precious metals.

Although SX was commercialized for uranium recovery in South Africa during the 1950s, it was not until 1968 that SX was used in base-metal hydrometallurgy, with the commissioning of the first copper plant (Bluebird Ranchers) in Arizona. Shortly thereafter, the Nchanga plant started operation in Zambia, and this remained the largest SX plant in the world for over a decade. For many years, copper and uranium remained the only applications of SX to the processing of mineral ores. During the last 15 years, however, the development of stable extractants with excellent selectivities for particular metal ions, coupled with advances in the engineering of this technology, increasing demands for higher purity products, and more environmentally friendly processing routes, led to the inclusion of SX in many flow sheets for the hydrometallurgical processing of a variety of precious and base metals. Large-scale copper plants treating over 550,000 m^3 of solution per day are found in Arizona and South America (Hopkins 2003; Kennedy et al. 1999), while precious metal refining processes may treat flow streams of only a few litres per hour (Mooiman 1993).

This chapter discusses hydrometallurgical applications that use SX for separation and purification in the primary recovery of metals from their ores. The approach taken is to examine the chemistry of the extractants in relation to the demands of the separation required and show how changing specifications for a commercial product or flow sheet configuration have led to improvements in the design of both reagents and equipment. The integration of SX operations with the up- and downstream operations is of critical importance in process applications; this aspect is addressed in discussions of flow sheet development and process design. Flow sheets for selected commodities are discussed, with the intention of demonstrating the capabilities and versatility of modern solvent-extraction systems. Practical operational issues, engineering aspects, and equipment design are also briefly covered.

5.2 CONSIDERATIONS FOR FLOW SHEET CONCEPTUALIZATION AND DESIGN

5.2.1 Nomenclature and Circuit Configurations

A generalised SX circuit is illustrated in Figure 5.1. Most commercial processes are operated in a continuous manner, although some batch processes are used, particularly when the kinetics of a reaction are very slow. SX circuits are usually

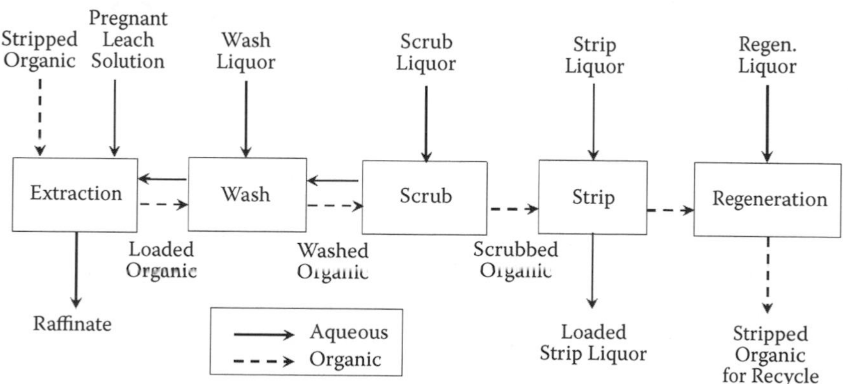

FIGURE 5.1 Generalized solvent extraction circuit.

operated in a *counter-current* manner, with the aqueous and organic streams flowing in opposite directions in order to maximise extraction (or stripping) efficiency and separation efficiency (Lloyd 2004). Some of the more widely used equipment for continuous operation is discussed in Section 5.4.

The aqueous stream requiring purification or concentration is the *feed* or, more typically for metallurgical applications, the *pregnant leach solution* (PLS). The feed is contacted with the stripped organic phase in the *extraction* circuit, and the extracted species is transferred across the interface from the aqueous phase to the *loaded organic phase*. The aqueous stream leaving the extraction circuit, depleted of the extracted species, is the *raffinate*. In the strip circuit, the *strip liquor* is contacted with the loaded organic phase and the extracted species is transferred back into an aqueous phase. The *stripped organic phase* is recycled to the extraction circuit. The concentrated or purified aqueous phase leaving the SX circuit is called the *loaded strip liquor* (LSL).

A *wash* circuit may be included to minimise the loss of organic phase by entrainment in the aqueous phases (e.g., contacting a versatic acid organic phase with strong acid to minimise aqueous solubility losses) or to minimise carry-over of contaminants to the LSL (e.g., Fe or Mn in Cu circuits) (Hein 2005). In this case, the aqueous wash liquor is contacted with the loaded organic phase and any residual aqueous-phase droplets are removed from the organic and leave the circuit in the spent wash liquor (or indirectly via the raffinate). The wash circuit could also be placed after the scrub, strip, or regeneration circuits, depending on the process chemistry or the purity requirements of the product.

In some flow sheets, a *scrub* circuit is introduced between the extraction and stripping circuits. A *scrub liquor* is introduced to "scrub off" unwanted coextracted species from the loaded organic phase by displacing the impurities with the more strongly complexed main element. This produces a *scrubbed organic* phase. For systems in which the extractant is completely selective for the species of interest or when high selectivity of extraction is not required (such as for bulk extractions), a scrub circuit is not necessary. Usually, however, some coextraction of other species occurs and a scrub circuit is employed to reduce the transfer of these species to the LSL.

The terms *wash* and *scrub* are sometimes used interchangeably. It should be noted that this is not consistent with the recommendations of the International Union of Pure and Applied Chemistry (IUPAC) for SX nomenclature (Rice et al. 1993). *Washing* refers to the *physical removal* of impurities from the organic phase, while *scrubbing* refers to the *chemical removal* of impurities from the organic phase.

Some flow sheets also include *regeneration* circuits. A regeneration circuit is used to convert the organic phase to the appropriate chemical form required for extraction. For example, stripping with strong acid may convert an extractant to its protonated form; extraction may, however, require the extractant to be in the sodium form so that pH control is facilitated. A regeneration step could be used for the conversion of the H^+ form of the extractant to the Na^+ form. An application requiring a different type of regeneration stage is the extraction of base metals by di(2-ethylhexyl)phosphoric acid (D2EHPA): any residual Fe(III) present in the feed solution loads onto this organic phase, but is not stripped by the conventional sulfuric acid strip liquor (see Section 5.3.4). Typically a separate HCl re-strip (regeneration) stage is included to control the level of iron build-up on the organic phase. Amine circuits also often require a regeneration circuit to protonate the extractant ahead of the extraction process. Other amine circuits may use a strong caustic wash to remove very strongly bound impurity species from the stripped organic phase to avoid loss of capacity on recycle of the organic to the extraction circuit.

5.2.2 Factors Influencing Composition of Organic Phase

Today, the wide range of available extractants includes *cation and anion exchangers*, *chelating* reagents, and reagents that complex the species of interest by *solvation*. Other components can also be added to the organic phase as *synergists* (to improve extraction or separation factors), as *phase modifiers* (to enhance phase separation or to increase the solubility of certain species), as *antioxidants* (to retard or prevent degradation of components of the organic phase), or as *phase-transfer catalysts* (to improve the reaction kinetics).

The viscosity of most extractants is high, so they are typically dissolved in *diluents*—usually inexpensive hydrocarbon mixtures selected to have flashpoints above 60°C (to minimize fire hazards and losses by volatilization) and a specific gravity ~0.8 to aid phase separation. Most petrochemical suppliers produce hydrocarbon mixtures with varying degrees of aromaticity and purity that are suitable for use as diluents (see Section 5.5.7). The modern trend is generally toward the phasing out of diluents containing aromatic components, because of their higher volatility, carcinogenic properties, and greater flammability, and to limit carbon emissions in compliance with environmental legislation. Certain applications, however, such as the processing of precious metals, require aromatic components to allow higher organic-phase loadings and greater solubility of the extracted complexes in the organic phase because of their better solvating power.

In choosing an optimum solvent composition for a commercial operation, chemical, physical, technical, and economic factors need to be considered. *Inter alia*, these include (Flett et al. 1983):

- Selectivity for the species of interest over other components in the aqueous feed.
- Loading capacity of the organic phase for the species of interest.
- Need to control pH and consumption of neutralizing reagents in particular applications.
- Ability to strip the extracted species with relatively inexpensive reagents
- Rates of extraction and stripping.
- Chemical stability of the organic phase in contact with the particular aqueous phase.
- Aqueous-phase solubility of the organic components.
- Organic-phase density and viscosity.
- Interfacial properties of the aqueous–organic system.
- Safety factors, such as toxicity, volatility, and flammability.
- Cost of initial fill and of regular top-ups.

The choice of organic-phase components favouring any one of these properties invariably involves poorer performance in other properties. It is often necessary to make compromises in choosing an optimum solvent for a specific application. For example, in the refining of a valuable metal such as gold or platinum, a fairly expensive reagent may be used without a significant impact on operating costs; however, in the refining of low-cost, high-tonnage metals such as copper or zinc, the reagent cost and losses become important factors in the choice of extractant. Similarly, the transport and availability of reagents to remote areas may impact as much on the choice of reagent as technical considerations. In pharmaceutical and food applications, the overriding factors are the product purity and product safety that can be attained.

5.2.3 Integration with Upstream and Downstream Operations

In integrated flow sheets, both the up- and downstream unit operations can impose constraints on the design of the SX operation. Some of these are considered below, with specific attention to the inclusion of SX in hydrometallurgical circuits (Flett et al. 1983; Fisher and Notebaart 1983).

5.2.3.1 Feed Solution

The feed solution to a metallurgical SX circuit usually originates from the leaching of an ore, concentrate, or secondary material. Although the leach liquor may have passed through a solid–liquid (S–L) separation step, residual solids are often present in solution. Different circuits can handle differing quantities of suspended solids; for example, fewer than 10 mgle of solids are recommended for copper extraction from sulfate circuits (Evans 1975), while up to 300 mgle can be tolerated in uranium circuits that use alkylphosphate extractants (Merritt 1971). The choice of contacting equipment may also be influenced by the solids' content of the clarified liquor. Pulsed columns can operate efficiently with much higher levels of entrained solids and crud than can conventional mixer–settlers (Movsowitz et al. 1997). High levels of silica in solution are often responsible for crud formation (see Section 5.5.6).

A feed solution may contain organic molecules such as humic acids from decomposed plant materials in a leach solution originating from a surface-mined ore. Surface-active agents (surfactants) such as flotation, flocculation, or coagulation reagents from upstream mineral processing operations, or brighteners, hardening, or smoothing agents in electrowinning applications, may be deliberately added. These organic species may have detrimental effects on one or more components of the solvent (Evans 1975).

It is important that the feed solution does not degrade the solvent. For example, the nitrate ion is a strong oxidising agent. Removal of species from this medium by SX may be difficult due to oxidation of the organic components. Leach solutions that contain dissolved chlorine and/or nitrates may cause oxidative degradation when in contact with certain organic phases, as periodically occurs at Rössing Uranium, Namibia (Munyangano 2007), and at certain South American copper plants (Virnig et al. 2003; Hurtado-Guzmán and Menacho 2003).

5.2.3.2 Raffinate

In many hydrometallurgical flow sheets, the SX raffinate is returned to an upstream leaching or milling circuit, either as a lixiviant or to maintain the water balance. Traces of the organic phase in the raffinate, either chemically dissolved or physically entrained, not only represent a loss of organic inventory, but may cause problems. Certain dissolved organics may be detrimental to bacteria in a bacterial leaching circuit, interfere with the action of flocculating agents during S–L separation of the leach solution, or adversely affect the efficiency of a flotation circuit.

Although most circuits are designed to achieve maximum recovery of the species of interest in a single pass, in certain flow sheets this may not be desirable. In the recovery of copper from scrap metal in an ammonia leach, for example, a relatively high concentration of copper in the raffinate is necessary for the optimal operation of the leaching circuit, while in other circuits the concentration of acid produced in the raffinate may be more critical than the corresponding metal extraction.

One feature of modern hydrometallurgical flow sheets is the inclusion of sequential SX operations. It is important that cross-contamination of the different organic systems is not introduced. For example, in a flow sheet developed for the recovery of copper and cobalt from tailings material (Feather et al. 2001; Dry et al. 1998), copper is extracted first using ACORGA®* M5640. This raffinate is treated for removal of manganese using D2EHPA. The raffinate from the second SX circuit is treated for cobalt recovery using CYANEX®** 272 (di(2,4,4-trimethylpentyl)phosphinic acid). The presence of D2EHPA in a CYANEX 272 circuit would completely upset the cobalt selectivity of the latter because of the wide difference in the relative pH-dependences of the two reagents (see Section 5.3.2).

5.2.3.3 Loaded Strip Liquor (LSL)

The purified LSL leaving the SX circuit is typically processed further, for example, by a second SX circuit, ion exchange (IX), crystallisation, precipitation, electrowinning (EW), pyrohydrolysis, or hydrogen reduction. When SX is coupled with an EW circuit for the recovery of the metal from the purified solution, the *advance electrolyte* to the EW circuit is the LSL, while the depleted *spent electrolyte* is recycled as the strip liquor. The relative solution concentrations of the strip liquor and the LSL (and the levels of various impurities) must be compatible with both SX and EW. Figure 5.2 shows a typical copper hydrometallurgical flow sheet in which the sulfuric acid balance of the circuit is maintained by the integration of leaching, SX, and EW. The ore shown is malachite, but similar reactions can be written for other acid-consuming copper minerals.

As with the raffinate stream, the organic content of the LSL exiting the SX circuit should be minimised as far as possible, both to reduce the operating costs of reagent top-up and to avoid introducing problems in the downstream operation. The organic content of the LSL is particularly important when the purified solution is to be treated by electrowinning (see Section 5.5.8). Copper EW is able to tolerate up to

* ACORGA is a registered trademark of Cytec Industries Inc.

** CYANEX is a registered trademark of Cytec Industries Inc.

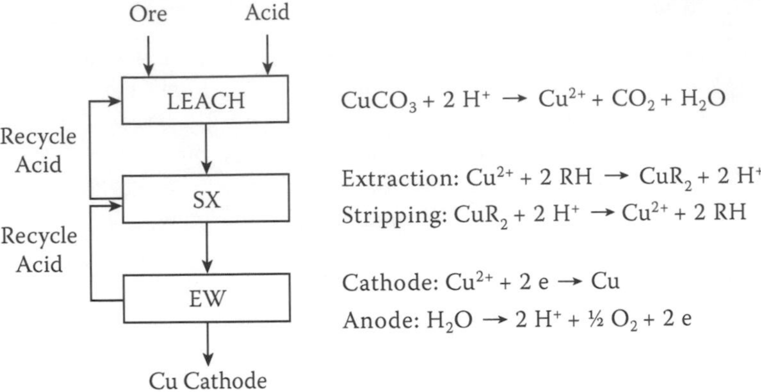

FIGURE 5.2 Flow sheet for recovery of copper by leaching, SX, and EW. RH = organic extractant; see Section 5.3.1.

30 mg/l organics in the electrolyte without adverse effects on the quality of the cathode; in contrast, however, zinc or cobalt EW requires electrolytes that are completely free of any organic species.

5.3 SELECTED APPLICATIONS OF SOLVENT EXTRACTION IN MODERN HYDROMETALLURGICAL FLOW SHEETS

5.3.1 Copper: A Mature but Expanding Technology

5.3.1.1 Copper Recovery from Sulfate Leach Liquors

One of the most remarkable success stories in the commercial application of SX occurred in the copper industry. The use of SX for the primary processing of copper has enjoyed spectacular growth over the past 35 years. The production of high-purity copper by a combination of leaching in sulfuric acid, upgrading and purification of the copper by SX, and recovery of the metal by EW has increased steadily, now approaching 30% of total copper production (Figure 5.3) (Kordosky 2002; Wallis and Chlumsky 1999). Almost 50 copper SX plants operate worldwide. The largest, El Abra in Chile, treats 5,000 m^3/h of leach solution and produces 225,000 t/a copper cathode (Kennedy et al. 1999; Taylor 1997), and Morenci in Arizona treats 23,000 m^3/h in four trains, producing 380,000 t/a cathode.

The operating cost of this hydrometallurgical route is typically 30% less per kilogram of cathode copper than conventional pyrometallurgical production routes that involve concentrating, smelting, and refining (Carter 1997). Other reasons for the growth and acceptance of SX in this industry include (1) the ability to treat low-grade and poor-quality raw materials cost effectively; (2) increased supplies of cheap sulfuric acid due to environmental pressure on smelters; (3) success in consistently producing a high-quality product; (4) the ability to survive low copper prices due to low operating costs and modest labor requirements; and (5) the ability to produce an electrolyte suitable for the EW of high-purity copper (>99.99%).

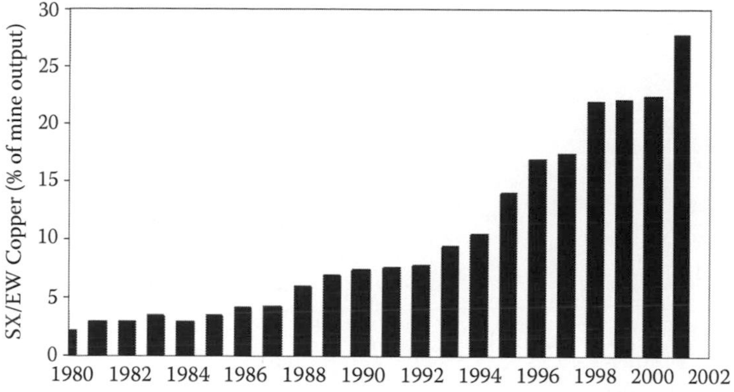

FIGURE 5.3 Contribution of SX/EW to world copper production. (From Wallis, T. L. and Chlumsky, G. F. 1999. In *Copper leaching, solvent extraction, and electrowinning technology*, Littleton, CO: Society for Mining, Metallurgy and Exploration. With permission.)

5.3.1.1.1 *Chemistry of Copper Extraction*

The copper-specific extractants used to extract copper from other base metals in acidic sulfate liquors are known as *hydroxyoximes*, and the interaction is known as *chelation*. The basic structures of modern oxime extractants are shown in Figure 5.4, where A = H for *aldoximes* and A = CH_3 for *ketoximes*. The R-chain is usually C_9H_{19} (occasionally $C_{12}H_{25}$). The complexation of copper by these extractants is shown schematically in Equation (5.1). Extraction of each copper ion releases two protons, thereby providing a useful source of acid for further leaching (see Figure 5.2). pH is an important parameter controlling the equilibrium position of this reaction.

$$Cu^{2+} + 2\ (\text{O-H, N-OH, H, } C_9H_{19}) \rightleftharpoons (C_9H_{19},\ \text{H, N, O, OH, } Cu^{+2},\ \text{O, N, OH, H, } C_9H_{19}) + 2H^{+} \tag{5.1}$$

Since the complexation of copper with the extractant molecules is an equilibrium reaction, copper can be stripped from the organic phase by invoking Le Chatelier's principle: the reaction is reversed by contacting the loaded organic phase with strong acid. This generates a purified, concentrated copper LSL from which the metal can be recovered by EW, precipitation, or crystallisation. The EW reactions also generate acid (see Figure 5.2). The spent electrolyte from the EW circuit is recycled as the strip liquor to the SX circuit, providing a closed loop for the acid requirements of the process.

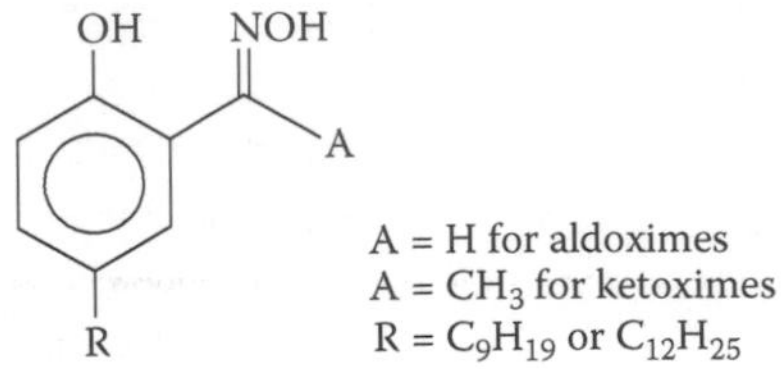

FIGURE 5.4 Modern oxime extractants.

5.3.1.1.2 Functionality of Copper Extractants

LIX®* 63, an aliphatic α-hydroxyoxime, was the first commercial reagent developed for copper. Although it was very selective for copper, it operated at pH values of 5 to 8. Since most leach solutions have pH values between 0.7 and 2.2, it was necessary to optimise the pH dependence of the extractants for use under more acidic conditions. The first reagent used in a full-scale plant was LIX 64N, a β-hydroxybenzophenone oxime developed by General Mills (now Cognis) and comprising a mixture of LIX 63 (2% by mass) and LIX 65N (40%) (Arbiter and Fletcher 1994; House 1985; Dasher and Power 1971).

Table 5.1 illustrates how the introductions of various changes to the chemical structures of these first-generation extractants improved their extraction characteristics, lowering the pH at which copper could be extracted and improving their selectivity for copper over iron. Compared with the aliphatic side chains of LIX 63, the introduction of aromatic rings into the oxime structure (LIX 65) provided an electron-withdrawing effect that increased the acidity of the exchangeable proton and allowed copper extraction at lower pH values. LIX 65 comprised two isomers, but steric constraints permit only the *anti*-isomer to act as a bidentate ligand and to complex the Cu^{2+} ion. Subsequently, LIX 64N and LIX 65N contained only the active isomer. The 8-hydroxyquinoline structure of KELEX®** 100 provided bidentate chelation at low pH, but significant co-extraction of acid also occurred. The introduction of a chlorine atom to the aromatic ring (LIX 70) provided a stronger electron-withdrawing effect and further decreased the pH at which copper extraction could be carried out. Interesting anecdotal reminiscences of the early development of the LIX reagents are given by House (1989, 1985).

5.3.1.1.3 Modern Copper Extractants

The first commercial reagents were all based on *ketoxime* functionality and were used exclusively for copper extraction for over a decade after the first full-scale application at Bluebird Ranchers Mine, Arizona, in 1968 (Arbiter and Fletcher 1994). Today, ketoximes are still successfully used in niche applications for the recovery of copper from dilute leach liquors and also find applications in nickel SX from ammoniacal solutions and in precious metal refining (see Sections 5.3.3.3 and 5.3.6.2). Particular applications of ketoximes in copper production are at El Tesoro and Lomas Bayas in the Atacama Desert of Chile, where the leach liquors of circuits often contain high

* LIX® is a registered trademark of Cognis Corporation.

** KELEX® is a registered trademark of Schering Berlin.

TABLE 5.1
First-Generation Copper Extractants: Influence of Chemical Structure on pH Dependence of Copper Extraction

Reagent	pH_{50}*	Structure
LIX 63	4.8	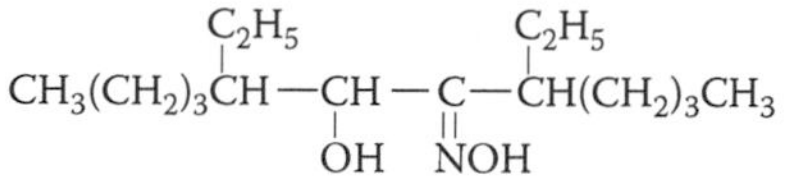 $CH_3(CH_2)_3CH(C_2H_5)-CH(OH)-C(=NOH)-CH(C_2H_5)(CH_2)_3CH_3$
LIX 65	3.3	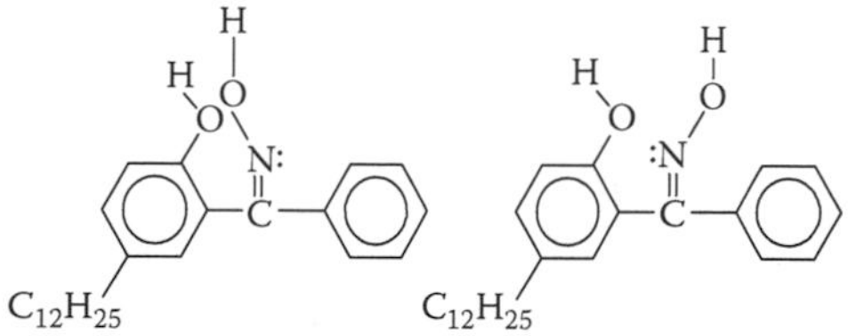 H, O, H, O, N:, C, $C_{12}H_{25}$; H, O, H, O, :N, C, $C_{12}H_{25}$
KELEX 100	1.8	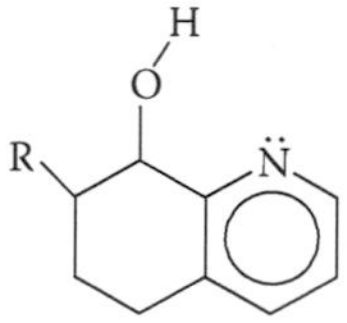 H, O, R, N
LIX 65N	2.9	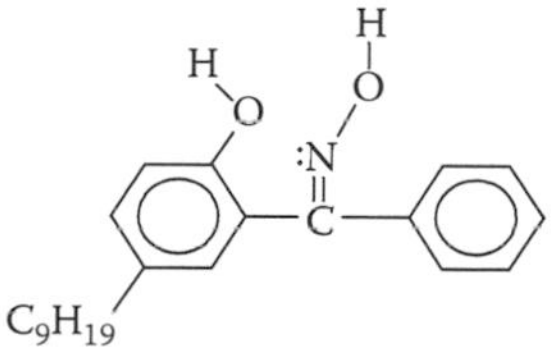 H, O, H, O, :N, C, C_9H_{19}
LIX 70	2.6	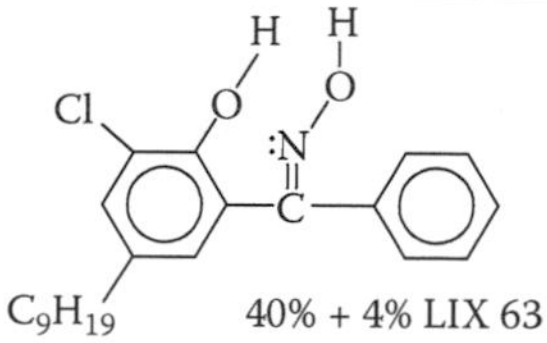 H, H, O, O, Cl, :N, C, C_9H_{19}; 40% + 4% LIX 63
SME 529		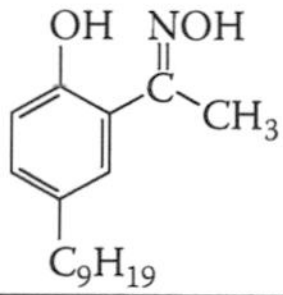 OH, NOH, C, CH_3, C_9H_{19}

* pH_{50} is defined as the equilibrium pH at which 50% of the copper in solution is extracted. pH_{50} is specific for particular operating conditions.

Source: Adapted from R. L. Atwood and J. D. Miller, paper presented at Annual Meeting of the American Institute of Mining, Metallurgical and Petroleum Engineers, San Francisco, 1972; and D. S. Flett, *Trans. Inst. Min. Metall.*, Sect. C, 83, no. 3, C30, 1974. (With permission.)

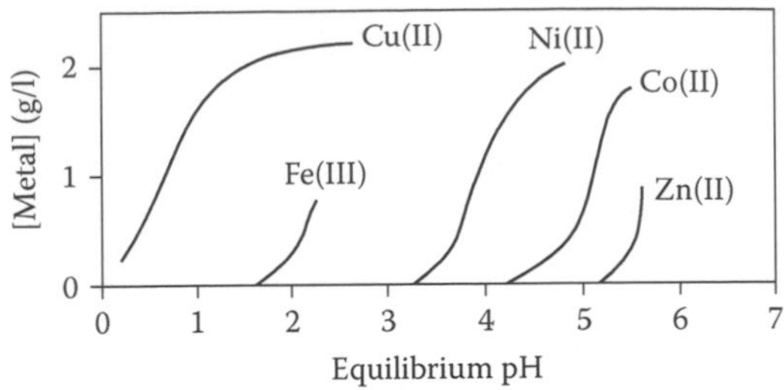

FIGURE 5.5 pH dependence of various cations by LIX 84-I ketoxime extractant. (Adapted from Cognis Corp. Undated. *MID Bluebook*. With permission.)

levels of nitrates and chlorides; ketoximes are stable to oxidation under these conditions (Kordosky 2003; Virnig et al. 2003).

Ketoximes also operate efficiently in circuits in which colloidal silica is present (they have low tendencies to form crud), have low aqueous solubility (low losses to the raffinate), and good phase separation. However, ketoximes do not generally allow the extraction of copper at pH values below 1.8, and have slow extraction kinetics (particularly in colder climates). Ketoximes exhibit copper selectivity over iron of about 300:1. Their selectivity as a function of pH for other cations commonly found in sulfate leach liquors is shown in Figure 5.5.

The second-generation *aldoxime* extractants were developed to overcome the shortcomings of the ketoximes (Maes et al. 2003; Dalton et al. 1986). Aldoxime extractants exhibit very fast extraction kinetics, high selectivity (Cu:Fe ≈ 2000:1), and high loading capacity. Because they extract at pH levels below 1, the stripping of copper is difficult. Aldoximes are therefore combined with polar organic compounds known as *modifiers* so that loaded copper can be stripped at the typical acid strength of spent electrolyte from the EW tankhouse (~180 g/l H_2SO_4) (Readett and Townson 1997). Figure 5.6 shows the influence of a stripping modifier on the copper loading of an aldoxime extractant; despite a small loss of loading capacity, a marked improvement in stripping efficiency can be attained. Nonylphenol, tridecanol, long-chain esters, and other proprietary formulations have been used as stripping modifiers.

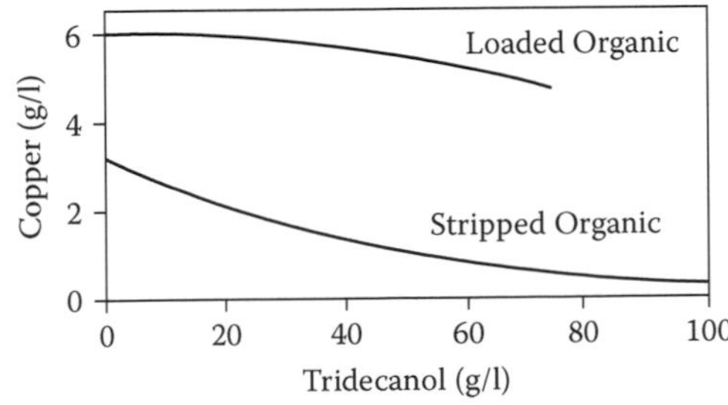

FIGURE 5.6 Copper loading and stripping of aldoxime as a function of modifier (tridecanol) concentration. (Adapted from Townsend, B. and Severs, K. J. 1990. *Min. Mag.* 162:26–35. With permission.)

TABLE 5.2
Comparison of Properties of Copper SX Reagents Based on Ester-Modified Aldoximes and Ketoxime–Aldoxime Mixtures

Property	Ester-Modified Aldoximes	Aldoxime–Ketoxime Mixtures
Strength of extractant	Strong/customised	Customized
Stripping	Customised	Customized
Cu/Fe selectivity	Excellent	Moderate
Extraction kinetics	Very fast	Fast
Phase separation	Very good	Very good
Stability	Very good	Very good
Crud formation	Dependent on circuit	Dependent on circuit
Entrainment losses	Dependent on operation and diluent	Dependent on operation and diluent

Aldoximes are less hydrolytically stable than ketoximes, but modern reagents use an ester modifier that increases stability. Solubility losses to the aqueous phase can also be higher. Ester-modified aldoximes are more selective for Cu over Fe than aldoxime–ketoxime mixtures that are in turn more selective than tridecanol-modified aldoximes. More recently, low viscosity-modified extractants have been introduced and offer the advantages of aldoximes, but with lower density and viscosity, and therefore lower entrainment losses (Sole and Feather 2003). A third class of reagents, mixtures of aldoximes and ketoximes, gives combinations of chemical and physical characteristics suited to a wide range of applications (Kordosky and Virnig 2003; Kordosky 1992; Kordosky et al. 1987).

Today modified aldoximes and aldoxime–ketoxime mixtures are the most widely used copper extractant systems. With advances in chemistry and manufacturing processes, these reagents now have high purity and many of the limitations of earlier reagents have been overcome. They have faster reaction kinetics, greater selectivity for copper over other base metals at low pH values, better extraction performance (because of a steep distribution isotherm), and more rapid phase disengagement (resulting in smaller settling requirements and reduced organic losses). Tables 5.2 and 5.3 summarise the characteristics of the main copper SX reagent systems in commercial use today.

During the past few years, vendors from China (the Chemorex* range of extractants) and India (MEX** extractants) have also started to penetrate the Western copper extractant market. This competition has led to improved technical support and service delivery, competitive pricing, and the introduction of new products (Sole et al. 2007a; Soderstrom 2006; Sole and Feather 1993) from the established vendors—all of which augur well for the future of this industry.

With copper concentrations in leach liquors ranging from 1 to 5 g/l (dump or heap leaching) to 30 to 50 g/l (vat or bioleaching), to 50 to 90 g/l Cu for pressure leaching applications, each process flow sheet will have different requirements for

* Chemorex is a registered trademark of Longlight International.

** MEX is a registered trademark of Atma International, LLC.

TABLE 5.3
Functionalities of Common Copper Extractants

Supplier	Name	Functionality
Cognis	LIX 984N	50:50 (vol.) mixture of aldoxime:ketoxime
	LIX 973N	70:30 (vol.) mixture of aldoxime:ketoxime
	LIX 622N	Tridecanol-modified aldoxime
	LIX 612N-LV	Low viscosity modified aldoxime
	LIX 84-I	Ketoxime
Cytec	ACORGA M5640	Ester-modified aldoxime
	ACORGA M5774	Ester-modified aldoxime with high modifier concentration
	ACORGA M5850	Ester-modified aldoxime
	ACORGA PT5050	Tridecanol-modified aldoxime

the recovery and purification of copper from the leach liquor. Today, the most sophisticated approach to extractant selection is based on customising the extractant composition for each plant, depending on the PLS, solids' concentrations, the presence of species such as nitrates or chlorides, engineering limitations, and geographical, climatic, or operational characteristics (Kordosky 2002). Advanced modeling capabilities available from major extractant vendors can speed up process development considerably, minimize the experimental testing required, and enable good integration with upstream and downstream operations.

As a consequence of significant advances in reagent customization during the past 30 years, large reductions in capital costs for copper SX plants have been possible. For example, the typical number of extraction stages has been reduced from four to two, the number of stripping stages has decreased from two to one, and mixer retention times have decreased from 3 to 4 minutes to 2 to 3 minutes. Low-profile units have cut capital costs. Copper production per unit area of plant size has also increased dramatically, from about 4 t/a Cu/m^2 of settler area in the early plants to 16 t/a/m^2 (Carter 1997; Hopkins 1994). Some of the most innovative advances in equipment design and circuit configurations have also been achieved in copper SX (Carter 1997; Hopkins 1994). Earlier plants were designed using only series configurations of the mixer–settler trains. The trend in recent years leans toward the use of series-parallel configurations, especially in very large plants, and from series-parallel to all-parallel (Rojas et al. 2005). This increases overall throughput and reduces capital costs.

5.3.1.2 Copper Recovery from Ammonia Leach Liquors

Copper SX has also been successfully achieved using leach systems other than sulfuric acid. The Arbiter plant in Anaconda, Montana, treated an ammoniacal leach liquor from a sulfide concentrate using LIX 64N (Anon 1973). More recently, BHP Minerals (now BHPBilliton) developed the Escondida process in which copper concentrates are leached with ammonia and ammonium sulfate, and then purified by SX with LIX 54 (Duyvestuyn and Subacky 1995; Duyvestuyn and Subacky 1993). This β-diketone extractant has the structure:

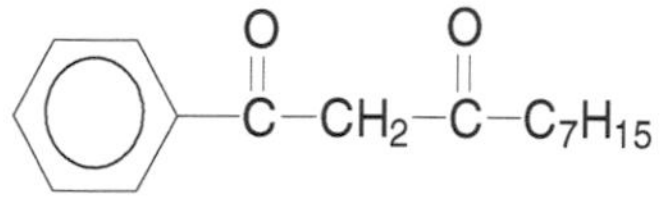

$$C_6H_5-C(=O)-CH_2-C(=O)-C_7H_{15}$$

The extraction of copper from solution is given by the reaction:

$$Cu(NH_3)_4{}^{2+}{}_{(aq)} + 2\ OH^-{}_{(aq)} + 2\ H_2O + 2\ HA_{(org)} \rightarrow CuA_{2\ (org)} + 4\ NH_4OH_{(aq)} \quad (5.2)$$

where copper is extracted as the simple cation, and ammonium hydroxide is regenerated for recycle to the leach. LIX 54 has a high net copper transfer (~30 g/l) compared with ketoximes and is readily stripped using relatively low concentrations of sulfuric acid (Kordosky 1992). A plant built and commissioned in 1994 in Escondida, Chile, (Duyvestuyn and Subacky 1995; Anon 1995), was subsequently closed.

5.3.1.3 Copper Recovery from Chloride Leach Liquors

Copper recovery by chloride hydrometallurgy has been piloted in several systems (Suttill 1989; Schweitzer and Livingston 1982; Demarthe et al. 1976), but has never been commercialised, mainly due to problems with electrowinning copper from chloride solution. Acorga CLX-50 (Dalton et al. 1991, 1983), specifically designed for the extraction of copper from chloride media, is interesting as it uses elegant coordination chemistry in a hydrometallurgical application. The extraction and stripping reactions are governed by chloride ion concentration rather than by pH as in conventional copper SX:

$$Cu^{2+}{}_{(aq)} + 2\ Cl^-{}_{(aq)} + 2\ L_{(org)} \rightarrow CuL_2Cl_{2\ (org)}, \quad (5.3)$$

where L is a diester of pyridine 3,5 dicarboxylic acid.

As research intensifies for economically attractive processes for the treatment of difficult ores and materials of construction become capable of handling more extreme environments, chloride-based processing routes are receiving renewed attention (Galbraith et al. 2007). It is possible that these technologies or their successors may form the bases of new flow sheets in the future.

5.3.2 Separation of Cobalt from Nickel: Neat Coordination Chemistry

The separation of cobalt from nickel by SX has been practised for many years, and innovative flow sheets have been developed. Several commercial processes using SX for the recovery of cobalt and nickel have been reviewed (Bautista 1993). Separations in both chloride and sulfate media are practiced commercially.

5.3.2.1 Chloride Leach Liquors

When cobalt is in solution as the chloride, it forms tetrahedral anionic complexes, $CoCl_4{}^{2}$, that can be extracted by amine-type extractants (Cox 2004; Cole 1994; Flett 1987):

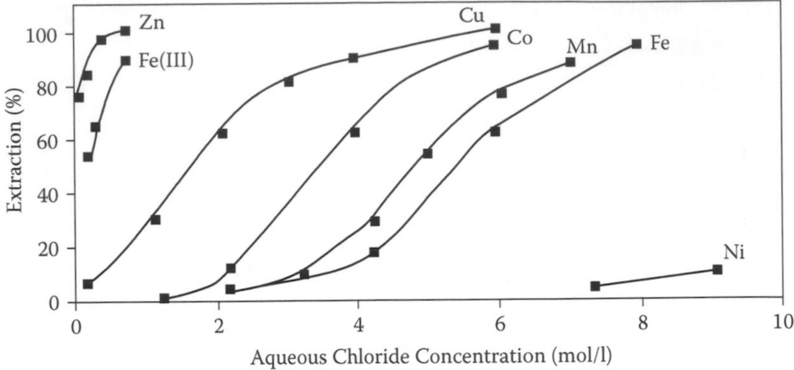

FIGURE 5.7 Extraction of some divalent base metal cations and Fe(III) from chloride solution by Alamine 336 hydrochloride. (From Nicol, M. J. et al. 1987. In *Comprehensive coordination chemistry*, vol. 6, Oxford: Pergamon. With permission.)

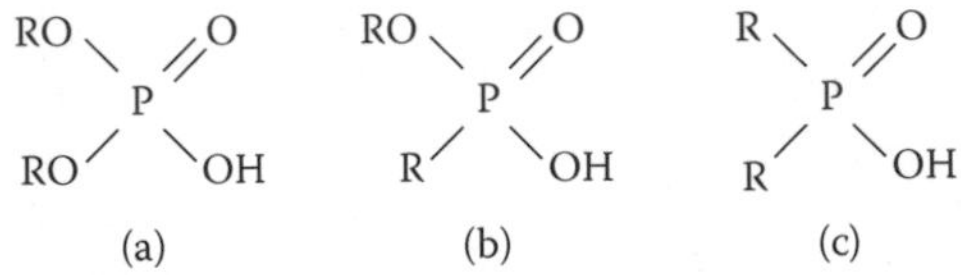

RO, RO, P, O, OH (a) — RO, R, P, O, OH (b) — R, R, P, O, OH (c)

FIGURE 5.8 Comparison of (a) dialkylphosphoric, (b) dialkylphosphonic, and (c) dialkylphosphinic acid extractants.

$$CoCl_4^{2-}{}_{(aq)} + 2\ R_4N^+Y^-{}_{(org)} \rightarrow (R_4N^+)_2CoCl_4^{2-}{}_{(org)} + 2\ Y^-{}_{(aq)} \tag{5.4}$$

where R denotes an alkyl group or hydrogen, and Y is a univalent anion. Nickel remains behind in the aqueous phase because the corresponding nickel chloride species is neutral ($NiCl_2$). Figure 5.7 shows the relative selectivities of base-metal extractions by a tertiary amine.

This separation principle has been applied commercially at Falconbridge Nikkelverk in Norway (Wigstol and Froyland 1972), at Sumitomo Metal Mining Co. in Japan (Suetsuna et al. 1980), and at the Société de Nickel refinery in France (Bozec et al. 1974). More recently, this chemistry has been included in the flow sheet for the processing of a nickel laterite in the Goro process in New Caledonia (Mihaylov et al. 2000) (see Section 5.3.3.2).

5.3.2.2 Sulfate Leach Liquors

In sulfate media, cobalt is separated from nickel using dialkylphosphorus extractants (Figure 5.8). The basicity of these reagents increases with decreasing distance of the alkyl chain from the central phosphorus atom, from the phosphoric acid structure

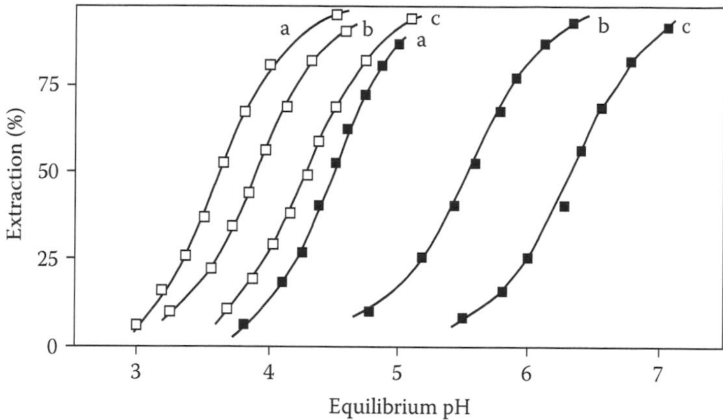

FIGURE 5.9 pH dependence of Co (□) and Ni (■) extraction by di(2-ethylhexyl)phosphorus extractants (a = phosphoric acid; b = phosphonic acid; c = phosphinic acid). (From Preston, J. S. and du Preez, A. C. 1988. *Mintek Report M378*. With permission.)

(exemplified by D2EHPA, DP-8R*) to the phosphonic acid (PC-88A**, Ionquest*** 801), to the phosphinic acid (CYANEX 272, Ionquest 290, LIX 272). This results in a corresponding improvement of the Co–Ni separation factors from 14 to 280 to 7000 (Rickelton and Nucciarone 1997).

The influence of the basicity of the ligands is demonstrated in Figure 5.9, showing the extraction of nickel and cobalt by three organophosphorus analogues. The extraction reactions are (Preston and du Preez 1988; Preston 1982):

$$Co^{2+}{}_{(aq)} + 2\ H_2A_{2\ (org)} \rightarrow CoA_2{\cdot}H_2A_{2\ (org)} + 2\ H^+{}_{(aq)} \tag{5.5}$$

$$Ni^{2+}{}_{(aq)} + 3\ H_2A_{2\ (org)} + H_2O \rightarrow NiA_2{\cdot}H_2A_2{\cdot}H_2O_{(org)} + 2\ H^+{}_{(aq)}, \tag{5.6}$$

where the extractant has been represented as the dimeric hydrogen-bonded species (Figure 5.10).

The selective extraction of cobalt over nickel is possible for the following reasons (Golding and Barclay 1988; Danesi et al. 1985; Preston 1982):

- Formation of the tetrahedrally coordinated cobalt species is more favourable in the organic phase than the octahedrally coordinated nickel species (Figure 5.10).
- The cobalt complex is hydrophobic, while the nickel complex can contain one or two coordinated water molecules in its inner sphere and is therefore more hydrophilic.

---

* DP-8R is a product of Diahachi Chemical Co. Ltd.

** PC-88A is a product of Diahachi Chemical Co. Ltd.

*** Ionquest® is a registered trademark of Rhodia Inc.

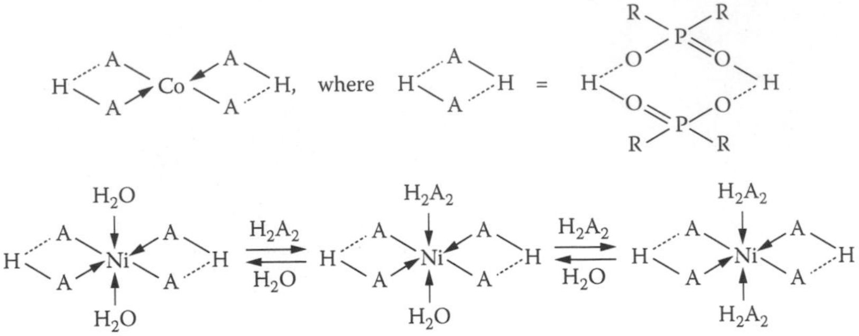

FIGURE 5.10 Coordination of Co(II) and Ni(II) by dialkylphosphinic acid. (From Preston, J. S. 1982. *Hydrometallurgy* 9:115–123. With permission.)

Figure 5.11 shows the pH dependence of extraction of organophosphorus acids for common base metal ions. In general, the extraction of a particular ion shifts to higher pH in the order phosphoric → phosphonic → phosphinic (i.e., the extractants are weaker). It is also significant that the order of extraction changes as well. In particular, the positions of Ca and Mg relative to base-metal cations should be noted.

Some older plants still use D2EHPA (Anglo Platinum Base Metal Refiners, South Africa) (Sole 1999a) or PC-88A (Nippon Mining Co., Japan) for the separation of cobalt from nickel (Flett 1987). Today, however, the preferred reagent is di(2,4,4-trimethylpentyl)phosphinic acid (commercially available as CYANEX 272, Ionquest 290, or LIX 272), used to produce some 40% of the world's cobalt (Rickelton and Nucciarone 1997). Not only are the Co–Ni separation factors higher, but the phosphinic acid is more stable with respect to the oxidative degradation of the diluent by Co(III), which can be a problem in these systems (Maxwell et al. 1999; Rickelton et al. 1991; Flett and West 1986). CYANEX 272 also has the advantage of being selective for cobalt over calcium, so gypsum crud formation is minimised in these systems. Any coextracted Fe(III) is also easily stripped by sulfuric acid (Cole 2002). Indicative of the improvements in performance possible is a comparison of the staging required for a typical Co–Ni separation: the Anglo Platinum D2EHPA circuit (Rustenburg, South Africa) uses 19 stages to purify a solution that contains essentially only Co and Ni, while the Outokumpu CYANEX 272 process requires only 11 stages to purify cobalt from a far more complex feed liquor.

Although Cytec has dominated this market for the past two decades with CYANEX 272, the patents on this reagent have now expired and similar reagents have recently been produced by Rhodia (Ionquest 290) and Cognis (LIX 272). Chinese and other vendors are also considering the production of this chemical. This competition will open up new possibilities for the cobalt SX market in the near future.

5.3.3 Nickel Laterite Processing: A Variety of Solvent Extraction (SX) Options

While high-grade nickel sulfides are processed by smelting, low-grade laterites (0.5 to 3% Ni), until relatively recently, could not be economically treated with available

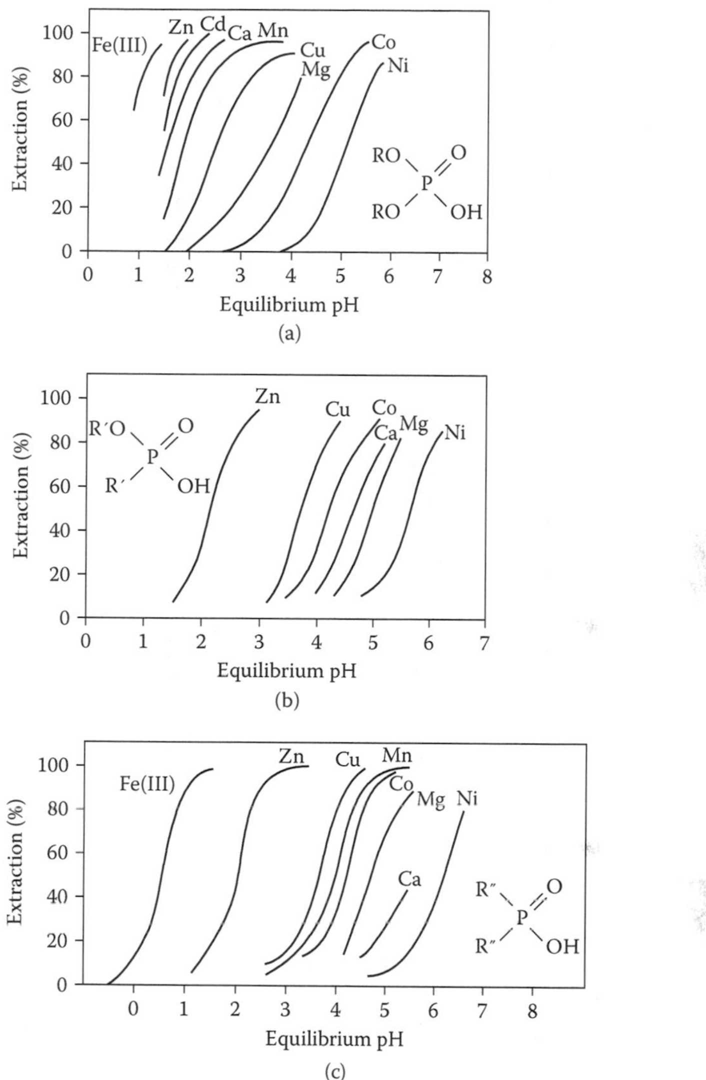

FIGURE 5.11 Extraction of base metals as function of pH by (a) D2EHPA (dialkylphosphoric acid), (b) Ionquest 801 (dialkylphosphonic acid), and (c) CYANEX 272 (dialkylphosphinic acid). (From Sole, K. C. and Cole, P. M. 2001. In *Ion exchange and solvent extraction*, vol. 15, New York: Marcel Dekker. With permission.)

technologies. Since 1998, several new laterite flow sheets have been commissioned or are under advanced development. These processes all use pressure acid leaching (PAL) to solubilize the metals of interest, but the downstream flow sheets have significant differences. They all involve SX as one or more of the unit operations, either for the removal of cobalt from the nickel-rich leach liquor or for the purification of nickel liquor. For a more detailed review of nickel SX applications, see Sole and Cole (2001).

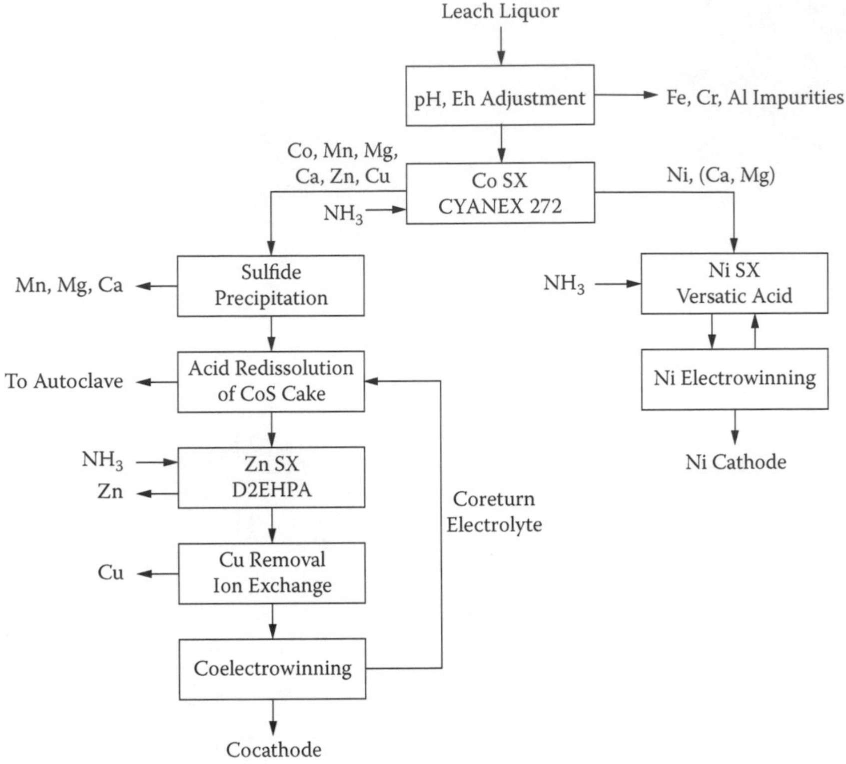

FIGURE 5.12 Bulong flow sheet for purification of nickel and cobalt from laterite leach liquor. (From Sole, K. C. and Cole, P. M. 2001. In *Ion exchange and solvent extraction*, vol. 15, New York: Marcel Dekker. With permission.)

The flow sheets discussed in this section provide a good overview of the modern versatility of SX chemistries. Among the extractant classes used for nickel purification are the dialkylphosphorus acids, sulfur-substituted dialkylphosphinic acids, carboxylic acids, and tertiary amines. The Bulong and Goro flow sheets are examples of two very different processes that make extensive use of SX to produce high-grade nickel products from the sulfate leaching of lateritic ores. The Queensland Nickel and Cawse flow sheets are examples of nickel recovery from ammoniacal solutions.

5.3.3.1 Bulong Process

The Bulong nickel operation near Kalgoorlie, Western Australia, was commissioned in 1998 with a design capacity of 9000 t/a Ni and 720 t/a Co (Donegan 2006; Ritcey et al. 1996). This plant closed in 2003, but later reopened as Avalon Nickel to treat nickel sulfide material. The original flow sheet remains of interest, particularly since it has many similarities with those planned for Tati Nickel, Botswana (Sole et al. 2005), and the Nkomati project in South Africa (Feather et al. 2002a).

Figure 5.12 shows a simplified flow sheet of the Bulong downstream purification process. Following dissolution of the ore at 4500 kPa and 250°C, the autoclave

TABLE 5.4
Compositions of Process Streams in Bulong Cobalt SX Circuit

	Concentration (mg/l)			
Element	**Feed**	**Raffinate**	**Loaded Organic**	**Loaded Strip Liquor**
Co	280	5.3	530	7200
Ni	3500	2800	10	120
Ca	500	500	5.8	180
Mn	990	0.4	2300	32 000
Zn	29	<0.2	66	950
Mg	15 000	14 000	3100	37 000
Co:Ni ratio	0.08	—	53	1440

Source: Updated from Soldenhoff, K. et al. 1998. *EPD Congress*, Warrendale, PA: The Minerals, Metals and Materials Society. (With permission.)

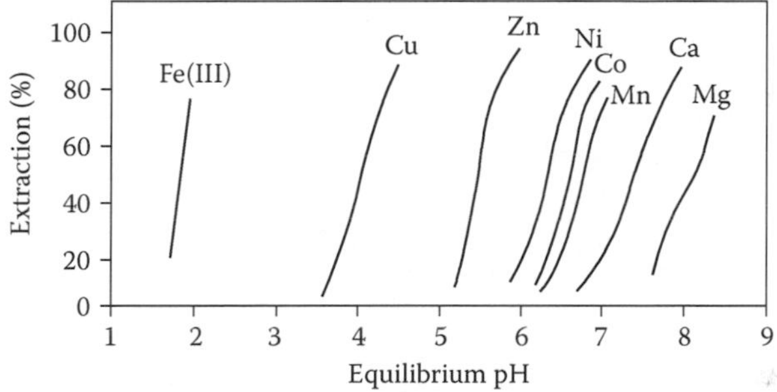

FIGURE 5.13 Selectivity of versatic acid for nickel and other cations. (Adapted from Preston, J. S. 1985. *Hydrometallurgy* 14:171–188. With permission.)

discharge was flashed down and thickened. The leach liquor was then processed directly for the recovery of the valuable metals. Fe(III), Al, and Cr(III) were removed to <1 mg/l by precipitation with limestone. In the first of three SX operations, Co, Cu, Zn, and Mn were separated from Ni by CYANEX 272. The relative selectivity of this extractant for the various base-metal cations is illustrated in Figure 5.11, while Table 5.4 shows illustrative compositions of the various process streams in the cobalt SX circuit. Extractions of Co, Mn, and Zn are excellent, with minimum co-extraction of nickel (Soldenhoff et al. 1998).

The raffinate (cobalt-barren stream) from the cobalt SX circuit was refined by nickel SX to produce a high-grade solution suitable for the production of nickel cathode. This extraction was carried out using *versatic acid*, a long-chain carboxylic acid, using pH control to limit the co-extraction of Mg and Ca (Figure 5.13). Although carboxylic acids are inexpensive, among their disadvantages is their high aqueous solubility (Nagel and Feather 2001). The consequent extractant losses to the raffinate are high and a solvent-recovery stage was needed. The relatively high pH (6 to 7) at which nickel is extracted requires tight control to avoid problems with the extraction of calcium and consequent gypsum formation. The nickel-containing LSL from this process had a purity of >99%, from which nickel cathode (>99.5% purity) was produced by EW. The acid balance between the nickel SX and EW circuits was maintained by recycling the spent electrolyte as the strip liquor.

A third SX operation removed zinc from the cobalt circuit. Cobalt was precipitated as a sulfide from the cobalt SX LSL to reject Mn and some Mg, before the filter cake was dissolved in dilute acid. Zinc was removed from the resulting liquor by SX using D2EHPA. As shown in Figure 5.11, this extractant achieves good selectivity for Zn over Co by appropriate pH control of the extraction reactions.

Ion exchange was used to remove residual copper from the advance electrolyte prior to the recovery of cobalt as high-purity cathode. Aminophosphonic acid resins exhibit the selectivity sequence Fe(III) > Pb > Cu > Zn ~ Al > Mg > Ca > Cd > Ni > Co > Sr > Ba > Na for metal cations, and are therefore suited to the removal of small quantities of copper and zinc from nickel or cobalt electrolytes. At Bulong, copper was reduced from 360 to <0.5 mg/l (>99.8% removal) while limiting the associated cobalt loss to 0.4% (Pavlides and Wyethe 2000).

As both the leaching and downstream processes were unproven technologies, Bulong's selection of this flow sheet represented a high-risk venture, and subsequent plant performance attracted much interest in the industry. Bulong was unfortunately plagued by engineering and operational problems after start-up and eventually was forced to close. It is nevertheless believed that the downstream process chemistry is inherently sound, and that this flow sheet has some advantages over other nickel processes that use a series of precipitation and redissolution steps rather than treating the leach liquor directly (Sole and Cole 2001).

5.3.3.2 Goro Process

The Goro process (Figure 5.14) developed by INCO (now CVRD-Inco) for the treatment of a nickel laterite deposit in New Caledonia (Bacon and Mihaylov 2002; Mihaylov et al. 2000; Mihaylov et al. 1995) also treats the laterite leach liquor directly but uses a completely different approach. The flow sheet involves the co-extraction of nickel and cobalt by SX with the sulfur-substituted organophosphinic acid, CYANEX 301, followed by their separation in chloride medium by SX using an amine extract.

The lateritic ore is acid leached under pressure at 270°C. The clarified autoclave discharge liquor is partially neutralised to precipitate Al, Cr(III), Cu, Fe, and Si. Trace quantities of residual copper are reduced to levels of <0.04 mg/l by ion exchange using a chelating resin with iminodiacetate functionality. Copper must be removed effectively because it poisons the extractant used in the subsequent SX operation and catalyses its oxidation (Sole and Hiskey 1995).

Ni, Co, and Zn are quantitatively and selectively extracted from other metals in solution by CYANEX 301. CYANEX 301 and CYANEX 302 are, respectively, the di- and mono-substituted sulfur analogues of CYANEX 272. Their structures and pH dependencies for the extraction of base metals are shown in Figure 5.15. CYANEX 301 has the unique capability of extracting Ni and Co selectively from Mn, while also being very selective over Mg and Ca. Since these three elements

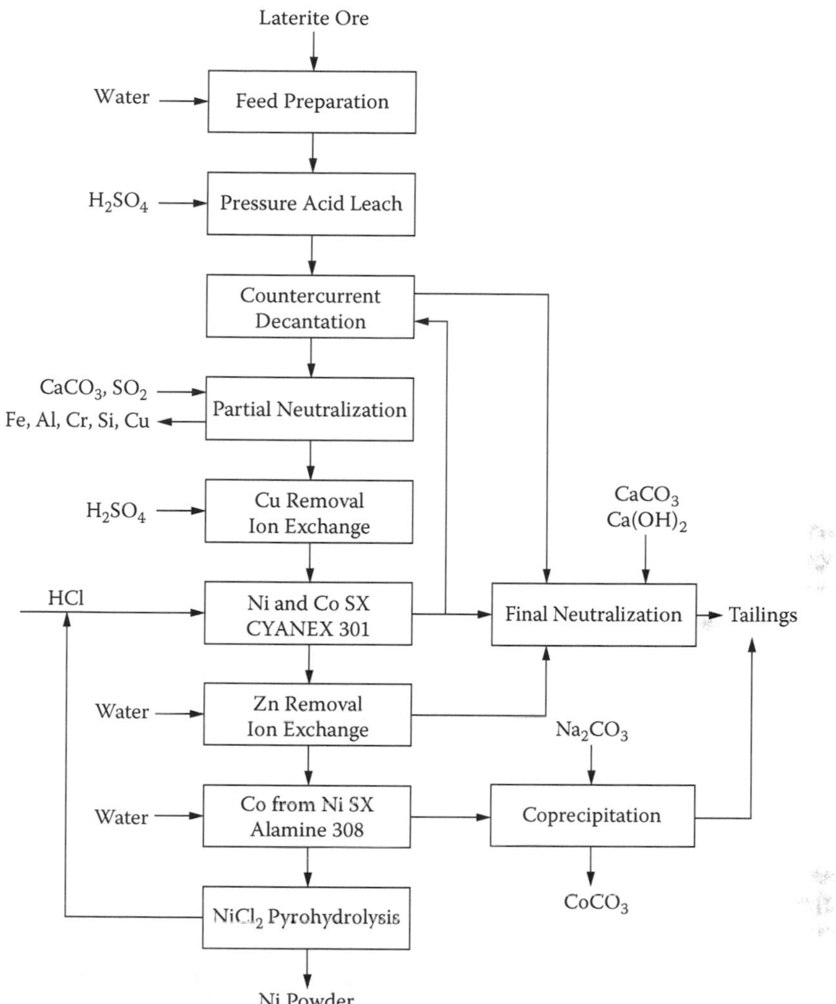

FIGURE 5.14 Goro process for recovery of nickel from laterites. (Adapted from Mihaylov, I. et al. 2000. *CIM Bull.* 93:124–130. With permission.)

constitute the main impurities in the ore, this is a tailor-made application for this extractant. Excellent extraction is also achieved under acidic conditions (pH 1 to 2), so no pH adjustment is required in the SX circuit and the waste streams can be neutralised with inexpensive limestone. (CYANEX 302 has not been commercially implemented to date, but is currently under development for use in nickel processing flow sheets.)

CYANEX 301 cannot be stripped with H_2SO_4, so HCl is used. The conversion from a sulfate to a chloride matrix serves a useful purpose in that cobalt and zinc exist as anionic MCl_4^{2-} species in solution, while nickel is present as the neutral $NiCl_2$ species. Zinc is removed using an anion exchange resin. Cobalt and nickel are

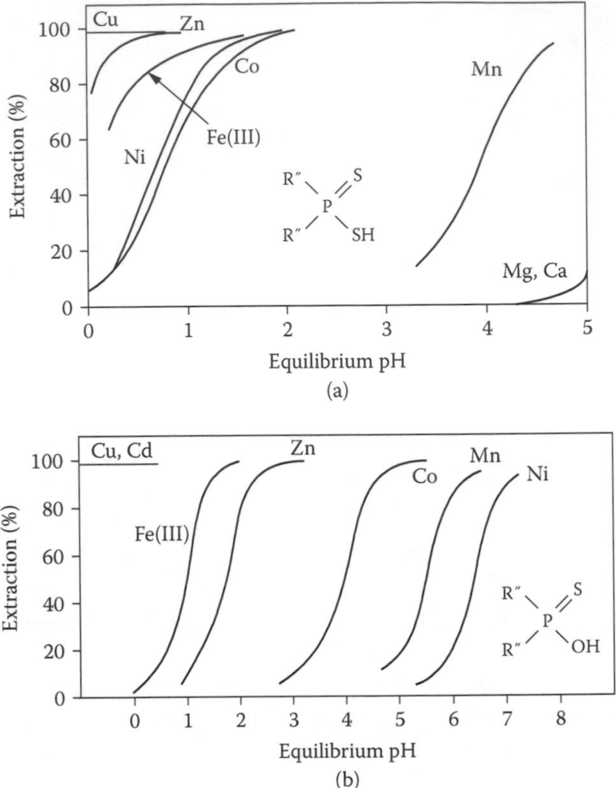

FIGURE 5.15 Extraction of base metals as function of pH by (a) CYANEX 301 (di(2,4,4-trimethylpentyl)dithiophosphinic acid and (b) CYANEX 302 (di(2,4,4-trimethylpentyl)thiophosphinic acid. (From Sole, K. C. and Cole, P. M. 2001. In *Ion exchange and solvent extraction*, vol. 15, New York: Marcel Dekker. With permission.)

then separated using an inexpensive tertiary amine extractant (Alamine®* 308) (see Figure 5.7). Nickel is recovered by pyrohydrolysis as NiO, regenerating HCl for recycle to the CYANEX 301 strip circuit, while cobalt is recovered as a carbonate salt.

The construction of Goro was underway in 2007 and the project will be the first commercial application of CYANEX 301. The SX contacting equipment to be used will be pulsed columns rather than the traditional mixer–settlers used in most base-metal plants (see Section 5.4.2). On completion, Goro is predicted to become one of the largest and most cost-effective nickel projects in the world (Bacon and Mihaylov 2002).

5.3.3.3 Queensland Nickel Process

The Yabulu refinery of Queensland Nickel Industries (QNI) in Townsville, Australia, is the largest nickel SX plant in the world. The flow sheet (Figure 5.16) comprises

* Alamine is a registered trademark of Cognis Corporation.

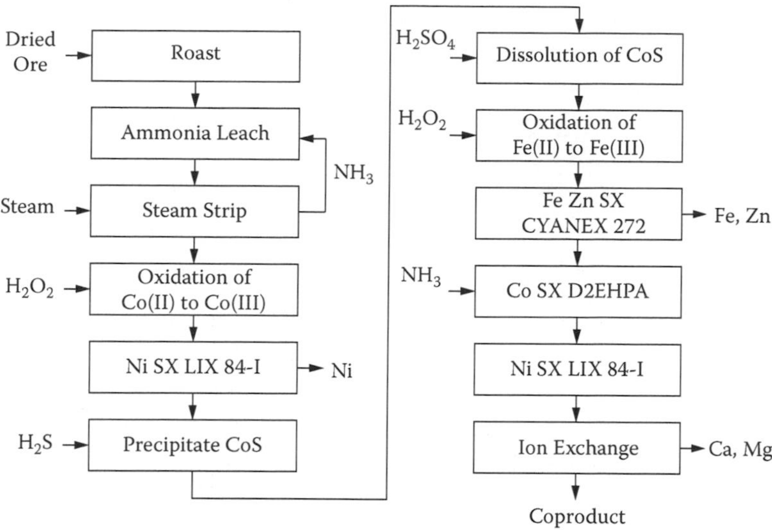

FIGURE 5.16 Simplified flow sheet of Queensland Nickel Industries process.

an ammonia leach circuit for the recovery of nickel (Mackenzie et al. 1998; Price and Reid 1993). The bulk of the nickel is extracted using the ketoxime extractant LIX 84-I (Mackenzie et al. 2005; Mackenzie and Virnig 2004; Mackenzie et al. 1998). The extraction of nickel from ammoniacal solutions depends on the particular nickel amine complex that is present in solution:

$$Ni(NH_3)_4{}^{2+}{}_{(aq)} + 2\ RH_{(org)} \rightarrow R_2Ni_{(org)} + 2\ NH_{3\ (aq)} + 2\ NH_4{}^+{}_{(aq)} \tag{5.7}$$

or

$$Ni(NH_3)_6{}^{2+}{}_{(aq)} + 2\ RH_{(org)} \rightarrow R_2Ni_{(org)} + 4\ NH_{3\ (aq)} + 2\ NH_4{}^+{}_{(aq)}. \tag{5.8}$$

As seen from the position of equilibrium in the reactions above, the extraction of nickel depends on the concentration of free ammonia. Stripping is achieved with a solution of strong ammonia and ammonium sulfate. Cobalt is recovered from the raffinate of the Ni SX circuit by precipitation as the sulfide. This material is then redissolved for further purification. Fe(III) and Zn are removed by SX with D2EHPA, Co is purified by CYANEX 272, and then any remaining Ni in the cobalt electrolyte is recovered by second the LIX 84-I SX step.

One of the main disadvantages of this flow sheet is that it requires cobalt in the feed solution to be oxidised from Co(II) to Co(III) prior to nickel SX. Divalent cobalt loads onto the organic phase, but does not strip. Co(III) also causes oxidation of the oxime extractant to a ketone, thereby resulting in loss of extractant capacity. Because some residual Co(II) is always present in the SX feed solution, it is necessary to include an oxime regeneration process and a reductive strip of the organic phase to remove cobalt (Skepper and Fittock 1996).

5.3.3.4 Cawse Process

The Cawse plant in Western Australia also recovered lateritic nickel from an ammoniacal liquor by SX (Burvill 1999; Kyle and Furfaro 1997). The flow sheet includes PAL, iron oxidation and removal, and precipitation of cobalt/nickel hydroxide. Today, the processing stops here and the mixed hydroxide precipitate is shipped to Scandinavia for refining. However, for several years, this intermediate product was subjected to an ammonia re-leach, followed by SX to produce a pure nickel sulfate for nickel EW while cobalt was precipitated as the sulfide.

LIX 84-I was also used as the extractant in this process. Nickel was selectively and quantitatively extracted, leaving <1 mg/l Ni in the raffinate. Cobalt(III) ammine was rejected to the raffinate, along with most of the other impurities in solution. In contrast to the QNI circuit, this flow sheet employed sulfuric acid for stripping:

$$R_2Ni_{(org)} + 2\,H^+_{(aq)} + SO_4^{2-}{}_{(aq)} \rightarrow 2\,RH_{(org)} + Ni^{2+}{}_{(aq)} + SO_4^{2-}{}_{(aq)}. \tag{5.9}$$

The concentration and flow rate of the acid stream are controlled to produce a LSL at pH of 3 to 3.5.

5.3.4 Zinc: First Application of SX to Primary Processing

Traditional processing routes for zinc sulfide concentrates produce high-tenor purified zinc solutions suitable for direct EW of the metal. Such schemes are not cost-effective for lower-grade zinc deposits. Several zinc projects are now under development in which the preferred flow sheets incorporate SX–EW technology (Cole and Sole 2003; Cole and Sole 2002). The Skorpion Zinc operation in Namibia (Figure 5.17) is an interesting example of the use of leaching and SX–EW to recover zinc from an ore previously considered untreatable from both the processing and economic points of view. Since producing first metal in May 2003, Skorpion Zinc has consistently met the stringent specification for the advance electrolyte to produce special high-grade (SHG) zinc (>99.995% purity) (Fuls et al. 2005; Gnoinski et al. 2005). This is the first plant to use direct SX on the mainstream leach liquor in primary zinc processing and is one of the lowest cost zinc producers.

The Skorpion ore is a low-grade silicate containing 10 to 40% Zn and ~ 26% Si. Following an atmospheric leach in sulfuric acid, Fe, Al, and Si are rejected from the leach liquor by precipitation. Zinc is then purified by SX with D2EHPA ahead of EW. Zinc EW typically requires an extremely pure electrolyte containing ~100 g/l Zn. Because of the low zinc content of this feed material, leaching conditions that produce a high-grade liquor would cause unacceptably high soluble losses of zinc in the subsequent filtration step. The Skorpion leach liquor also contains significant quantities of fluoride and chloride that are detrimental to zinc EW.

The selection of SX as the purification route serves several purposes in this flow sheet. Sulfuric acid leaching of silicates yields silicic acid, $Si(OH)_4$, in solution which, if treated incorrectly, may form silica gel that impairs S–L separation. Leaching can be carried out so that a fairly dilute liquor is obtained, thereby minimising soluble losses of the valuable metal and avoiding silica gel formation (Matthew and

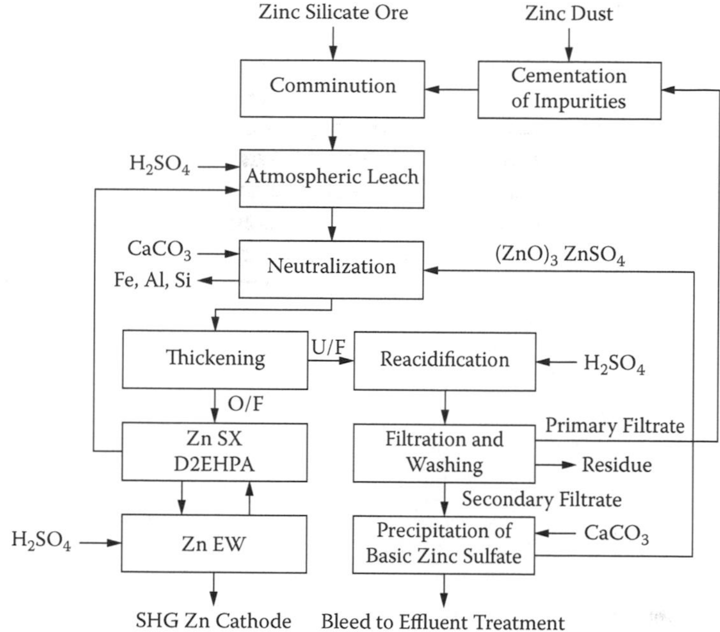

FIGURE 5.17 Simplified flow sheet of Skorpion Zinc process. (From Cole, P. M. and Sole, K. C. 2003. *Miner. Proc. Extr. Metall. Rev.* 24:91–137. With permission.)

Elsner 1977; Wood et al. 1977). In addition to rejecting other base-metal impurities (except for Fe(III)) (Figure 5.11), the halides can be effectively rejected from the circuit by appropriate control of the SX operating conditions. The process stream is upgraded from 30 to 100 g/l Zn, providing a solution suitable for EW. The selectivity of D2EHPA for Zn over Ca eliminates gypsum precipitation in the EW circuit, which is a major contributor to cellhouse downtime in conventional circuits.

The Skorpion plant is the largest yet built for zinc SX, with a PLS flow rate of 960 m^3/h (Fuls et al. 2005). The circuit (Figure 5.18) comprises three extraction, three washing, and two strip stages. The first two washing stages use demineralised water to remove physically entrained impurities (specifically halides); diluted spent electrolyte is employed as a scrub liquor in the third wash stage to remove coextracted base metal impurities from the loaded organic phase. Spent electrolyte is employed as the strip liquor, and the advance electrolyte generated is fed forward to EW for zinc recovery as cathode. Iron build-up on the organic phase is controlled by treating a bleed stream with 6 *M* HCl in the regeneration circuit.

The success of the SX process chemistry can be assessed with reference to the typical quality of the zinc cathode produced, which consistently exceeds SHG grade due to the high electrolyte purity. Table 5.5 shows the specification for SHG cathode and a typical analysis of the Skorpion Zinc product.

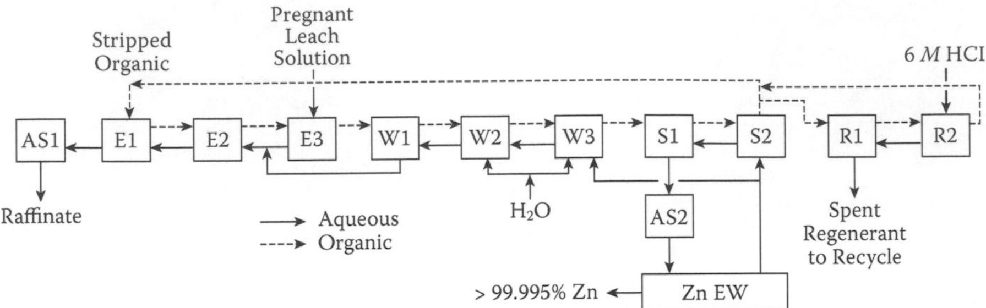

FIGURE 5.18 Skorpion Zinc SX circuit configuration. (From Cole, P. M. and Sole, K. C. 2003. *Miner. Proc. Extr. Metall. Rev.* 24:91–137. With permission.)

5.3.5 Uranium: Old Technologies and New Applications

Although uranium SX has not been of major commercial importance for several decades, it is receiving renewed interest as a "clean" energy source. A large number of projects presently under development involve SX. The technology has not changed significantly during the past three decades, and it remains of interest as one of the first full-scale applications of SX. Much of the knowledge applied to uranium SX was developed during the Manhattan Project, and several interesting accounts of this research have been published (Seaborg 1972, 1963).

The recovery of uranium from ores uses SX to reject impurities and concentrate the uranium in solution so that it can be economically recovered (Gupta and Singh 2003; Lloyd 1983). The choice of extractant depends on the lixiviant used in the upstream leaching operation, which, in turn, depends on the type of ore in which the uranium is found. Most uranium-bearing ores are readily leached in sulfuric acid and the uranium is recovered by SX using amines or dialkylorganophosphorus acids. Phosphate ores (such as those in Florida) are leached in a mixture of sulfuric and phosphoric acids or in phosphoric acid alone. Hot nitric acid has also been used as a lixiviant for uranium ores (as at Phalaborwa, South Africa). The two common extraction systems for the recovery of uranium(VI) from sulfate leach liquors are compared in Table 5.6.

TABLE 5.5
Comparison of British Standard for SHG Zinc with Typical Skorpion Zinc Cathode Analysis

Element	British Standard (%)	Skorpion Cathode (%)	Element	British Standard (%)	Skorpion Cathode (%)
Zn	99.995	99.9956	Cu	0.0010	0.0007
Cd	0.0030	0.0000	Sn	0.0010	0.0000
Fe	0.0020	0.0015	Al	0.0011	0.0001
Pb	0.0030	0.0027			

Source: From Cole, P. M. and Sole, K. C. 2003. *Miner. Proc. Extr. Metall. Rev.* 24:91–137. (With permission.)

TABLE 5.6
Comparison of Amine and Alkylphosphoric Acid Extractants for the Recovery of Uranium from Sulfate Solution

Parameter	Tertiary Amine	D2EHPA
Extraction kinetics	Very rapid	Relatively slow
Selectivity	Sensitive to vanadates, molybdates, and simple anions	Sensitive to Fe(III), Th(IV), V(IV), Ti(IV), rare earths, and Mo
Stability	Stable	Slightly unstable
Cost	More expensive	Cheaper
Solids	Tend to be sensitive to suspended solids, flotation reagents, oils, etc.	Relatively insensitive to the presence of suspended solids
Phase separation	Relatively slow	Rapid
Stripping	Easy to strip under mild conditions	Usually stripped with Na_2CO_3; needs modifier to prevent third phase formation

Source: Adapted from Lloyd, P. J. D. 1983. In *Handbook of Solvent Extraction*, New York: John Wiley & Sons. (With permission.)

Uranium can be present in acid solution as either U^{4+} or UO_2^{2+}, depending on the redox potential. It is necessary to convert all uranium to one or other of these species to ensure efficient extraction. If the solution is oxidised, however, any iron present forms Fe(III) that may interfere in certain extraction processes, and it may be necessary to remove iron via a precipitation route prior to uranium SX.

5.3.5.1 Sulfuric Acid Leach Liquors

5.3.5.1.1 Organophosphorus Extractants

Organophosphorus reagents were among the first used for the commercial recovery of uranium from solutions obtained from the leaching of low-grade ores with sulfuric acid. In the DAPEX process (Lloyd 1983; Merritt 1971; Blake et al. 1958), D2EHPA is used to selectively extract uranium from vanadium and iron(III) under conditions of controlled pH and electrochemical potential (Nicol et al. 1987). Tri-*n*-butylphosphate (TBP) or isodecanol is added to the organic phase to prevent the formation of a third phase. The extraction of uranium(VI) by D2EHPA (represented as the dimeric H_2A_2 structure) occurs via the reaction:

$$UO_2^{2+}{}_{(aq)} + 2\,H_2A_{2\,(org)} \rightarrow UO_2(HA_2)_{2\,(org)} + 2\,H^+{}_{(aq)}. \tag{5.10}$$

The most widely accepted structure of the extracted complex is (Nicol et al. 1987; Baes 1962):

O
A A
H U H
A A
O

At low acidities, the dimeric complex has also been reported (Nicol et al. 1987, Sato 1982).

Uranium is stripped from the loaded organic phase with a solution of sodium carbonate and recovered as sodium uranyltricarbonate, while the extractant is regenerated in the sodium form:

$$UO_2(HA_2)_{2\ (org)} + 3\ Na_2CO_{3\ (aq)} \rightarrow 2\ NaHA_{2\ (org)} + Na_4UO_2(CO_3)_{3\ (aq)}. \quad (5.11)$$

5.3.5.1.2 Amine Extractants

Today, amines are more widely used for the recovery of uranium from sulfate leach liquors, and uranium recovery is one of the most important commercial uses of amines (Nicol et al. 1987). Plants in the United States (Merritt 1971), Canada (Ritcey and Ashbrook 1972), Australia (Bellingham 1961), South Africa (Boydell and Viljoen 1982; Finney 1977), and Namibia (Sole et al. 2005; Lewis and Kesler 1980) have been described.

Amine systems achieve higher uranium purity than organophosphorus systems (due to the greater selectivity of amines for uranium) and have lower extractant losses due to their lower aqueous-phase solubility. SX is applied either directly to the weakly acidic leach liquor (also known as the Purlex or Amex process, Vaal River West, South Africa) or to the strongly acidic eluate from an ion-exchange preconcentration treatment of the leach liquor (the Bufflex or Eluex process, Vaal River South, South Africa) (Nicol et al. 1987). A more modern variation (such as at Southern Cross Resources' Uranium One, South Africa) is to treat the ore by pressure leaching followed by SX.

Uranium(VI) is present in solution as the uranyl sulfate anion. The extraction of uranium(VI) by amines occurs in the order of tertiary > secondary > primary amine. The extraction of iron(III) occurs in the reverse order, so tertiary amines represent an obvious choice of extractant. The tertiary alkyl amine sold as Alamine 336 or Armeen* 380 is widely used, usually in conjunction with an alcohol phase modifier (such as isodecanol) to prevent the formation of a third phase and inhibit the formation of emulsions.

The structure of the tertiary amine can be represented as R_3N. In order for the extraction to occur, the extractant must be protonated. This is achieved in a single step during the extraction process due to the high acid content of the leach liquor. In sulfuric acid medium, some bisulfate extraction may also occur:

$$2\ R_3N_{\ (org)} + H_2SO_{4\ (aq)} \rightarrow (R_3NH^+)_2SO_4^{2-}{}_{(org)} \quad (5.12)$$

$$R_3N_{\ (org)} + H^+ + HSO_4^-{}_{(aq)} \rightarrow R_3NH^+HSO_4^-{}_{(org)} \quad (5.13)$$

Once the extractant is protonated, the uranium complex is extracted by an anion-exchange process:

* Armeen is a registered trademark of Akzo Nobel.

$$2\,(R_3NH^+)_2SO_4{}^{2-}{}_{(org)} + UO_2(SO_4)_3{}^{4-}{}_{(aq)} \rightarrow (R_3NH^+)_4UO_2(SO_4)_3{}^{4-}{}_{(org)} + 2\,SO_4{}^{2-}{}_{(aq)} \quad (5.14)$$

$$(R_3NH^+)_2SO_4{}^{2-}{}_{(org)} + UO_2(SO_4)_2{}^{2-}{}_{(aq)} \rightarrow (R_3NH^+)_2UO_2(SO_4)_2{}^{2-}{}_{(org)} + SO_4{}^{2-}{}_{(aq)} \quad (5.15)$$

The formation of adducts such as $[(R_3NH)_2SO_4]_2{\cdot}UO_2SO_4(H_2O)_3$ has also been postulated (Nicol et al. 1987). The mechanism of uranium extraction from sulfate media by amines has been comprehensively reviewed by Mackenzie (2005) and Cattrall and Slater (1973).

The loaded organic phase is washed with water or dilute H_2SO_4 to remove any physically entrained eluate from the organic phase because it will contribute to the carry-over of impurities from the feed liquor to the product.

Uranium can be stripped from the loaded organic phase using a variety of reagents, including NaCl, $(NH_4)_2SO_4$, Na_2CO_3, ammoniacal ammonium sulfate, or ammonia gas (Nicol et al. 1987). These stripping systems present few choices based on technical grounds (Lloyd 1983), so the choice of reagent is usually determined by economic factors. Stripping with ammonium sulfate is most commonly practiced in South Africa:

$$\begin{aligned}&(R_3NH^+)_4UO_2(SO_4)_3{}^{4-}{}_{(org)} + 4\,NH_4OH_{(aq)} \rightarrow \\ &4\,R_3N_{(org)}) + (NH_4)_2SO_{4\,(aq)} + (NH_4)_2UO_2(SO_4)_{2\,(aq)} + 4\,H_2O_{(aq)}\end{aligned} \quad (5.16)$$

The mechanism of stripping involves deprotonation of the amine by increasing the pH. It is important to control the pH of the reaction to avoid precipitation of solid ammonium diuranate (ADU) in the strip circuit:

$$2\,(NH_4)_2UO_2(SO_4)_2 + 6\,NH_4OH \rightarrow (NH_4)_2U_2O_7 + 4\,(NH_4)_2SO_4 + 3\,H_2O \quad (5.17)$$

The LSL (known in the industry as OK liquor) typically has a uranium concentration of 7 to 10 g/l U_3O_8 (2 to 3 g/l U). This is treated with ammonia to produce a slurry of ADU, which is then sent for further processing. The stripped organic phase is treated in a regeneration circuit using a mixture of Na_2CO_3 and NaOH to ensure the removal of a range of other impurities from the organic phase so that they do not build up in the circuit.

5.3.5.2 Phosphoric Acid Leach Liquors

Many phosphate rock deposits contain quantities of radioactive elements such as uranium and thorium. Selective leaching of uranium from raw phosphate ores is difficult because the U(VI) ion is incorporated into the crystal structure of apatite $(Ca_5(PO_4)_3(OH,F,Cl))$, rather than adsorbtively associated with it. Uranium is, therefore, typically recovered from phosphate rocks by recovering it from phosphoric acid produced by sulfuric acid leaching of phosphate ores. The radioactive species are also leached and must be removed during purification of the acid. Uranium in phosphoric acid is extracted as the UO_2^{2+} species using a mixture of D2EHPA and tri-*n*-octylphosphine oxide (TOPO) and forming an organic-phase adduct:

$$UO_2(HA_2)_{2\ (org)} + TOPO_{(org)} \rightarrow UO_2(HA_2)_2 \cdot TOPO_{(org)}. \tag{5.18}$$

The enhanced extraction observed by using a combination of two extractants compared with extraction achieved by either extractant independently is known as a *synergistic effect*. The mechanism of the synergistic effect in uranium SX has been discussed by many authors (Nicol et al. 1987). In the extraction of uranium, this effect increases with increasing basicity of the phosphoryl oxygen of the neutral additive. $(RO)_3PO < (RO)_2RPO < (RO)R_2PO < R_3PO$. TBP has also been used as a synergist, but TOPO typically gives high synergism and high selectivity. Di(nonylph enyl)phosphoric acid can be used in place of D2EHPA (Vijayalakshmi et al. 2005).

Uranium is stripped from the loaded organic phase with ammonium carbonate, and the uranium is precipitated as ammonium uranyltricarbonate:

$$\begin{aligned} &UO_2(HA_2)_2(TOPO)_{n\ (org)} + 3\ (NH_4)_2CO_{3\ (aq)} \rightarrow \\ &NH_4(HA_2)_{\ (org)} + (NH_4)_4UO_2(CO_3)_{3\ (s)} + n\ TOPO_{(org)}. \end{aligned} \tag{5.19}$$

A single-step process does not allow the economic production of high enough concentrations suitable for precipitation (~20 g/l U_3O_8), but a two-cycle countercurrent process appears to be a feasible route for the processing of such liquors (Gupta and Singh 2005; Girivalkar et al. 2004; Singh et al. 2003). This allows the stepwise upgrading of a feed solution that contains a low concentration of U (~200 mg/l) using extractants that have relatively low distribution ratios for uranium.

5.3.5.3 Nitric Acid Leach Liquors

High-grade uranium ores are sometimes leached in hot nitric acid, particularly when significant amounts of radioactive elements such as thorium are present in the rock (Tunley and Nel 1974). Uranium forms the uranyl nitrate species that can be extracted by a solvating mechanism using TBP, forming the species $UO_2(NO_3)_2(TBP)_2$ (Nicol et al. 1987). TBP extraction of uranium from nitric acid is also practised for the purification of uranium (Ashbrook and Lakashmanan 1983) and for its recovery from irradiated nuclear fuel (Chesne and Germain 1992; Naylor and Wilson 1983).

5.3.6 Refining of Precious Metals

The classical refining methods used for precious metals remained unchanged for several decades. Refining relied heavily on precipitation methods and was extremely tedious and inefficient because of the substantial and numerous recycles required, the large number of solid–liquid separations, and the consequent long residence times of the valuable metals in the circuit. In the past two decades, significant advances have been made in the processing of these elements, and SX is now widely used as the main separation technique.

TABLE 5.7
Precious Metal Oxidation States and Aqueous Chloro Complexes

Metal	Oxidation State	Coordination Number	Complex Geometry	Complexes Formed	
				Low [Cl⁻]	High [Cl⁻]
Au	3+	4	Square planar	$AuCl_4^-$	$AuCl_4^-$
Ag	1+	2	Linear	$AgCl_2^-$	$AgCl_2^-$
Ru	3+	6	Octahedral	$Ru_2OCl_8(H_2O)_2^{4-}$	$RuCl_6^{3-}$
	4+	6	Octahedral		$RuCl_6^{2-}$
Rh	3+	6	Octahedral	$Rh(H_2O)_2^{3+}$	$RhCl_6^{3-}$
Pd	2+	4	Square planar	$PdCl_4^{2-}$	$PdCl_4^{2-}$
	4+	6	Octahedral	$PdCl_6^{2-}$	$PdCl_6^{2-}$
Os	4+	6	Octahedral		$OsCl_6^{2-}$
Ir	3+	6	Octahedral	$Ir(H_2O)_xCl_{6-x}^{(3-x)-}$	$IrCl_6^{3-}$
	4+	6	Octahedral	$IrCl_6^{2-}$	$IrCl_6^{2-}$
Pt	2+	4	Square planar	$PtCl_4^{2-}$	$PtCl_4^{2-}$
	4+	6	Octahedral	$PtCl_6^{2-}$	$PtCl_6^{2-}$

Anglo Platinum Precious Metals Refiners has been in the forefront of SX developments for the refining of the so-called PGMs (platinum-group metals: Pt, Pd, Ir, Rh, Ru, and Os). The technologies now employed were largely developed by Johnson Matthey in the United Kingdom and production started at Rustenburg, South Africa, in 1989. Today, this is the largest PGM refinery in the world, with current annual production of 116,000 ozT/a Au, 2.3 million ozT/a Pt, and 1.2 million ozT/a Pd. While specific details of the SX refining steps remain closely guarded, outlines of some commercial flow sheets have been published. They provide interesting insights into the clever chemistry and novel ideas employed in the development of these processes (Demopoulos et al. 2002; Harris 1993; Mooiman 1993; Benner et al. 1991; Al-Bazi and Chow 1984; Charlesworth 1981; Cleare et al. 1979).

Precious metal feed materials are almost always leached in HCl, with chlorine gas as the oxidizing agent. The dominant aqueous species formed under these conditions are shown in Table 5.7. A variety of oxidation states and coordination structures exist. These differences are exploited in designing extractant systems that will be selective for one element over the others (Al-Bazi and Chow 1984).

The extractants used for PGMs fall into three main classes (Mooiman 1993):

- Solvating extractants (long-chain alcohols, ethers, and TBP)
- Anion-exchange extractants (long-chain alkylamines)
- Coordinating extractants (oximes or dialkylsulfides)

For the first two of these classes, selectivity is largely based on the charge-to-size ratio of the chloroanion (Table 5.7) (Harris 1993; Edwards and Te Riele 1983). Large anions with low charges are most easily extracted, and this accounts for the

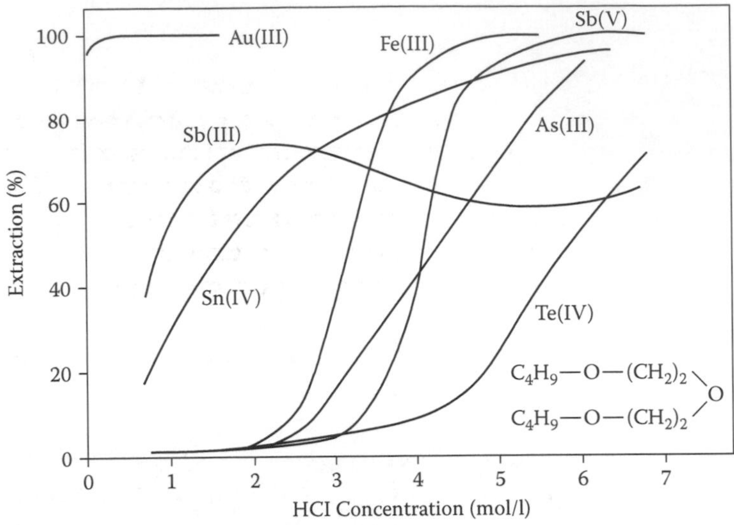

FIGURE 5.19 Selectivity of gold extraction with dibutylcarbitol. (Adapted from Amer, S. 1983a. *Rev. Metal. CENIM* 19:161–183.)

selectivity of extraction of PGM chloroanions over the corresponding base metal species. Charge considerations result in differences in the order of extraction within the PGMs. The lower the charge density, the more extractable is the species. In general, the order of extraction from strong HCl is $AuCl_4^- > PtCl_6^{2-} \approx IrCl_6^{2-} > PdCl_4^{2-} > RhCl_6^{3-} \approx IrCl_6^{3-}$.

In most processes that use SX, gold is removed from solution first, followed by palladium, and then platinum. The remaining PGMs are recovered in a variety of ways, either up-front or at the end of the overall flow sheet. In each case, a pure solution of the respective metal is obtained, which is then subjected to a reduction to produce the pure metal powder or sponge. The final products are obtained by melting to obtain ingots, granules, or delivery bars.

5.3.6.1 Gold

Since the $AuCl_4^-$ anion is the most readily extracted of all the precious metal species, it is important that it is removed first from the leach liquor so that it does not contaminate the downstream products. Gold can be selectively removed from solution using a weakly solvating reagent. Extractants include ketones, alcohols, TBP, and amines (Amer 1983a; Mooiman 1983).

INCO (United Kingdom) first used SX for the refining of gold from chloride solution (Grant and Drake 2002; Rimmer 1989; Barnes and Edwards 1982; Rimmer 1974). The extractant, dibutylcarbitol (DBC, also known as butyldiglyme), is a straight-chain, high molecular mass molecule that contains three oxygen atoms in ether positions (see Figure 5.19). The extraction mechanism has been shown to involve the formation of an ion pair between the solvated auric chloride anion and the oxygen-donor reagent (Sergeant and Rice 1977).

Although several PGM refiners around the world have used DBC, this reagent suffers from several disadvantages. Under the high acid conditions of the leach, it is not particularly selective over some of the metalloid elements often found in solutions of this nature (Figure 5.19). It is also very difficult to strip effectively, so recovery of gold from the loaded organic phase is by direct reduction with oxalic acid in a batch process. This necessitates good washing of the gold product to prevent contamination and organic losses. The extractant is also fairly soluble in the aqueous phase (~3 g/l) and dissolved organic losses are higher than are generally considered acceptable. Because of the relatively high cost of the reagent, it may be necessary to recover dissolved organic from the aqueous raffinate by distillation. Nevertheless, gold of 99.9% purity was successfully produced for many years using this technology.

Anglo Platinum (South Africa) recovers gold from PGM leach liquor by SX with methyl isobutylketone (MIBK), an oxygen-donor solvating reagent whose selectivity is largely based on the charge-size ratio of the chloroanion (Benner et al. 1991; Charlesworth 1981; Cleare et al. 1979). MIBK has an aqueous solubility of about 2%, so again organic losses are high.

The Minataur™ (acronym for Mintek Alternative Technology for Au Refining) process for the refining of gold was developed in South Africa (Mintek 2001; Sole et al. 1998). This technology uses SX as the main purification step and a variety feed materials ranging in gold content from 50 to 99% can be treated. The identity of the extractant remains proprietary; it is known to be inexpensive and does not present the disadvantages of DBC mentioned above. Gold loadings of >100 g/l on the organic phase can be achieved and stripping efficiency is high (Sole et al. 1998; Feather et al. 1997a, 1997b). The process produces gold of 99.99% purity. Advantages of the process are the reduced gold lock-ups and residence times in the circuit and the ability to produce high-purity gold with minimal recycles in the flow sheet. This process is presently operating in South Africa, Algeria, and Dubai (Scott and Matchett 2004; Feather et al. 2002b).

5.3.6.2 Palladium

Most research into the recovery of palladium by SX has centred on the use of sulfur-based extractants (Du Preez and Preston 2002; Amer 1983b). Organic sulfides are selective for palladium over all other precious metals except gold, so gold is removed from the HCl leach liquor ahead of palladium. The extraction reaction is:

$$PdCl_4^{2-}{}_{(aq)} + 2\ R_2S_{\ (org)} \rightarrow PdCl_2(R_2S)_{2\ (org)} + 2\ Cl^-{}_{(aq)}. \tag{5.20}$$

The extraction mechanism involves substitution of the inner sphere chloro ligands by the dialkyl sulfide. Palladium extraction with this reagent is extremely slow, taking several hours to reach equilibrium. The reaction goes to completion, however, leaving less than 1 mg/l Pd remaining in the raffinate. The reaction is carried out in batch mode. Strong ammonia solution is used for stripping:

$$PdCl_2(R_2S)_{2\ (org)} + 4\ NH_{3\ (aq)} \rightarrow Pd(NH_3)_4^{2+}{}_{(aq)} + 2\ R_2S_{\ (org)} + 2\ Cl^-{}_{(aq)}. \tag{5.21}$$

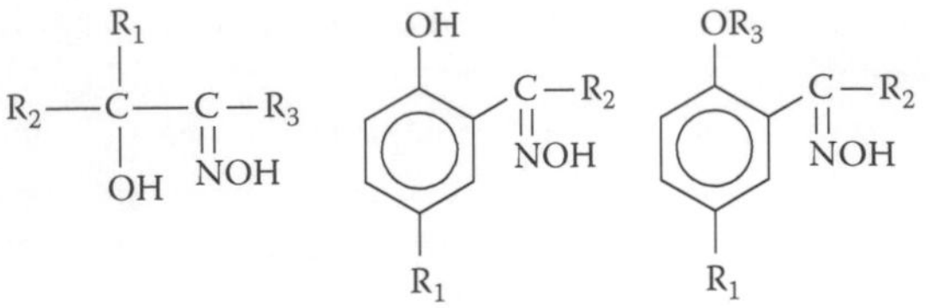

FIGURE 5.20 Structures of oximes that extract palladium.

Palladium is recovered from the LSL by acidification with HCl to precipitate $Pd(NH_3)_2Cl_2$. The salt is calcined to produce the metal. Alternatively, the salt can be reduced with NaCOOH or HCOOH and then calcined.

INCO used di-*n*-octylsulfide in a highly purified paraffin diluent to produce palladium of 99.95% purity (Rimmer 1989; Lea et al. 1983; Barnes and Edwards 1982). Degussa (Germany) was believed to use di-*n*-hexylsulfide for the extraction of palladium (Harris 1993; Renner 1985). The process is similar to the INCO route, but higher purities (99.9%) of palladium were possible.

Various oximes, similar to those used in copper extraction, have also been widely studied as palladium extractants. The general chemical structures of oximes that will extract palladium are shown in Figure 5.20. Johnson Matthey (United Kingdom) and Anglo Platinum (South Africa) recover palladium with a β-hydroxyoxime. Although the kinetics of this extraction are extremely slow, the extraction rate can be enhanced by the addition of amines that act as accelerators (phase-transfer catalysts). An advantage of these reagents is that they can be stripped with strong acid, avoiding the use of ammonia (Brits and Deglon 2007). Stripping is typically achieved using 6 *M* HCl. Recovery of the metal is by salt formation, followed by reduction or calcination, similar to that described above.

5.3.6.3 Platinum

Platinum is recovered after palladium in most flow sheets. TBP is the most widely used platinum extractant (Rimmer 1989; Lea et al. 1983; Barnes and Edwards 1982). Figure 5.21 shows the effect of HCl concentration on the extraction of PGMs by TBP. The process has a very narrow operating window between about 3 and 4 *M* HCl. It is also necessary to adjust the reduction potential of the solution to ensure that iridium(IV) is not coextracted.

Amines are good platinum extractants, and schemes in which primary through quaternary amines are employed have been proposed. The extraction occurs via an anion-exchange mechanism. The reaction for tertiary amines can be written:

$$2\ R_3N_{(org)} + 2\ HCl_{(aq)} \rightarrow 2\ [R_3NH^+]Cl^-_{(org)} \tag{5.22}$$

$$2\ [R_3NH^+]Cl^-_{(org)} + PtCl_6^{2-}{}_{(aq)} \rightarrow [R_3NH^+]_2PtCl_6^{2-}{}_{(org)} + 2\ Cl^-_{(aq)}. \tag{5.23}$$

In these systems, complete stripping can only be achieved by deprotonation of the extractant:

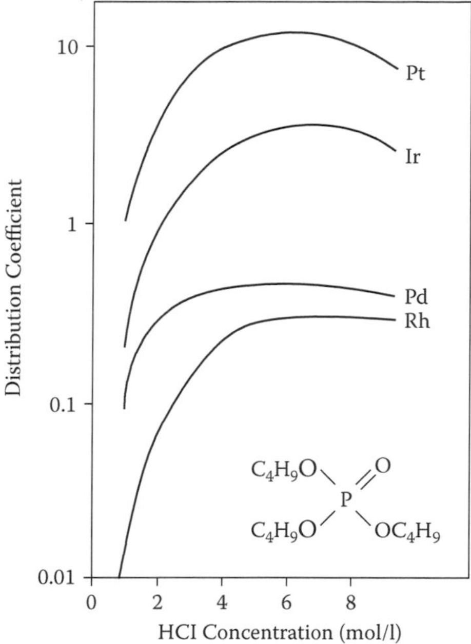

FIGURE 5.21 Effect of HCl concentration on extraction of PGMs by TBP. (Adapted from Amer, S. 1983a. *Rev. Metal. CENIM* 19:161–183.)

$$[R_3NH^+]_2PtCl_6^{2-}{}_{(org)} + 2\ Na_2CO_{3\ (aq)} \rightarrow$$
$$2\ R_3N_{(org)} + PtCl_6^{2-}{}_{(aq)} + 2\ Na^+{}_{(aq)} + H_2O + CO_{2\ (aq)}. \quad (5.24)$$

In addition to this anion-exchange extraction reaction, the irreversible formation of inner-sphere amine complexes can occur (Nicol et al. 1987):

$$2\ [R_3NH^+]Cl^-{}_{(org)} + PtCl_6^{2-}{}_{(aq)} \rightarrow Pt(R_3N)_2Cl_{4\ (org)} + 2\ Cl^-{}_{(aq)} + 2\ H^+{}_{(aq)}. \quad (5.25)$$

Inner-sphere complexation is minimised by the use of tertiary amines that have greater steric hindrance than primary or secondary amines. Amines are generally employed in combination with alcohols, phenols, or carboxylic acids that modify the pH dependence of extraction so that stripping is easier (Mooiman 1993). The relative extraction efficiencies of the amines as a function of chloride concentration are shown in Figure 5.22.

Johnson Matthey reported the use of tri-*n*-octylamine to recover platinum from the gold- and palladium-depleted leach liquor. It is necessary to use 12 *M* HCl for stripping with this reagent (Harris 1993; Al-Bazi and Chow 1984).

5.4 PROCESS EQUIPMENT

Although this chapter focuses mainly on commercial applications of SX, the type of equipment commonly used for carrying out these processes on an industrial scale is

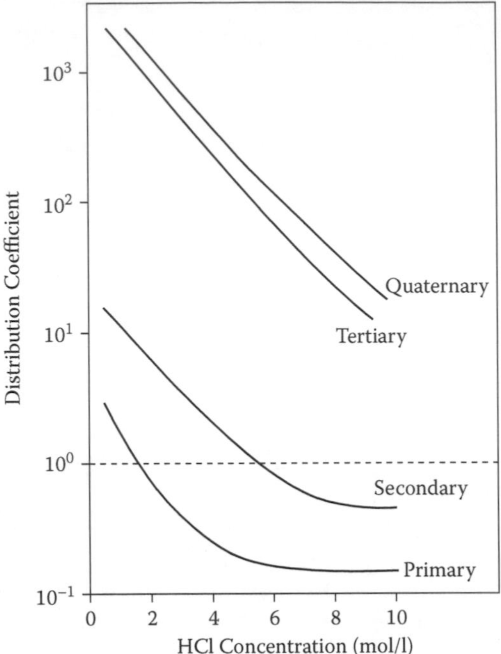

FIGURE 5.22 Extraction of $PtCl_6^{2-}$ complex by various amines as function of hydrochloric acid concentration. (Adapted from Cleare, M. J. et al. 1979. *J. Chem. Tech. Biotech.* 29:210–224. With permission.)

briefly mentioned. The discussion that follows is not intended to be comprehensive, but merely to provide an indication of the main SX contactors in use today. The interested reader is referred to excellent material in the bibliography (Blass 2004; Ritcey 2004; Nyman et al.; Slater 1996; Godfrey and Slater 1994; Bateman 1993; Nyman et al. 1992; Sonntag et al. 1989).

To ensure that good mass transfer of the extracted and stripped species across the organic–aqueous interface occurs, adequate contact between the two phases is necessary. Depending on various physical and chemical factors inherent in the extraction system, different contact systems may be appropriate. Some of the factors that influence the choice of contactor type are (Mooiman 1993):

- Organic volatility
- Volumetric throughput
- Kinetics of extraction and stripping and residence times required
- Relative organic to aqueous phase ratios
- Number of stages required for each section
- Corrosivity of aqueous phase
- Available space
- Value of product contained in solution inventory
- Need for pH control
- Stage efficiency requirements

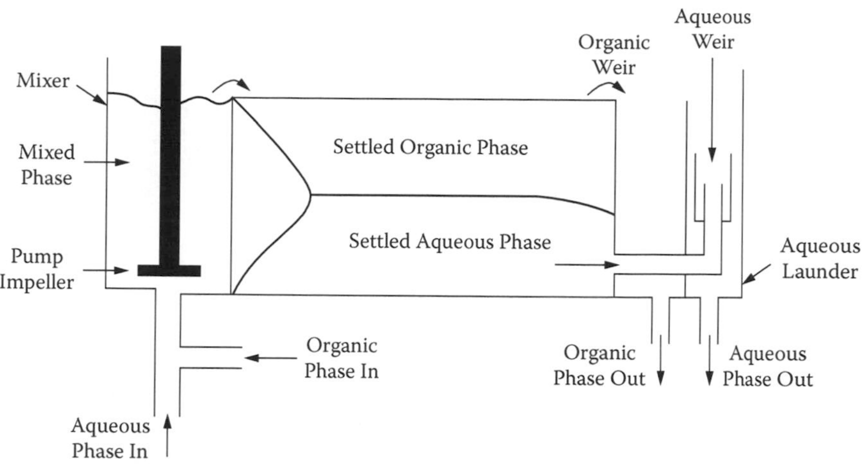

FIGURE 5.23 Conventional mixer–settler unit.

- Phase dispersion and disengagement rate
- Capital cost

5.4.1 Mixer–Settlers

In the hydrometallurgical industry, mixer–settler units have traditionally been used for commercial-scale SX. These are simple devices that are relatively inexpensive (Vancas 2005, 2003). Figure 5.23 shows a typical cross-section of a box-type mixer–settler. Mixer–settlers usually operate in a countercurrent configuration, with the organic and aqueous flows in opposite directions. For particular circumstances such as extraction systems with very slow kinetics, co- or cross-current flow configurations may be used. The contact between the phases is discontinuous, taking place in discrete steps, with separation of the phases between each contact. Each extraction or stripping unit is known as a *stage*.

The aqueous and organic phases are drawn into the mixing compartment from adjacent units by a pumping impeller. The use of the impeller obviates the need for separate pumps for interstage solution transfer. The two phases are mixed for sufficient time to allow mass transfer to take place. The mixed phases then overflow into the settling compartment, where the two phases disengage by gravity. The organic phase overflows the top weir, while the aqueous phase is removed via a barometric leg. The adjustable barometric leg allows the relative height of the interface in the settler to be varied (Amer 1980).

Mixer–settlers are usually preferred for systems that exhibit poor phase disengagement and require a considerable settling area. Stage efficiencies in this configuration are high. Interstage pH control (if necessary) is usually easier in mixer–settlers than in columns.

Several variations on this basic design have been developed by vendors of SX equipment such as Bateman, Krebs, Davy International, and Outokumpu (Sonntag et al. 1999; Hopkins 1994; Dilley et al. 1993; Nyman et al. 1992; Rowden et al. 1981). The modern trend, particularly with large volumetric inputs, is to use multiple mixers with three or more mixing compartments. This gives improved mass transfer per stage without overmixing (which causes excessive organic entrainment). For copper installations, for example, stage efficiency is quoted as improving from 80% with a single mixer to 95% with three mixers per stage (Cole 1994).

5.4.2 Columns

Column configurations allow contacts between the two phases to be continuous. Such contactors have been widely used in the nuclear, pharmaceutical, and petrochemical industries for many years.

A *reciprocating-plate column* consists of a stack of perforated plates and baffles supported by a central reciprocating shaft driven by an external mechanism above the column. The amplitude and reciprocating speed are adjustable, allowing for variation of the power input to the system. Various designs of baffles have been studied to maximise the interfacial contact area and minimise the residence time required (Benner et al. 1991). The open area of the plates is about 60%. The light phase is introduced at the base of the column and the heavy phase at the top. Columns are usually operated with the organic as the dispersed phase to minimise solvent inventory and for safety reasons.

A *packed column* is filled with an inert hydrophobic or hydrophilic packing material and the two phases are introduced as described above. No external mechanism is required for agitation of the phases and mass transfer results from the turbulence introduced to the flow patterns by the packing material.

Pulsed columns operate on a similar principle, but the agitation is provided by reciprocating pulsation of the solutions rather than by movement of the column internals. This arrangement allows the single moving part to be located outside the column and also offers the opportunity to change the nature of the internal packings of the column for different extraction requirements. Figure 5.24 illustrates a pulsed column designed by Bateman (Movsowitz et al. 1997; Bateman 1993).

Columns are useful for processing low flow rates and for systems that exhibit a tendency to form emulsions. An important benefit of a column contactor is the large number of possible theoretical stages and the ability to operate closer to the operating line rather than the equilibrium curve, thereby maximising mass-transfer kinetics. The settling volume is considerably lower than for the corresponding mixer–settler, so columns are preferred for systems in which solution lock-up and low solvent inventories are important (such as in precious metal extraction systems). Columns take up very little floor space, but require considerable headroom; mixer–settler requirements are the opposite (Movsowitz et al. 2001; Fox et al. 1998).

Another important advantage of columns is that the extraction and stripping processes occur in a closed system. This is crucial for systems in which toxicity is an issue, such as in the extraction of gold using MIBK or the separation of radioactive species in the nuclear industry. It is also much easier to maintain an inert atmosphere in the system such as for situations in which it is necessary to control the oxidation

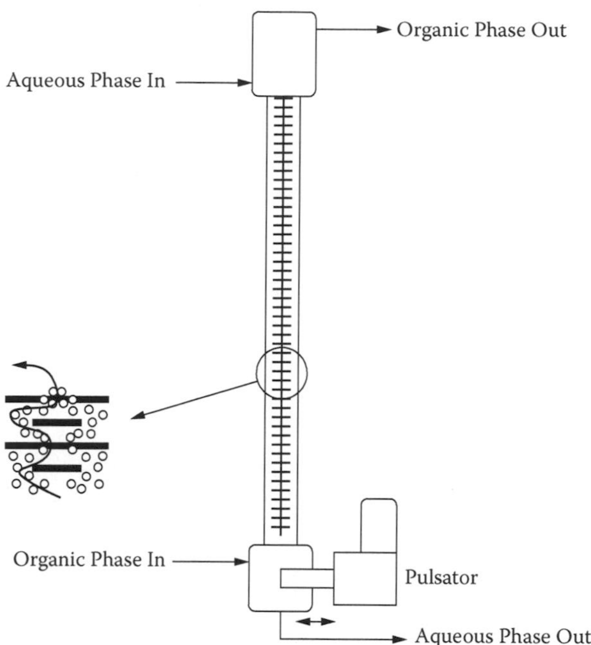

FIGURE 5.24 Bateman pulsed column. (From Bateman Corporation brochure. 1993. With permission.)

state of a particular species, as in the extraction of iridium. The column configuration also offers advantages in terms of reduced volatile losses and improved safety.

5.4.3 Centrifugal Contactors

A centrifugal contactor usually involves a combination of a high-speed mixer and centrifugal separator in a single unit. Several different designs are available from manufacturers such as Robatel, Centrek (Kuznetsov et al. 2002), Westfalia, and Podbielniak. Figure 5.25 illustrates the contacting and separation phases.

The main advantages of these units are their very fast kinetics and very low solution hold-up. The latter feature provides distinct benefits for systems in which low organic inventories are required, such as precious metal refineries. Centrifugal contactors are also widely used in the nuclear processing industry. They are not appropriate for systems in which long residence times are required or for low flow rates.

5.5 SELECTED SOLVENT EXTRACTION OPERATIONAL ISSUES

5.5.1 Mixer Efficiency

The crux of the SX process is the transfer of a particular species across an aqueous–organic interface. If mass transfer conditions are not adequate, then extraction and/or stripping efficiencies will be lower than expected. Optimized operation of the mass transfer process is crucial, whatever type of contacting equipment is used. For

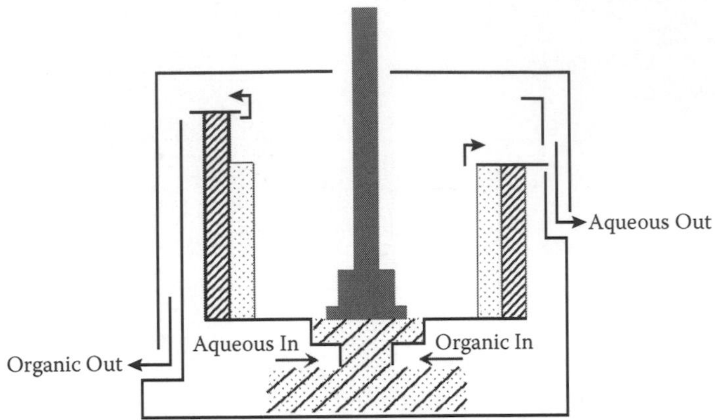

FIGURE 5.25 Aqueous and organic flow in a centrifugal extractor (based on Robatel BXP unit). (Adapted from Mooiman, M. B. 1983. *Precious metals.* Allentown, PA: International Precious Metals Institute. With permission.)

mixer–settlers, mixer efficiency is defined as the ratio of metal transfer in the mixer to metal transfer at equilibrium (expressed as a percentage). This can be calculated based on the aqueous and organic concentrations entering and leaving a particular mixer. Some factors that affect mixer efficiency are the following (Cognis 2000):

Mixer Residence Time—The longer the residence time, the closer the approach to equilibrium of the mixture.

Impeller Speed—The rule for tip speed in most SX circuits is 270 to 350 m/min. This is high enough to ensure good mass transfer, but not too high to cause excessive entrainment due to high shear and slow-breaking emulsions. Excessive impeller speeds can also cause vortex formation in the mixer and enhance air entrainment (see Section 5.5.6).

Organic–Aqueous (O:A) Ratio—An O:A volumetric ratio in the mixing compartment near unity produces the best mass transfer.

Temperature—Increasing temperature improves mixer efficiency.

Short Circuiting—This occurs when the residence time of a dispersion is lower than the average in the circuit. It is generally desired to have the extraction or stripping circuit operate as close as possible to plug-flow conditions.

Mixer and Impeller Design—Modern approaches employ the use of computational fluid dynamics (CFD) modelling to improve designs of mixer boxes, baffles, and impellers (Giralico et al. 2003; Gigas and Giralico 2002). The choice of whether or not to use mixer–settlers or columns can also influence the efficiencies achievable (Fox et al. 1998).

5.5.2 Phase Continuity

When aqueous and organic phases are mixed, two possible events will occur with respect to the emulsion formed (illustrated in Figure 5.26). If the aqueous phase forms the matrix and organic droplets are dispersed within it, then the mixing is

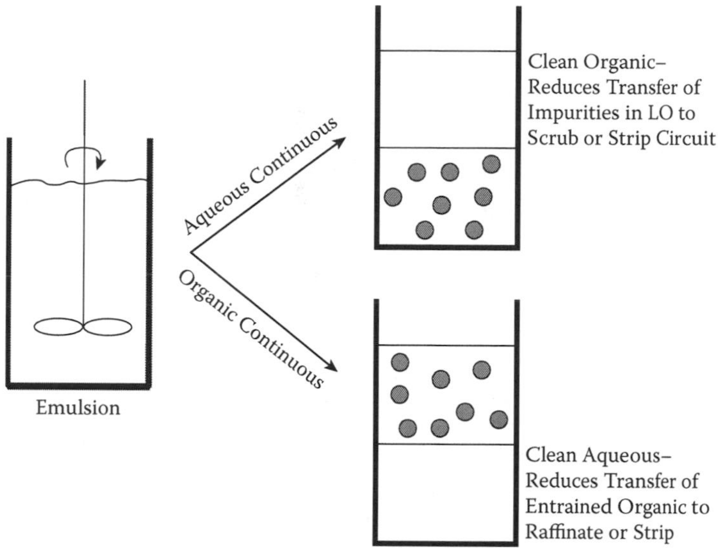

FIGURE 5.26 Organic and aqueous phase continuity in mixing systems.

aqueous continuous. The aqueous phase is the *continuous phase* and the organic phase will be the *dispersed phase*. The reverse situation is *organic continuity*.

When an aqueous continuous emulsion separates, the organic phase will be very clean and free of aqueous phase, while the aqueous phase will contain some entrained organic phase. Similarly, when an organic continuous emulsion separates, the aqueous phase will be clean, while the organic phase will contain an entrained aqueous phase.

It is possible to individually control the continuity of a particular mixer or column. This is an important operational parameter to ensure minimal contamination of product. A general rule is that the phase continuity should be the *opposite* of the most important phase leaving a particular stage. For example, where the LSL or raffinate leaves the SX, it is generally preferred to run organic continuous, as this will minimise the contamination of these phases by entrained organic. Organic present in the LSL could report to the EW tankhouse where it can produce very detrimental effects, especially in the cases of metals such as zinc and nickel (Sole et al. 2007b). Organics in the raffinate are often returned to the leach circuit where they could interfere with flocculant performance or have other detrimental effects.

In stages where the loaded organic phase transfers from the extraction to scrub or scrub to strip circuits, it is preferable to run aqueous continuous to avoid aqueous carry-over of impurities to the subsequent circuit. Certain circuits prefer to run in a specific continuity and will continually "flip" to the opposite continuity (Vancas 2005).

The rate of phase separation is often influenced by continuity, particularly with aging of the organic phase or in the presence of solids or other surfactants that have entered the circuit. Sometimes, it is also preferable to choose a particular continuity to control crud in a circuit (see Section 5.5.6). Crud may be compacted and stabilized at the interface by operating in a particular continuity, depending on the composition

TABLE 5.8
Factors Influencing Phase Disengagement in SX Systems

Aqueous Contaminants	Organic Contaminants
Suspended solids	Modifiers
Polymeric flocculants and coagulants	Degradation products
Precipitates (gypsum, hydroxides, jarosites)	Humates and other organic materials from leach
Bacteria, fungi, and algae	Surface-active contaminants such as oils, grease, cleaning fluids
Biological oxidation products of organic phase	
Aqueous-soluble polar organic molecules	

Source: Modified from Cognis Corp., 2000. Unpublished. (With permission.)

and hydrophobicity of the crud. This may make for more stable operation of the circuit, even at the expense of increased entrainment.

5.5.3 Phase Separation

Once the phases have been adequately mixed, they are required to separate. Because SX is an interfacial phenomenon, the characteristics of the interface are particularly important in ensuring good kinetics as well as good phase disengagement. The interfacial tension can be influenced by the presence of foreign materials in SX circuits, some possibilities of which are indicated in Table 5.8. In general, aqueous contaminants slow aqueous continuous phase separations, while organic contaminants retard separation of organic continuous emulsions.

It is important to routinely monitor both the chemical and physical integrity of the organic phase. Standard tests are available to monitor the rate of phase disengagement. Any deterioration should be interpreted as a sign of contamination of the circuit or degradation of the organic phase.

Coalescence of the organic phase can be enhanced by mechanical means. So-called picket fences are installed at various distances along a settler to force interruption of the flow patterns and permit coalescence on their surfaces. Other approaches use packings of hydrophobic materials in the settlers. While these can be helpful, they can also cause significant disruption in systems that produce excessive crud formation or are saturated in calcium and should therefore be used with caution.

5.5.4 Third Phase Formation

A mixed emulsion will generally separate into two phases: aqueous and organic. Under certain conditions, however, the undesirable situation of *third phase formation* may occur. The organic phase separates into two phases of different densities. This phenomenon generally occurs under conditions of high loading of the extractant and inadequate solvation of the complexed species by the diluent. This is commonly seen in amine systems such as uranium or vanadium SX in which the extracted complexes typically have very high molecular masses. This can be avoided by decreasing the

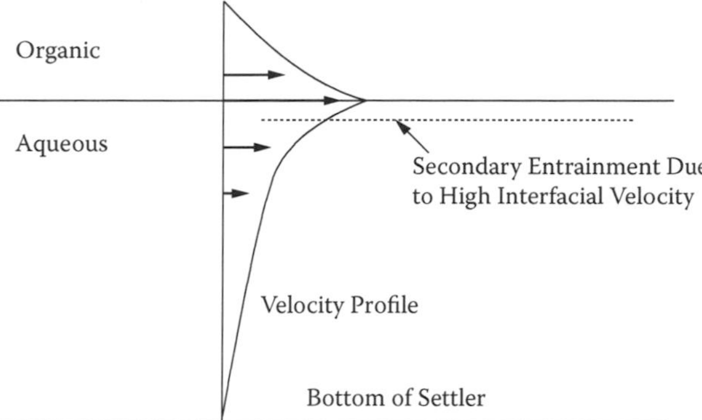

FIGURE 5.27 Effect of reduced organic height on linear velocity profile in settler. (Adapted from Cognis Corp., 2000. Unpublished. With permission.)

metal loading on the organic phase or, preferably, changing to a diluent that offers better solvation properties (see Section 5.5.7).

5.5.5 Organic Depth in Settlers

In mixer–settler units, the interface height in the settler is an important parameter in ensuring efficient phase separation and low entrainment. Irrespective of the flow rates treated or sizes of the settlers, the total height of a settler seldom exceeds 1 m. Because of the cost associated with the organic inventory, there is often a tendency to reduce the organic height in a settler, but other considerations are also important.

In general, the linear velocities of the aqueous and organic phases through the settler should be the same. The interface height will therefore be established from the advance O:A ratio. This means that the differential velocity at the interface is zero, which minimizes turbulence at the interface and hence minimizes entrainment. A narrow organic height in the settler also reduces the residence time of the organic phase, further contributing to reduced separation efficiency. This is shown schematically in Figure 5.27.

The optimum organic linear velocity in most SX systems falls in the range 3 to 6 cm/s. High velocities through a settler can result in inadequate separation of the dispersion band and carryover of unsettled aqueous and organic phases. It is generally recommended that an organic height of 25 to 35 cm be maintained in all SX circuits.

5.5.6 Crud Formation

The term *crud* is widely used in describing amorphous third phase solids that occur in SX circuits, but the nature of crud varies widely from circuit to circuit. It is generally a voluminous, low density mass, comprising a mixture of aqueous and organic phases, solids from various sources, and air (Cytec 2006; Wang 2005; Ritcey 1980). Although it looks ugly and is difficult to cure completely, it is seldom fatal and occurs to a greater or lesser extent in all SX circuits. Bottom crud occurs at the bottom of a settler and is of higher density than the aqueous phase. Other types are interfacial crud and floating crud (with density lower than that of the organic phase).

Bottom crud usually comprises solids that have entered the SX circuit in the PLS and have settled out due to gravity in the settler. This crud is a nuisance in that it occupies space and requires periodic removal from the circuit, but otherwise it does not generally interfere with the operation of the circuit. Floating crud is the most serious because it is transferred through the circuit with the organic phase, permitting impurities and solids to move through the circuit and contaminate product. Most commonly, crud is interfacial in nature, and its impact can generally be limited by controlling phase continuity, allowing it to compact, and restricting it to a particular stage. In modern circuit design, provision can be made for the continuous removal of crud via a suction mechanism that travels across a settler at the appropriate height.

The chemical composition of crud is circuit-specific. Commonly, however, crud will contain species such as polymeric (colloidal) silica (including dust) and insoluble species such as gypsum. It may also contain degradation products from the organic phase and precipitated salts from solution. It is often stabilized by air entrained by excessive mixing in the mixers. Depending on whether the crud is hydrophobic or hydrophilic, different species may adsorb to the crud. Its surface activity may influence phase separation and kinetics of extraction and stripping and permit transfer of adsorbed impurities through the circuit.

The most common approaches to reducing and/or avoiding crud formation are ensuring adequate clarification of the PLS (see Section 5.5.9) and avoiding overmixing and consequent air entrainment. Ensuring that operational conditions are such as to avoid gypsum precipitation is also important. Crud formation is not always detrimental, and some circuits (especially copper) prefer to operate with small amounts of crud at the interface as the crud promotes phase disengagement, reduces entrainment, and allows for smoother solution flow patterns through the settler.

Most circuits employ some means of treating crud for recovery of the aqueous and organic phases. These could be as simple as a decantation tank or plate-and-frame filter, or increasing in complexity to the use of centrifuges (Mukutuma et al. 2007). Oxime systems can be treated with clay to remove crud and organic degradation products.

5.5.7 Choice of Diluent

In many older SX textbooks, the authors considered the diluent to be an inert carrier for the extractant and assumed it did not play a role in the chemical or physical properties of the process. More recently, however, considerable attention is paid to the nature and choice of diluents because the choices can convey significant operational and cost advantages to a process.

The diluents most commonly employed in hydrometallurgical SX circuits are by-products of the petroleum industry. Depending on the distillation range under which they are collected, they will have differing aromatic and aliphatic contents. *Aromatic* diluents (based on unsaturated benzene-ring structures and/or naphthalenes) tend to have lower boiling and flash points, higher volatility and toxicity, increased tendency to oxidation, and higher aqueous-phase solubility, but better solvation characteristics. *Aliphatic* diluents (straight- or branched-chain saturated hydrocarbons), also known as *paraffins*, are more stable to oxidation, have lower aqueous solubility, and improved safety features because of higher points, lower flammability and volatility, but do not provide such good solvation of extracted species. Branched-chain aliphatic diluents are generally better than straight chains.

Most western countries have legislated reductions of hydrocarbon emissions. In SX plants, this means a tendency away from the use of aromatic components. In Africa, Australia, and South America, diluents containing 20 to 25% aromatic contents are still common, while in North America, diluents seldom contain more than 5% aromatics. Some processes, however, require 100% aromatic diluents because of their solvation properties. The uranium and precious metals industries often employ aromatic diluents, while cobalt, nickel, and zinc systems can employ 100% aliphatic diluents.

It is always recommended that pilot conditions include testing of diluents, and that a diluent not be introduced to a plant without adequate understanding of its chemical and physical properties and the process implications for the change.

5.5.8 Organic Losses and Recovery

Extractant and diluent replacements represent considerable operating costs in SX circuits. It is therefore preferable to operate a plant in such a manner as to minimize these losses. Organic losses can occur through a number of physical and chemical mechanisms including oxidative degradation, hydrolytic degradation, soluble losses to the aqueous phase, losses associated with crud formation, entrainment losses to raffinate and LSL, and evaporation and aerosol formation.

Suitable choice of process conditions (temperature, redox potential of liquors, nature of diluent, phase continuity, etc.) can reduce the chemical contributions to organic losses. Appropriate plant design and operation also play a role in minimizing physical losses. In many cases, appropriate engineering that minimizes organic losses will also lead to improvements in plant safety by avoiding mist and aerosol formation. Engineering also includes the decision whether to employ mixer–settlers or columns as contacting equipment (see Section 5.4). A detailed discussion of these factors is beyond the scope of this chapter, but the interested reader is referred to excellent reviews on this topic (Hearn 2005a, 2005b; Hopkins 2005; Miller 2005; Moore 2005; Haig et al. 2003).

Several mechanisms can be employed to recover organics lost from a circuit. An after-settler on the raffinate or LSL streams allows additional settling time for entrained organic phase to settle out of the aqueous phase. Flotation columns (Jameson cells) have been employed to bubble air through the solution, onto which organic droplets will attach themselves and can be floated to the surface and removed. Coalescers are typically packed beds of a hydrophobic or hydrophilic material that can encourage the coalescence of the dispersed phase on the surface of the packing material. A particular innovation in this area is the CoMatrix®* column (Minson et al. 2005; SpinTek Filtration 2005; Greene 2002) that uses a matrix packing followed by a garnet–anthracite dual media filtration system to remove organic phase.

* CoMatrix is a registered trademark of Spintek Filtration.

Activated carbon has also been employed successfully in applications that require very low levels of organic in the stream going forward (Sole et al. 2007b).

Most crud-treatment systems also allow for the recovery of the organic phase back to the SX circuit. A newer development is the use of a *diluent wash*— an additional stage added to the extraction circuit to allow the raffinate to come into contact with the diluent alone, facilitating the recovery of entrained and soluble organic back to the diluent. This is particularly useful for nickel systems that employ carboxylic acids that have high aqueous solubilities.

As indicated in previous sections, the physical properties of the organic phase are extremely important in allowing the smooth operation of a circuit. It is important to ensure that any recovered organic is not contaminated. Many plants use storage ponds to collect settled-out organic phase. These ponds typically process waste streams from the entire plant and so oils and greases from machinery and other sources can be transported to the organic phase. The organic should typically be treated through a regeneration or crud-treatment process before it is returned to the SX circuit.

5.5.9 Clarification of Pregnant Leach Solution (PLS)

One critically important factor in the design of an SX process is clarification of the feed liquor entering the circuit. This is often overlooked and can cause untold operational problems at a later stage. An SX circuit should have only two phases: aqueous and organic. A third phase allowed to enter the circuit can significantly interrupt the interfacial properties of the system, causing *inter alia* deterioration in phase disengagement, crud formation, gypsum formation, stabilisation of emulsions, loss of selectivity of the extractant, and reduction in circuit capacity. A general guideline is that the solids' content of the PLS should be <10 mg/l. Furthermore, solids in the circuit will also require periodic downtime of the circuit for their removal, leading to lower productivity.

Devices that can be used to improve S–L separation ahead of the SX circuit include thickeners, clarifiers, and filters. Of particular interest in recent designs are pinned-bed clarifiers (Langton et al. 2008), sand filters, Wakishimizu filters (Fuls and Nong 2007), and centrifuges (Mukutuma et al. 2007). All have advantages and disadvantages for specific systems, and testing of clarification equipment should be a vital component of any piloting program.

5.6 CONCLUSIONS AND FUTURE DEVELOPMENTS

Almost all base metals may be recovered or removed as impurities by SX, as can most precious metals and the rare earths (Cox 2004). Even some metalloid elements such as indium, germanium, and gallium are purified by SX for various high-tech applications. Considering that the first copper plant began operations only 40 years ago, this achievement reflects the confidence that a traditionally conservative industry has placed in this technology. In modern applications, the needs for higher purity products and more environmentally friendly operating conditions have led to the replacement by SX of many of the more traditional processing routes (Ritcey 2005,

1986). This trend is likely to continue as more complex and lower grade materials are processed. In particular, increasing applications are predicted for precious metals (Mooiman 1993; Harris 1993), actinides (Nash 2002), and rare earths (Gupta and Krishnamurthy 1992), all of which employ very complex circuits, a variety of solvents, and often sequential circuits.

For most of the last century, copper, nickel, and zinc were processed primarily by smelting and associated pyrometallurgical techniques. The few plants that employed hydrometallurgical processing relied heavily on precipitation methods to achieve the separations required. The flow sheets involved numerous recycles and repurification steps to achieve adequate product purities, often at the expense of high operating costs or poor single-pass product recoveries. Advances in equipment design, developments in specialised SX extractants, and innovative process chemistry today allow myriad process options for the recovery of base metals. Modern flow sheets show a trend toward the use of multiple SX steps using several different extractants, as opposed to the "safe" one-step copper SX of the past. This presents engineering challenges in ensuring process isolation and the integration of SX with the upstream and downstream operations.

The more widespread use of PAL and halide-based processes has also introduced new opportunities for SX. Despite the many "teething" problems of the new nickel plants, these commissioning experiences have led to a better understanding and acceptance of the capabilities and reliability of SX by the base-metal industry. With most of the materials, engineering, and operational issues associated with autoclave leaching now better understood, new projects are able to concentrate on optimising the downstream processing routes, enabling more harmonious integration of the leaching and purification circuits. Several hydrometallurgical flow sheets for the treatment of both oxide and low-grade sulfide materials are under advanced feasibility studies and pilot-plant development for copper, nickel, and zinc. Considerable variation exists in the approaches taken to solution purification, but all of the processes employ at least one SX circuit. While some of these flow sheets remain to be proved in long-term operation, there is no doubt that the adoption of this technology for large, capital-intensive projects augurs well for the future of SX in hydrometallurgical applications.

ACKNOWLEDGMENTS

Sincere appreciation is expressed to my colleagues and friends who assisted in various ways with the preparation of this chapter. In particular, I thank Angus Feather and Gary Kordosky (Cognis), Matthew Soderstrom and Owen Tinkler (Cytec), Peter Cole (TWP Matomo Projects, South Africa), and Dave Robinson (CSIRO, Australia) for their critical comments. I am also indebted to the many expert reviews cited in the bibliography. This chapter is based on material presented at various professional continuing education courses (Sole 1999a, 1999b, 2001, 2003, 2006).

REFERENCES

Al-Bazi, S. J. and Chow, A. 1984. Platinum metals: solution chemistry and separation methods (ion exchange and solvent extraction). *Talanta* 31:815–836.

Amer, S. 1980. Aplicaciones de la extracción con disolventes a la hidrometalurgia. I Parte. *Generalidades. Rev. Metal. CENIM* 16:291–303.

Amer, S. 1983a. Aplicaciones de la extracción con disolventes a la hidrometalurgia, XI parte. Metales preciosos (I). *Rev. Metal. CENIM* 19: 96–112.

Amer, S. 1983b. Aplicaciones de la extracción con disolventes a la hidrometalurgia, XI parte. Metales preciosos (II). *Rev. Metal. CENIM* 19:161–183.

Anon. 1973. In clean-air copper production Arbiter process is first off the mark. *Eng. Min. J.* 174:74–75.

Anon. 1995. Escondida's new cathode plant. *Eng. Min. J.* 196(5):24–28.

Arbiter, N. and Fletcher, A. W. 1994. Copper hydrometallurgy: evolution and milestones. *Min. Eng.* 46:118–123.

Ashbrook, A. W. and Lakshmanan, V. I. 1983. Uranium purification. In *Handbook of solvent extraction*, eds. T. C. Lo, M. H. I. Baird, and C. Hanson, 799–804. New York: John Wiley & Sons.

Atwood, R. L. and Miller, J. D. 1972. Structure and composition of commercial copper chelate extractants. In Proceedings Annual Meeting of the American Institute of Mining, Metallurgical and Petroleum Engineers, San Francisco, February 20–24.

Bacon, G. and Mihaylov, I. 2002. Solvent extraction as an enabling technology in the nickel industry. In Proceedings international solvent extraction conference, vol. 1, eds. K. C. Sole, P. M. Cole, J. S. Preston, and D. J. Robinson, 1–13. Johannesburg: South African Institute of Mining and Metallurgy.

Baes, C. F. 1962. The extraction of metallic species by dialkylphosphoric acids. *J. Inorg. Nucl. Chem.* 24:707–720.

Barnes, J. E. and Edwards, J. D. 1982. Solvent extraction at Inco's Acton precious metal refinery. *Chem. Ind.* (5):151–155.

Bateman Corporation. 1993. Bateman Pulse Column brochure, Bateman Israel.

Bautista, R. G. 1993. The solvent extraction of nickel, cobalt and their associated metals. In *Extractive metallurgy of copper, nickel and cobalt*, vol. I: Fundamental aspects, eds. R. G. Reddy and R. N. Weizenbach, 827–852. Warrendale, PA: The Minerals, Metals & Materials Society.

Bellingham, A. I. 1961. Application of solvent extraction to the recovery of uranium from El Sherana ore. *Austral. Inst. Min. Metall. Proc.* (198):85–112.

Benner, L. S., Suzuki, T., Meguro, K., and Tanaka, S. (eds.). 1991. *Precious metals science and technology*, 375–398. Allentown, PA: International Precious Metals Institute.

Blake, C. A., Baes, C. F. and Brown, K. B. 1958. Solvent extraction with alkylphosphoric compounds. *Ind. Eng. Chem.* 50:1763–1767.

Blass, E. 2004. Engineering design and calculation of extractors for liquid–liquid systems. In *Solvent extraction principles and practice*, 2nd ed., eds. J. Rydberg, M. Cox, C. Musikas, and G. R. Choppin, 367–414. New York: Marcel Dekker.

Boydell, D. W. and Viljoen, E. B. 1982. Uranium processing in South Africa from 1961 to 1981. In Proceedings 12th Council of Mining and Metallurgical Institutions congress, ed. H. W. Glen, 575–582. Johannesburg: South African Institute of Mining and Metallurgy.

Bozec, C., Demarthe, J. M., and Gandon, L. 1974. Recovery of nickel and cobalt from metallurgical wastes by solvent extraction. In Proceedings international solvent extraction conference, 1201–1229. London: Society of Chemical Industry.

Brits, J. H. and Deglon, D. A. 2007. Palladium stripping rates in PGM refining. *Hydrometallurgy* 89:253–259.

Burvill, D. 1999. Engineering aspects of the Cawse nickel/cobalt laterite project. In Proceedings ALTA 1999 nickel/cobalt pressure leaching and hydrometallurgy forum. Melbourne: ALTA Metallurgical Services.

Carter, R. A. 1997. Copper hydromet enters the mainstream. *Eng. Min. J.* 198(9):26–30.

Cattrall, R. W. and Slater, S. J. E. 1973. The extraction of metal ions from aqueous sulfate media by alkylamines. *Coord. Chem. Rev.* 11:227–245.

Charlesworth, P. 1981. Separating the platinum group metals by liquid–liquid extraction. *Platinum Met. Rev.* 25:106–112.

Chesne, A. and Germain, M. 1992. The use of solvent extraction in the nuclear fuel cycle: forty years of progress. In Proceedings international solvent extraction conference, ed. T. Sekine, 539–548. Essex, UK: Elsevier Science.

Cleare, M. J., Charlesworth, P., and Bryson, D. J. 1979. Solvent extraction in platinum group metal processing. *J. Chem. Tech. Biotech.* 29:210–224.

Cognis Corp. Undated. The chemistry of metals recovery using LIX® reagents, *MID Bluebook*.

Cognis. 2000. Operation of copper solvent extraction plants: a Cognis overview. Unpublished.

Cole, P. M. 1994. Review of and advances in solvent extraction, In *Hydrometallurgy school*. Johannesburg: South African Institute of Mining and Metallurgy.

Cole, P. M. 2002. The introduction of solvent-extraction steps during the upgrading of a cobalt refinery. *Hydrometallurgy* 64:69–77.

Cole, P. M. and Sole, K. C. 2002. Solvent extraction in the primary and secondary processing of zinc. In Proceedings international solvent extraction conference, vol. 2, eds. K. C. Sole, P. M. Cole, J. S. Preston, and D. J. Robinson, 863–870. Johannesburg: South African Institute of Mining and Metallurgy.

Cole, P. M. and Sole, K. C. 2003. Zinc solvent extraction in the process industries. *Miner. Proc. Extr. Metall. Rev.* 24:91–137.

Cox, M. 2004. Solvent extraction in hydrometallurgy. In *Solvent extraction principles and practice*, 2nd ed., eds. J. Rydberg, M. Cox, C. Musikas, and G. R. Choppin, 455–506. New York: Marcel Dekker.

Cytec. 2006. Crud: how it forms and techniques for controlling it, Cytec brochure MCT-1102. http://www.acorga.com.

Dalton, R. F., Severs, K. J., and Stephens, G. 1986. Advances in solvent extraction for copper with optimised use of modifiers. In Proceedings mining Latin America, Santiago, Nov. 17–19.

Dalton, R. F., Price, R., and Quan, P. M. 1983. Novel extractants for chloride leach systems. In Proceedings international solvent extraction conference, 189–190. Denver: American Institute of Chemical Engineers.

Dalton, R. F., Diaz, G., Price, R., and Zunkel, A. D. 1991. The Cuprex metal extraction process: recovering copper from sulphide ores. *JOM* 43(8):51–56.

Danesi, P. R., Reichley-Yinger, L., Mason, G., Kaplan, L., Horwitz, E. P., and Diamond, H. 1985. Selectivity-structure trends in the extraction of Co(II) and Ni(II) by dialkyl phosphoric, alkyl alkylphosphonic, and dialkylphosphinic acids. *Solvent Extr. Ion Exch.* 3:435–452.

Dasher, J. and Power, K. 1971. Copper solvent-extraction process: from pilot study to full-scale plant. *Eng. Min. J.* 172:111–115.

Demarthe, J. M., Gandon, L., and Georgeaux, A. 1976. A new hydrometallurgical process for copper. In *Extractive metallurgy of copper*, eds. J. C. Yannopoulos and J. C. Agarwal, 825–848. Warrendale, PA: The Minerals, Metals and Materials Society.

Demopoulos, G. P., Chang, Y., Benguerel, E., and Riddle, M. 2002. Opportunities and challenges in developing a solvent extraction process for rhodium based on the use of stannous chloride. In Proceedings of the international solvent extraction conference, eds. K. C. Sole, P. M. Cole, J. S. Preston, and D. J. Robinson, vol. 2, 908–915. Johannesburg: South African Institute of Mining and Metallurgy.

Dilley, M., Errington, M. T., and Naden, D. 1993. Application of the Davy combined mixer electrostatically assisted settler (CMAS) to copper stripping from organic SX solutions. In Proceedings international solvent extraction conference, 140–147. London: Society of Chemical Industry.

Donegan, S. 2006. Direct solvent extraction of nickel at Bulong operations. *Miner. Eng.* 19:1234–1245.

Dry, M. J., Iorio, G., Jacobs, D. F., Cole, P. M., Feather, A. M., Sole, K. C., Engelbrecht, J., Matchett, K. C., Cilliers, P. J., O'Kane, P. T., and Dreisinger, D. B. 1998. Cu/Co tailings treatment project, Democratic Republic of Congo. In Proceedings ALTA 1998 nickel/cobalt pressure leaching and hydrometallurgy forum. Melbourne: ALTA Metallurgical Services.

Du Preez, A. C. and Preston, J. S. 2002. The solvent extraction properties of di-n-hexyl sulphoxide in relation to the refining of platinum-group metals. In Proceedings international solvent extraction conference, eds. K. C. Sole, P. M. Cole, J. S. Preston, and D. J. Robinson, vol. 2, 896–901. Johannesburg: South African Institute of Mining and Metallurgy.

Duyversteyn, W. C. and Subacky, B. J. 1993. The Escondida process for copper concentrates. In The Paul E. Queneau international symposium extractive metallurgy of copper, nickel and cobalt, vol. 1, Fundamental aspects, eds. R. G. Reddy and R. N. Weizenbach, 881–910. Warrendale, PA: The Minerals, Metals and Materials Society.

Duyvesteyn, W. P. C. and Sabacky, B. J. 1995. Ammonia leaching process for Escondida concentrates. *Trans. Inst. Min. Metall. C* 104:C117–178.

Edwards, R. I. and Te Riele, W. A. M. 1983. Commercial processes for precious metals. In *Handbook of solvent extraction*, eds. T. C. Lo, M. H. I. Baird, and C. Hanson, 725–732. New York: John Wiley & Sons.

Evans, J. P. 1975. Process design of hydrometallurgical copper flow sheets. *Min. Mag.* 133: 271–276.

Feather, A., Sole, K. C., and Bryson, L. J. 1997a. Gold refining by solvent extraction: the Minataur™ Process. J. S. *Afr. Inst. Min. Metall.* 97:169–173.

Feather, A., Sole, K. C., and Bryson, L. 1997b. Refinacíon del oro por medio de la extraccíon con disolventes: el proceso Minataur (MR). In XXII Convencíon nacional de minería. Acapulco, Mexico: Asociacíon de Ingenieros de Minas, Metalurgistas y Geólogos de México.

Feather, A. M., Sole, K. C., and Dreisinger, D. B. 2001. Pilot-plant evaluation of manganese removal and cobalt purification by solvent extraction. In Proceedings international solvent extraction conference, vol. 2, eds. M. Cox, M. Hidalgo, and M. Valiente, 1443–1448. London: Society of Chemical Industry.

Feather, A. M., Bouwer, W., Swarts, A., and Nagel, V. 2002a. Pilot-plant solvent extraction of cobalt and nickel for Avmin's Nkomati project. In Proceedings international solvent extraction conference, vol. 2, eds. K. C. Sole, P. M. Cole, J. S. Preston, and D. J. Robinson, 946–951. Johannesburg: South African Institute of Mining and Metallurgy.

Feather, A., Bouwer, W., O'Connell, R., and Roux, J. 2002b. Commissioning of the new Harmony gold refinery. In Proceedings international solvent extraction conference, eds. K. C. Sole, P. M. Cole, J. S. Preston, and D. J. Robinson, vol. 2, 922–927. Johannesburg: South African Institute of Mining and Metallurgy.

Finney, S. A. 1977. A review of the progress in the application of solvent extraction for the recovery of uranium from ores treated by the South African gold mining industry. In Proceedings international solvent extraction conference, vol. 2, 567–576. Montreal: Canadian Institute of Mining and Metallurgy.

Fisher, J. F. C. and Notebaart, C. W. 1983. Commercial processes for copper. In Handbook of solvent extraction, eds. T. C. Lo, M. H. I. Baird, and C. Hanson, 649–671. New York: John Wiley & Sons.

Flett, D. S. 1974. Solvent extraction in copper hydrometallurgy: A review. *Trans. Inst. Min. Metall. C* 83:C30–C38.

Flett, D. S. 1987. Chemistry of nickel–cobalt separation. In *Extractive metallurgy of nickel*, 76–97. New York: John Wiley & Sons.

Flett, D. S. and West, D. W. 1986. The cobalt-catalysed oxidation of solvent extraction diluents. In Proceedings international solvent extraction conference, vol. II, 3–10. Frankfurt: Dechema.

Flett, D. S., Melling, J., and Cox, M. 1983. Commercial solvent systems for inorganic processes. In Handbook of solvent extraction, ed. T. C. Lo, M. H. I. Baird, and C. Hanson, 629–647. New York: John Wiley & Sons.

Fox, M. H., Ralph, S. J., Sithebe, N. P., Buchalter, E.M., and Riordan, J. J. 1998. Comparison of some aspects of the performance of a Bateman pulsed column and conventional mixer/settlers. Proceedings ALTA 1998 nickel cobalt pressure leaching and hydrometallurgy forum. Melbourne: ALTA Metallurgical Services.

Fuls, H. and Nong, K. 2007. Piloting of the Ishigaki Fibre-Wakishimizu (fibre media rapid filtration equipment) at Skorpion Zinc. In Africa's base metals resurgence, 4th Southern Africa base metals conference, 223–240. Johannesburg: Southern African Institute for Mining and Metallurgy.

Fuls, H., Sole, K. C., and Bachmann, T. 2005. Solvent extraction of zinc from a primary source: the Skorpion Zinc experience. In proceedings international solvent extraction conference, Beijing: China Academic Journal Electronic Publishing House.

Galbraith, G. G., Wang, Q., Li, L., Blake, A. J., Wilson, C., Collinson, S. R., Lindoy, L. F., Plieger, P. G., Schröder. M., and Tasker, P. 2007. Anion selectivity in zwitterionic amide-functionalised metal salt extractants. *Chem. Eur. J.* 13:6091–6107.

Gigas, B. and Giralico, M. 2002. Advanced methods for designing today's optimum solvent extraction mixer settler unit. In Proceedings international solvent extraction conference, eds. K. C. Sole, P. M. Cole, J. S. Preston, and D. J. Robinson, vol. 2, 1388–1395. Johannesburg: South African Institute of Mining and Metallurgy.

Giralico, M., Gigas, B., and Preston, M. 2003. Optimised mixer settler designs for tomorrow's large flow production requirements. In Proceedings Copper 2003, Hydrometallurgy of copper: modelling, impurity control and solvent extraction, vol. VI, book 2, eds. P.A. Rivieros, D. G. Dixon, D. B. Dreisinger and J. H. Menacho, 775–794. Montreal: Canadian Institute of Mining, Metallurgy and Petroleum.

Girivalkar, A., Kotekar, M. K., Mallavarapu, A., Mishra, S. L., Mukherjee, T. K., Ravishankar, V., and Singh, H. 2004. Recovery of high purity uranium from fertilizer grade weak phosphoric acid by two cycles of extractions using selective extracting solvent of synergistic extractant system of di(2-ethylhexyl) and tri-n-butylphosphate in refined kerosene. Patent WO 200487971 A1.

Gnoinski, J., Bachmann, T., and Holtzhausen, S. 2005. Skorpion Zinc: defining the edge of zinc refining. In Lead–Zinc'05 Proceedings international symposium lead zinc processing, ed. T. Fujisawa, 1315–1325. Kyoto: Mining and Materials Processing Institute of Japan.

Godfrey, J. C. and Slater, M. J., eds. (1994), *Liquid–liquid extraction equipment.* Chichester: John Wiley & Sons.

Golding, J. A. and Barclay, C. D. 1988. Equilibrium characteristics for the extraction of cobalt and nickel into di(2-ethylhexyl) phosphoric acid. *Can. J. Chem. Eng.* 66:970–979.

Grant, R. A. and Drake, V. A. 2002. The application of solvent extraction to the refining of gold. In Proceedings international solvent extraction conference, vol. 2, eds. K. C. Sole, P. M. Cole, J. S. Preston, and D. J. Robinson, 940–945. Johannesburg: South African Institute of Mining and Metallurgy.

Greene, W. A. 2002. Organic recovery with CoMatrix filtration. In Proceedings international solvent extraction conference, vol. 2, eds. K. C. Sole, P. M. Cole, J. S. Preston, and D. J. Robinson, 1428–1433. Johannesburg: South African Institute of Mining and Metallurgy.

Gupta, C. K. and Krishnamurthy, N. 1992. Extractive metallurgy of rare earths. *Int. Mater. Rev.* 37:197–248.

Gupta, C. K. and Singh, H. 2003. *Uranium resource processing: secondary resources.* Berlin: Springer-Verlag.

Gupta, C. K. and Singh, H. 2005. Uranium resource processing: secondary resources, India. In Proceedings technical committee meeting, June, 15–18 1999, 73–79. Vienna: International Atomic Energy Agency.

Haig, P., Maxwell, J., and Koenen, T. 2003. Electrostatic hazards in solvent extraction plants. In Proceedings ALTA SX/IX world summit on SX fire protection, Melbourne: ALTA Metallurgical Services.

Harris, G. B. 1993. A review of precious metals refining. In *Precious Metals* 1993, 351–374. Allentown, PA: International Precious Metals Institute.

Hearn, G. 2005a. What sparks danger in solvent extraction? *Min. Mag.* (March):32–33.

Hearn, G. 2005b. Static electricity: danger inherent in the solvent extraction process. In Proceedings ALTA SX/IX world summit on SX fire protection, Melbourne: ALTA Metallurgical Services.

Hein, H. 2005. The importance of a wash stage in copper solvent extraction. In HydroCopper 2005, Proceedings III international copper hydrometallurgy workshop, eds. J. M. Penacho and J. Casas de Prada, 425–436. Santiago: Universidad de Chile.

Hopkins, W. R. 1994. SXEW: a mature but expanding technology. *Min. Mag.* 170:259–265.

Hopkins, W. 2003. Modern SX-EW: very large to very small. In Proceedings 16th international copper conference. Portugal, Feb. 17–19.

Hopkins, W. 2005. Fire hazards in SX plant design: an update. In Proceedings ALTA SX/IX world summit on SX fire protection. Melbourne: ALTA Metallurgical Services.

House, J. E. 1985. Success stories in speciality chemicals. *SRI Newsletter* (June):1–8.

House, J. E. 1989. The development of the LIX® reagents. *Miner. Metall. Proc.* 6(2):1–6.

Hurtado-Guzmán, C. and Menacho, J. M. 2003. Oxime degradation chemistry in copper solvent extraction plants. In Proceedings hydrometallurgy of copper: modelling, impurity control and solvent extraction, vol. 6, book 2, eds. P.A. Rivieros, D. G. Dixon, D. B. Dreisinger, and J. H. Menacho, 719– 734. Montreal: Canadian Institute of Mining, Metallurgy and Petroleum.

Jenkins, I. J. 1979. Solvent extraction in the atomic energy industry: a review. *Hydrometallurgy* 4:1–20.

Kennedy, B., Davenport, W. G., Jenkins, J., and Robinson, T. 1999. Electrolytic copper electrowinning and solvent extraction: world operating data. In *Copper leaching, solvent extraction, and electrowinning technology*, ed. G. V. Jergensen II, 41–87. Littleton, CO: Society for Mining, Metallurgy and Exploration.

Kordosky, G. 1992. Copper solvent extraction: state of the art. *JOM* 44(5):40–45.

Kordosky, G. A. 2002. Copper recovery using leach/solvent extraction/electrowinning technology: forty years of innovation, 2.2 million tones of copper annually. In Proceedings international solvent extraction conference, vol. 2, eds. K. C. Sole, P. M. Cole, J. S. Preston, and D. J. Robinson, 853–862. Johannesburg: South African Institute of Mining and Metallurgy.

Kordosky, G. A. 2003. Copper SX circuit design and operation: current advances and future possibilities. In Proceedings ALTA copper 2003. Melbourne: ALTA Metallurgical Services.

Kordosky, G. and Virnig, M. 2003. Equilibrium modifiers in copper solvent extraction reagents: friend or foe? In *Hydrometallurgy* 2003, vol. 1, Leaching and solution purification, eds. C. A. Young, A. M. Alfantasi, C. G. Anderson, D. B. Dreisinger, G. B. Harris, and A. James, 905–916. Warrendale, PA: The Minerals, Metals and Materials Society.

Kordosky, G. A., Olafson, S. M., Lewis, R. G., and Defner, V. L. 1987. A state-of-the-art discussion on the solvent extraction reagents used for the recovery of copper from dilute sulfuric acid solutions. *Sep. Sci. Tech.* 22:215–232.

Kusnetsov, G. I., Pushkov, A. A., and Kosogorov, A. V. 2002. Industrial applications of CENTREK centrifugal extractors. In Proceedings international solvent extraction conference, eds. K. C. Sole, P. M. Cole, J. S. Preston, and D. J. Robinson, vol. 2, 1322–1327. Johannesburg: South African Institute of Mining and Metallurgy.

Kyle, J. H. and Furfaro, D. 1997. The Cawse nickel/cobalt laterite project metallurgical process development. In *Hydrometallurgy and refining of nickel and cobalt*, eds. W. C. Cooper and I. Mihayov, 379–389. Montreal: Canadian Institute of Mining, Metallurgy and Petroleum.

Langton, M., Fuls, H. F., Sole, K. C., Rampersod, A., Blaauw, C., and Martinson, B. 2008. Innovations in the configuration of a pinned-bed clarifier for clarification of a zinc leach liquor. In Proceedings International Solvent Extraction Conference, Tucson, AZ, in press.

Lea, R. K., Edwards, J. D., and Colton, D. F. 1983. Process for the extraction of precious metals from solutions thereof. U.S. Patent 4,390,366.

Lewis, I. E. and Kesler, S. 1980. A case study: optimising the performance of the uranium SX plant at Rossing Uranium Limited in Namibia. In Proceedings international solvent extraction conference, vol. 3, Liege, Belgium: Association des Ingenieurs Sortis de l'Universite de Liege.

Lloyd, P. J. D. 1983. Commercial processes for uranium from ore. In *Handbook of solvent extraction*, eds. T. C. Lo, M. H. I. Baird, and C. Hanson, 763–782. New York: John Wiley & Sons.

Lloyd, P. J. D. 2004. Principles of industrial solvent extraction. In *Solvent extraction principles and practice*, 2nd ed., eds. J. Rydberg, M. Cox, C. Musikas, and G. R. Choppin, 339–366. New York: Marcel Dekker.

Mackenzie, M. 2005. Cognis amines in mining chemical applications. *Cognis Tech. Bull.*

Mackenzie, J. M. W., Virnig, M. J., Boley, B. D., and Wolfe, G. A. 1998. Extraction of nickel from ammoniacal leach solutions: extractant and solution chemistry issues. In Proceedings ALTA nickel/cobalt pressure leaching and hydrometallurgy forum. Melbourne: ALTA Metallurgical Services.

Mackenzie, J. M. W. and Virnig, M. J. 2004. Solvent extraction technology for the extraction of nickel using LIX 84-INS: update and circuit comparisons. In Proceedings international laterite nickel symposium, 457–475. Warrendale, PA: The Minerals, Metals and Materials Society.

Mackenzie, M., Virnig, M., and Feather, A. 2005. The recovery of nickel from HPAL laterite solutions using LIX 84-INS and some possible impurity control strategies. In Proceedings 10th ALTA 2005 nickel/cobalt conference. Melbourne: ALTA Metallurgical Services.

Maes, C., Tinkler, O., Moore, T., and Mejías, J. 2003. The evolution of modified aldoxime copper extractants. In Proceedings copper 2003 hydrometallurgy of copper: modelling, impurity control and solvent extraction, vol. 6, book 2, eds. P. A. Riviero, D. G. Dixon, D. B. Dreisinger, and J. H. Menacho, 753–760. Montreal: Canadian Institute of Mining, Metallurgy and Petroleum.

Matthew, I. G. and Elsner, D. 1977. The hydrometallurgical treatment of zinc silicate ores. *Met. Trans. B* 8B:73–83.

Maxwell, B., Rasdell, S., and Carlin, P. 1999. Oxidative stability of diluents in Co/Ni solvent extraction. In Proceedings ALTA 1999 nickel/cobalt pressure leaching and hydrometallurgy forum. Melbourne: ALTA Metallurgical Services.

Merritt, R. C. 1971. The extractive metallurgy of uranium. Golden, CO: Colorado School of Mines Research Institute.

Mihaylov, I. O., Krause, E., Laundry, S., and Luong, C. 1995. Process for the extraction and separation of nickel and/or cobalt. U.S. Patent 5,378,262.

Mihaylov, I., Krause, E., Colton, D. F., Okita, Y., Duterque, J.-P., and Perraud, J.J. 2000. The development of a novel hydrometallurgical process for nickel and cobalt recovery from Goro laterite ore. *CIM Bull.* 93:124–130.

Miller, G. 2005. Engineering design for lowering fire risk. In Proceedings ALTA SX/IX world summit on SX fire protection, Melbourne: ALTA Metallurgical Services.

Minson, D., Jiménez, C., and Gilmour, J. 2005. Filtros Co-Matrix: resultados de una plata pilota y sus applicaciones en las plantas de extracción por solventes. In HydroCopper: proceedings III international copper hydrometallurgy workshop, eds. J. M. Menacho and J. Casas de Prada, 437–448. Santiago: Universidad de Chile.

Mintek. 2001. Annual Review 2000: Product introduction. http://www.mintek.ac.za/pubs/ar00/marketing.htm.

Mooiman, M. B. 1993. The solvent extraction of precious metals: a review. In *Precious metals* 411–434. Allentown, PA: International Precious Metals Institute.

Moore, L. 2005. Using principles of inherent safety for design of hydrometallurgical solvent extraction plants. In Proceedings ALTA SX/IX world summit on SX fire protection, Melbourne: ALTA Metallurgical Services.

Movsowitz, R. L., Kleinberger, L., and Buchalter, E. M. 1997. Application of Bateman pulsed columns for uranium SX: from pilot to industrial columns. In Extraction Metallurgy Africa 1997. Johannesburg: South African Institute of Mining and Metallurgy.

Movsowitz, R. L., Kleinberger, R., Buchalter, E. M., Grinbaum, B., and Hall, S. 2001. Comparison of the performance of full-scale pulsed columns versus mixer–settlers for uranium solvent extraction. In Proceedings international solvent extraction conference, vol. 2, eds. M. Cox, M. Hidalgo, and M. Valiente, 1455–1460. London: Society of Chemical Industry.

Mukutuma, A., Schwartz, N., Chisakuta, G., Mbao, B., and Feather, A. 2007. A case study of the operation of a Flottweg Tricanter® centrifuge for solvent-extraction crud treatment at Bwana Mkubwa, Ndola, Zambia. In Africa's base metals resurgence, 4th Southern African conference on base metals, 393–403. Johannesburg: The Southern African Institute of Mining and Metallurgy.

Munyangano, B. M. 2007. Solvent degradation-Rossing Uranium Mine. *J. S. Afr. Inst. Min. Metall.* 107:415–417.

Nagel, V. and Feather, A. M. 2001. The recovery of nickel and cobalt from a bioleach liquor saturated in calcium using versatic acid in a synergistic mixture with 4-nonylpyridine. In Proceedings ALTA nickel/cobalt-7. Melbourne: ALTA Metallurgical Services.

Nash, K. L. 1993. A review of the basic chemistry and recent developments in trivalent f-element separations. *Solvent Extr. Ion Exch.* 11:729–768.

Nash, K. L. 2002. Virtues and vices of reagents for actinide partitioning in the twenty-first century. In Proceedings international solvent extraction conference, vol. 2, eds. K. C. Sole, P. M. Cole, J. S. Preston, and D. J. Robinson, 1109–1117. Johannesburg: South African Institute of Mining and Metallurgy.

Naylor, A. and Wilson, P. D. 1983. Recovery of uranium and plutonium from irradiated nuclear fuel. In *Handbook of solvent extraction*, eds. T. C. Lo, M. H. I. Baird, and C. Hanson, 768–798. New York: John Wiley & Sons.

Nicol, M. J., Fleming, C. A., and Preston, J. S. 1987. Applications to extractive metallurgy. In *Comprehensive coordination chemistry: synthesis, reactions, properties and applications of coordination compounds*, vol. 6, eds. G. Wilkinson, R. D. Gillard, and J. A. McCleverty, 779–842. Oxford: Pergamon.

Nyman, B., Aaltonen, A., Hulthom, S. E., and Karpale, K. 1992. Application of new hydrometallurgical developments in the Outokumpu HIKO process. *Hydrometallurgy* 29:461–478.

Nyman, B., Ekman, R., Kuusisto, R., and Pekkala, P. 2003. The OutoCompact SX technology: ideal approach to copper solvent extraction. In Proceedings copper 2003 hydrometallurgy of copper: modelling, impurity control and solvent extraction, vol. 6, book 2, eds. P.A. Rivieros, D. G. Dixon, D. B. Dreisinger, and J. H. Menacho, 761–774. Montreal: Canadian Institute of Mining, Metallurgy and Petroleum.

Pavlides, A. G. and Wyethe, J. 2000. Ion exchange column design for separation of nickel traces from cobalt electrolyte. In Proceedings ALTA SX/IX-1. Melbourne: ALTA Metallurgical Services.

Preston, J. S. 1982. Solvent extraction of cobalt and nickel by organophosphorus acids: comparison of phosphoric, phosphonic, and phosphinic acid systems. *Hydrometallurgy* 9:115–133.

Preston, J. S. 1985. Solvent extraction of metals by carboxylic acids. *Hydrometallurgy* 14:171–188.

Preston, J. S. and du Preez, A. C. 1988. The solvent extraction of cobalt, nickel, zinc, copper, calcium, magnesium, and the rare earth metals by organophosphorus acids. Mintek Report M378.

Price, M. J. and Reid, J. G. 1993. Separation and recovery of nickel and cobalt in ammoniacal systems: process development. In Proceedings international solvent extraction conference, 159–166. London: Society of Chemical Industry.

Readett, D. and Townson, R. 1997. Practical aspects of copper solvent extraction from acidic leach liquors. Proceedings ALTA 1997 copper hydrometallurgy forum. Melbourne: ALTA Metallurgical Services.

Renner, H. 1985. The selective solvent extraction of palladium by the use of di-normal-hexyl sulphide. Mintek Report M217.

Rice, N. M., Irving, H. M. N. H., and Leonard, M. A. 1993. Nomenclature for liquid–liquid distribution (solvent extraction). *Pure Appl. Chem.* 65:2373–2396.

Rickelton, W. A. and Nucciarone, D. 1997. The treatment of cobalt/nickel solutions using Cyanex® extractants. In Hydrometallurgy and refining of nickel and cobalt, eds. W. C. Cooper and I. Mihaylov, 275–291. Montreal: Canadian Institute of Mining, Metallurgy and Petroleum.

Rickelton, W. A., Robertson, A. J., and Hillhouse, J. H. 1991. The significance of diluent oxidation in cobalt–nickel separation. *Solvent Extr. Ion Exch.* 9:73–84.

Rimmer, B. F. 1974. Refining of gold from precious metal concentrates by liquid–liquid extraction. *Chem. Ind.* (2):63–66.

Rimmer, B. F. 1989. Refining of platinum group metals by solvent extraction. In *Precious metals* 1989, ed. B. Harris, 217–226. Allentown, PA: International Precious Metals Institute.

Ritcey, G. M. 1980. Crud in solvent extraction processing: review of causes and treatment. *Hydrometallurgy* 5:97–107.

Ritcey, G. M. 1986. The process and the environment: is solvent extraction the answer? In Proceedings international solvent extraction conference, vol. 1, 51–62. Frankfurt: Dechema.

Ritcey, G. M. 2004. Development of industrial solvent extraction processes. In Solvent extraction: principles and practice, 2nd ed., eds. J. Rydberg, M. Cox, C. Musikas, and G. R. Choppin, 277–338. New York: Marcel Dekker.

Ritcey, G. M. 2005. Solvent extraction in hydrometallurgy: present and future. In Solvent extraction for sustainable development, proceedings of the international solvent extraction conference. Beijing: China Academic Journal Electronic Publishing House.

Ritcey, G. M. and Ashbrook, A. W. 1972. Solvent extraction: principles and applications to process metallurgy, part 2, 282. Amsterdam: Elsevier.

Ritcey, G. M., Hayward, N. L., and Salinovich, T. 1996. The recovery of nickel and cobalt from lateritic ores. Australian Patent PN0441,

Rojas, E., Araya, G., Picardo, J., and Hein, H. 2005. Serie paralelo óptimo: un neuvo concepto de configuracíon en extraccíon por solventes. In HydroCopper 2005, Proceedings III international copper hydrometallurgy workshop, ed. J. M. Menacho and J. Casas de Prada, 449–458. Santiago: Universidad de Chile.

Rowden, G. A., Dilley, M., Bonney, C. F., and Gillett, G. A. 1981. Davy McKee's CMS contactor: its development and applicability. In Proceedings hydrometallurgy, ed. C. F. Bonney and G. A. Rowden, paper F2. London: Society of Chemical Industry.

Sato, T. 1962. The extraction of uranium from sulfuric acid solutions by di(2ethylhexyl)phos phoric acid. *J. Inorg. Nucl. Chem.* 24:699–706.

Schweitzer, F. W. and Livingston, R. 1982. Duval's CLEAR hydrometallurgical process. In Proceedings AIME annual meeting, Dallas, TX, Feb. 14–18.

Scott, S. A. and Matchett, K. 2004. MINATAUR™: the Mintek alternative technology to gold refining. J. S. *Afr. Inst. Min. Metall.* 104:339–343.

Seaborg, G. T. 1963. Man-made transuranium elements. Englewood Cliffs: Prentice-Hall.

Seaborg, G. T. 1972. Nuclear milestones: A collection of speeches. San Francisco: W. H. Freeman.

Sergeant, H. C. and Rice, N. M. 1977. The mechanism of uptake of gold by dibutyl carbitol. In Proceedings international symposium hydrometallurgy, 385–404. Belgium: Benelux Metallurgie.

Singh, H., Mishra, S. L., Mallavarapu, A., Ravishankar, V., Girivalkar, A., Kotekar, M. K., and Mukherjee, T. K. 2003. Process for recovery of high purity uranium from fertilizer grade weak phosphoric acid. Patent Class 423008000 (U.S.Patent and Trademark Office).

Skepper, I. G. and Fittock, J. E. 1996. Nickel cobalt separation by ammoniacal solvent extraction: the operating experience. In Proceedings international solvent extraction conference, 777–782. Melbourne: The University of Melbourne Press.

Slater, M. J. 1996. Liquid–liquid extraction equipment: progress and problems. In Proceedings international solvent extraction conference, 35–42. Melbourne: The University of Melbourne Press.

Soderstrom, M. 2006. New reagent developments in copper SX. In Proceedings ALTA Copper 2006. Melbourne: ALTA Metallurgical Services.

Soldenhoff, K., Hayward, N., and Wilkins, D. 1998. Direct solvent extraction of cobalt and nickel from laterite-acid pressure leach liquors. In EPD Congress 1998, ed. B. Mishra, 153–165. Warrendale, PA: The Minerals, Metals and Materials Society.

Sole, K. C. 1999a. Separation and purification of cobalt and nickel by solvent extraction. In Solvent extraction and electrowinning: basic principles and applications. Johannesburg: South African Institute of Mining and Metallurgy.

Sole, K. C. 1999b. Process experience: chemical and industrial applications. International Solvent Extraction School Lecture 3, Barcelona, July 12–16.

Sole, K. C. 2001. Solvent extraction in modern base metal processing. In Proceedings 6th world congress of chemical engineering. Melbourne, Australia.

Sole, K. C. 2003. Solvent extraction and ion exchange. In Hydrometallurgy short course. Montreal: Canadian Institute of Mining, Metallurgy and Petroleum.

Sole, K. C. 2006. Solvent extraction and ion exchange short course. Johannesburg: Anglo Research and Matomo Projects.

Sole, K. C. and Hiskey, J. B. 1995. Solvent extraction of copper by Cyanex 272, Cyanex 302 and Cyanex 301. *Hydrometallurgy* 37:129–147.

Sole, K. C. and Cole, P. M. 2001. Purification of nickel by solvent extraction. In Ion exchange and solvent extraction, vol. 15, eds. Y. Marcus and A. SenGupta, 143–195. New York: Marcel Dekker.

Sole, K. C. and Feather, A. M. 2003. Solvent extraction of copper from high-tenor pressure leach solutions using new modified aldoximes. In Proceedings copper 2003, vol. 6, book 2, eds. P. A. Rivieros, D. Dixon, D. B. Dreisinger, and J. Menacho, 691–706. Santiago, Chile.

Sole, K. C., Feather, A., Watt, J., Bryson, L. J., and Sorensen, P. F. 1998. Commercialisation of the Minataur™ process: commissioning of Harmony Gold Refinery. In EPD congress, ed. B. Mishra, 175–186. Warrendale, PA: The Minerals, Metals and Materials Society.

Sole, K. C., Cole, P. M., and Feather, A. M. 2005. Solvent extraction in southern Africa: update of recent developments. *Hydrometallurgy* 78:52–78.

Sole, K. C., Viljoen, K., Ferreira, B., Soderstrom, M., Tinkler, O. and Hoffman, L. 2007a. Customising copper–iron selectivity in modified aldoxime extractants: pilot plant evaluation. In Proceedings copper 2007, vol. IV, book 2, 3–14. Montreal: Canadian Institute of Mining, Metallurgy and Petroleum.

Sole, K. C., Stewart, R. J., Maluleke, R. F., Rampersad, A., and Mavhungu, A. E. 2007b. Removal of entrained organic phase from zinc electrolyte: pilot plant comparison of CoMatrix and carbon filtration. *Hydrometallurgy*, 89:11–20.

Sonntag, A. A., Szantz, I., and Goodman, D. A. 1989. New developments in Krebs mixer–settler technology. *Trans. Soc. Min. Metall. Expl.* 290:13–19.

SpinTek Filtration. 2005. CoMatrix tower. http://www.spintek.com/co-matrix_towers.htm.

Suetsuna, A., Ono, N., Iio, T. I., and Yamada, K. 1980. Sumitomo's new process for nickel and cobalt recovery from sulfide mixture. Metall. Soc. AIME, TMS Paper Selection A80-2.

Suttill, K. R. 1989. Solvent extraction, a key in maintaining copper production. *Eng. Min. J.* 190(9):24–26.

Taylor, A. 1997. Copper SX/EW seminar: Basic principles, alternatives, plant design. Melbourne: ALTA Metallurgical Services.

Townsend, B. and Severs, K. J. 1990. The solvent extraction of copper: a perspective. *Min. Mag.* 162:26–35.

Tunley, T. H. and Nel, V. W. 1974. The recovery of uranium from uranthorianite at the Palabora Mining Company. In Proceedings international solvent extraction conference, 1519–1533. London: Society of Chemical Industry.

Vancas, M. F. 2003. Solvent extraction settlers: a comparison of various designs. In Proceedings copper 2003 hydrometallurgy of copper: modelling, impurity control and solvent extraction, vol. 6, book 2, eds. P. A. Rivieros, D. G. Dixon, D. B. Dreisinger, and J. H. Menacho, 707– 718. Montreal: Canadian Institute of Mining, Metallurgy and Petroleum.

Vancas, M. F. 2005. A practical guide to SX plant operation. In Proceedings international solvent extraction conference. Beijing: China Academic Journal Electronic Publishing House.

Vijayalakshmi, R., Mishra, S. L., and Singh, H. 2005. Separation of uranium from weak phosphoric acid using di-nonylphenylphosphoric acid in combination with tri-n-butylphosphoric acid. *Int. J. Nucl. Energy Sci. Tech.* 1:178–183.

Virnig, M., Eyzaguirre, D., Jo. M., and Calderon, J. 2003. Effects of nitrates on copper SX circuits: a case study. In Proceedings copper 2003 hydrometallurgy of copper: modelling, impurity control and solvent extraction, vol. 6, book 2, eds. P.A. Rivieros, D. G. Dixon, D. B. Dreisinger and J. H. Menacho, 795–810. Montreal: Canadian Institute of Mining, Metallurgy and Petroleum.

Wallis, T. L. and Chlumsky, G. F. 1999. Economic considerations for SX/EW operations. In *Copper leaching, solvent extraction, and electrowinning technology*, ed. G. V. Jergensen II, 89–91. Littleton, CO: Society for Mining, Metallurgy and Exploration.

Wang, C. Y. 2005. Crud formation and its control in solvent extraction. In Proceedings international solvent extraction conference. Beijing: China Academic Journal Electronic Publishing House.

Wigstol, E. and Froyland, K. 1972. Solvent extraction in nickel metallurgy: the Falconbridge matte leach process. *Het Ingenieursblad* 41:476–486.

Wood, J. T., Kern, P. L., and Ashdown, N. C. 1977. Electrolytic recovery of zinc from oxidised ores. *J. Met.* (11):157–171.

6 Modeling and Optimization in Solvent Extraction and Liquid Membrane Processes

Inmaculada Ortiz Uribe and J. Angel Irabien Gulias

CONTENTS

6.1 INTRODUCTION

The development of new and more efficient extractants together with the use of more advantageous contactors have led to a continuous rise in the applications of solvent extraction (SX) and liquid membrane (LM) processes. However, rigorous mathematical models together with the aplication of optimization tools that enable the prediction of the operation optimum conditions are still needed to promote industrial application.

Among the principal applications of SX and LM processes, the separation and recovery of metallic components from aqueous solutions presents great relevance and is considered in what follows. Selective separation relies on the chemical reaction of the removed species with the selected extractant. Starting with the proper description of the chemical equilibrium between extractant and the extracted species the mathematical model requires an expression that allows determination of the extraction rate as a function of the operation variables. This task usually requires solving for the coupling between chemical reaction rates and mass transfer rates, leading to multiparametric expressions.

In this chapter the methodology needed for the mathematical modeling of SX and LM processes is presented.

6.2 MODELING OF THE CHEMICAL EQUILIBRIA

Considering again the extraction of metallic compounds with a selective extractant, it is usually thought that the reactions proceed fast enough to consider that the species involved are in equilibrium everywhere. For the extraction of a cationic species with an organic extractant HX, the stoichiometric equation would be represented by the following expression:

$$M^{m+} + nHX \leftrightarrow M(HX)_{n-m}X_m + mH^+ \quad (6.1)$$

Using commercial extractants at medium-high concentrations and eliminating purification processes usually imply a nonideal behavior of the liquid phases taking part in the extraction reaction. Thus, the thermodynamic equilibrium constant will be

$$K_a = \frac{\left(a_{\overline{[Me_iX_n(HX)_{m-n}]}}\right)\left(a_{(H^+)}\right)^n}{\left(a_{[Me_i]}\right)\left(a_{\overline{(HX)}_m}\right)^n} \quad (6.2)$$

Nonideality of the aqueous phase is taken into account by refering to the components activity rather than to the components concentration, $a_i = \gamma_i.C_i$. Nonideality of the organic phases is related to their ability to form aggregates such as dimer molecules that decrease the extraction capacity. Determination of the nonideal behavior of organic phases is usually a more complex task than for aqueous phases, and several ways have been proposed for this purpose: determination of the activity coefficients [1]; calculation of the aggregation number [2]; or description of the nonideal behavior as a function of the composition of the organic phase [3,4].

Accounting for the nonideal behavior of both fluid phases through the activity coefficients of the involved species, the following expression is obtained:

$$K_a = k_c \cdot K_\gamma = \frac{\left|\overline{M(HX)}_{n-m}\overline{X}_m\right|\left|H^+\right|^m}{\left|M^{m+}\right|\left|\overline{HX}\right|^n} \cdot \frac{\overline{\gamma}_{\overline{M(HX)}_{n-m}\overline{X}_m} \cdot \gamma^m_{H+}}{\gamma^{m+} \cdot \gamma^n_{\overline{HX}}} \quad (6.3)$$

Linearizing Equation (6.3) by taking logarithms it transforms into

$$\ln K_a = \ln k_c + \ln K_\gamma \quad (6.4)$$

with

$$\ln K_\gamma = \ln \frac{\gamma_{H+}^m}{\gamma^{m+}} + \ln \frac{\overline{\gamma}\,\overline{|M(HX)_{n-m}X_m|}}{\gamma_{\overline{HX}}^n} \tag{6.5}$$

Activity coefficients of ionic species are usually dependant on the ionic strength of the solution, $\gamma_i = \gamma_i$ (I); I, ionic strength of the solution, the Debye-Hückel equation being the simplest one in order to predict the values of the activity coefficients. Working with dilute metallic solutions it is likely to assume that ln

$$\frac{\gamma_{H+}^m}{\gamma^{m+}}$$

≈ constant at infinite dilution, and including this term in a new parameter, K_γ' the following equation is obtained:

$$\ln K_\gamma' = \ln \frac{\overline{\gamma}_{\overline{|M(HX)_{n-m}X_m|}}}{\gamma_{\overline{HX}}^n} \tag{6.6}$$

Equation (6.6) expresses nonideality of the organic phase, and the major difficulty lies in the lack of available information on the activity coefficients of the organic species. Assuming that the behavior of the organic extractant depends linearly on its concentration, as expressed by Equation (6.7),

$$\ln \frac{\overline{\gamma}_{\overline{|M(HX)_{n-m}X_m|}}}{\gamma_{\overline{HX}}^n} = A \cdot \frac{C_E}{C_T} \tag{6.7}$$

A being an empirical parameter that needs experimental finding. After introduction of Equation (6.7) into Equation (6.3) the latter transforms into

$$K_a = K_c \cdot C_T^n \tag{6.8}$$

Equation (6.8) permits the determination of the equilibrium constant of a liquid–liquid extraction reaction working with commercial extractants and providing that the empirical parameter "n" is known.

6.3 KINETICS OF SOLVENT EXTRACTION PROCESSES

The kinetics of SX and LM processes are in general a complex function of both the chemical reaction kinetics and the mass transfer rates [5]. Great attention has been devoted in the literature to the kinetic modeling of solvent extraction systems. Two extreme situations may be found regarding the rate controlling mechanisms: (1) the diffusional regime in which mass transfer controls the overall kinetics and the chemical reaction is assumed to be at equilibrium; and (2) the chemical regime where the extraction rate is independent of the mass transfer kinetics depending only on the velocity of the chemical reaction. A third situation where the kinetic control is shared between both regimes is usually found under experimental conditions. Therefore, a proper description of real systems would require the knowledge of mass transfer coefficients and intrinsic kinetic coefficients. The latter have been determined through the years from experimental analysis under the assumption of chemical regime [6–10]. However, some literature has pointed out the effect of wrongly assuming chemical control in the interpretation of experimental results [5,11,12]. As an example, the work of Pérez de Ortiz et al. [11] on the extraction of copper by LIX 65N reported that a variety of conflicting chemical kinetics could be satisfied by varying the range of concentrations and the hydrodynamic conditions, concluding that criteria for the nature of the rate controlling mechanism or location of the reaction do not always provide an unequivocal answer concerning the type of control if experiments are not carefully designed to distinguish between possible models.

Next, a methodology is presented for the experimental determination of the kinetics of liquid–liquid extraction systems.

The extraction system consists of an aqueous solution of metal ions in contact with an organic phase containing the extractant. At a suitable pH value, one of the species reacts with the extractant, forming a complex species soluble in the organic phase. The reaction is reversible, and the metal can be stripped from the organic phase. However, under extraction conditions and at initial contact times—that is, when the concentration of the complex species in the organic phase is low—the reverse reaction can be neglected. For a reaction represented by the stoichiometric Equation (6.1):

$$M^{m+} + nHX \leftrightarrow M(HX)_{n\text{-}m}X_m + mH^+$$

The rate of forward reaction will be

$$r_+ = k_+ C_M^a C_X^b C_H^d \tag{6.9}$$

and for the reverse reaction

$$r_- = K_- C_{MX}^e C_H^f \tag{6.10}$$

where M denotes the metal, X the anionic part of the complexing agent, MX the complex species, and H the hydrogen ion.

The low solubility in the aqueous phase of metal organic extractants reduces the possible locations of the rate controlling reaction to (1) the diffusional aqueous film and (2) the interfacial plane itself. Figure 6.1 gives a representation of the concentration profiles corresponding to an interfacial reaction. A model that considers a reaction region of variable thickness has been proposed by Hughes and Rod [13].

The steady state conservation equation for the species j involved in reaction (Equation 6.1) is

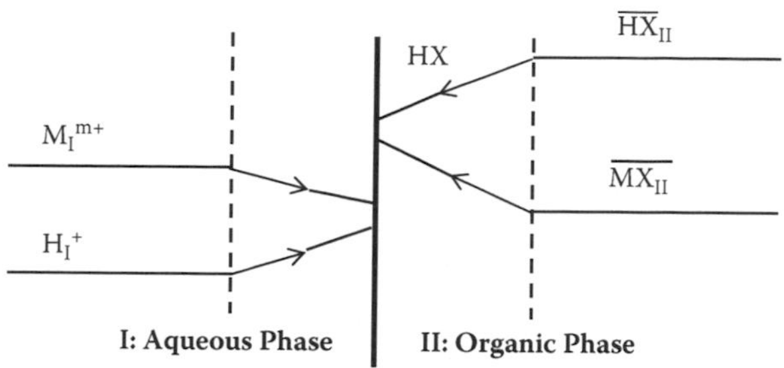

FIGURE 6.1 Representation of the concentration profiles corresponding to an interfacial reaction.

$$D_j \frac{d^2 C_j}{dy^2} - r_j = 0 \tag{6.11}$$

If the reaction takes place at the interface, Equation (6.11) reduces to

$$D_j \frac{d^2 C_j}{dy^2} = 0 \tag{6.12}$$

Integration of Equations (6.11) and (6.12) with the appropriate boundary conditions and with the rate of reaction given by Equations (6.9) and (6.10) leads to two different models. Equation (6.11) cannot be solved analytically for the rate of reaction given by Equation (6.10), but Hughes and Rod [13] integrated it analytically for a reaction of pseudo-first order in the extractant. Equation (6.12) can be solved analytically for the general case stated by Equations (6.9) and (6.10).

The following expressions represent the integrated models in dimensionless form:

$$\frac{mR}{(KD_x C_M)^{1/2} C_x / p_x} = \left(1 - \frac{R}{K_M . C_M}\right)^{1/2} \left(1 - \frac{mR}{K_x . C_x}\right) \tag{6.13}$$

for a pseudo-first-order reaction in the aqueous diffusional film, and

$$\begin{aligned} \frac{R.a_s}{r_M} &= \left(1 - \frac{R}{k_a . C_M}\right)^a \left(1 - \frac{nR}{k_x . C_x}\right)^b \left(1 + \frac{mR}{k_a C_H}\right)^d \\ &\quad - \frac{1}{k_e C_M^a . C_x^{b-e} . C_H^{d-f}} \left(\frac{C_{MX}}{C_x} + \frac{R}{k_x C_x}\right)^e \left(1 + \frac{mR}{k_a C_H}\right)^f \end{aligned} \tag{6.14}$$

for the interfacial reaction, where R is the extraction flux, a_s is the ratio of the interfacial area to aqueous phase volume, K_M and K_a are mass transfer coefficients in the aqueous phase, K_X is the mass transfer coefficient in the organic phase, and K_e is the chemical equilibrium parameter.

The denominators on the left-hand side of Equations (6.13) and (6.14) are the maximum possible interfacial fluxes as given by the chemical regime. Calling these fluxes R_{max}, it can be seen that in both equations $R/R_{max} \rightarrow 1$ as $R/K_j C_j \rightarrow 0$. Thus, the left-hand side of these equations represents the kinetic efficiency of extraction with respect to the chemical regime.

Defining the following dimensionless groups

- $\rho = R/R_{max}$, extraction efficiency
- $Da_1 = R_{max}/k_x C_x$, Damkhöler number of the organic phase
- $Da_2 = R_{max}/k_M C_M$, Damkhöler number of the aqueous phase
- $P = C_M/C_H$, dimensionless concentration
- $Q = KeC_M^a C_X^{b-e} C_H^{d-f}$, chemical equilibrium
- $q = C_{MX}/C_X$, dimensionless concentration

results in the dimensionless equations

$$\rho = (1 - \rho Da_2)^{1/2}(1 - n\rho\, Da_1), \tag{6.15}$$

and

$$\rho = (1 - \rho Da_2)^a(1 - n\rho\, Da_1)(1 + m\rho\, Da_2 P)^d - (1/Q)(q + \rho\, Da_1)^e(1 + m\rho\, Da_2 P)^f \tag{6.16}$$

Under extraction conditions and initial reaction rates, $1/Q \to 0$ and $q \to 0$, so that Equation (6.16) can be simplified to

$$\rho = (1 - \rho Da_2)^a(1 - n\rho\, Da_1)(1 + m\rho\, Da_2 P)^d \tag{6.17}$$

Da_1 and Da_2, Damkhöler numbers of the organic and aqueous phases represent the ratio between the maximum possible rate of extraction—that is, that given by the chemical kinetics for the bulk concentrations of the species involved and the rate of transport of metal and extractant, respectively, for the maximum concentration driving forces that are theoretically possible. Equations (6.15) and (6.17) are just two examples of extraction models.

6.3.1 Effect of Hydrodynamic Conditions and Concentrations on the Kinetic Efficiency

A numerical study of Equation (6.15) and Equation (6.17) will yield a quantitative evaluation of the effect on the extraction rate of mass transfer coefficients and concentrations. Given that Equation (6.15) is restricted to pseudo-first-order reactions and does not include C_H, Equation (6.17) will be used as an example. To simplify the analysis reaction orders a and b are made equal to 1 and d equal to –1.

Results are shown graphically as log ρ versus log Da_2 plots for different values of p and Da_1 in Figure 6.2 through Figure 6.4.

At a given value of the concentration of metal to that of hydrogen ion, P, changes in Da_1 can only be brought about by changing the value of the mass transfer coefficient of the extractant, whereas variations of Da_2 can be obtained by changing either the concentration ratio C_x/C_H or the mass transfer coefficient of the metal ion. The curves therefore represent the effect of hydrodynamic conditions and concentration

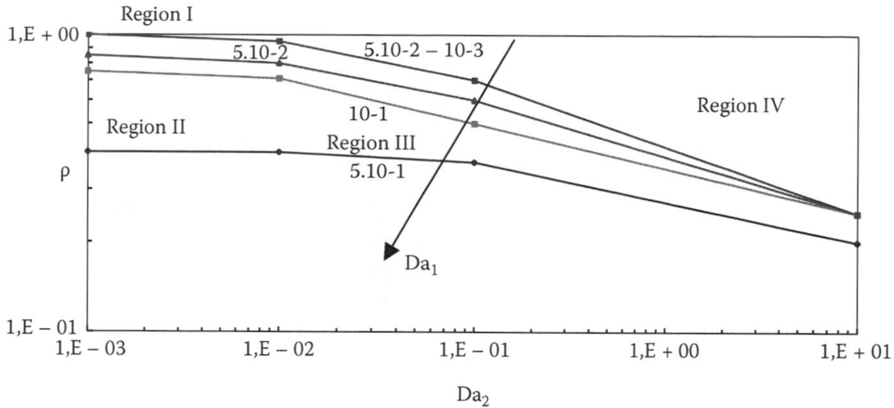

FIGURE 6.2 Representation of the effect of individual mass transfer coefficients (D_a) on dimensionless extraction efficiency ρ.

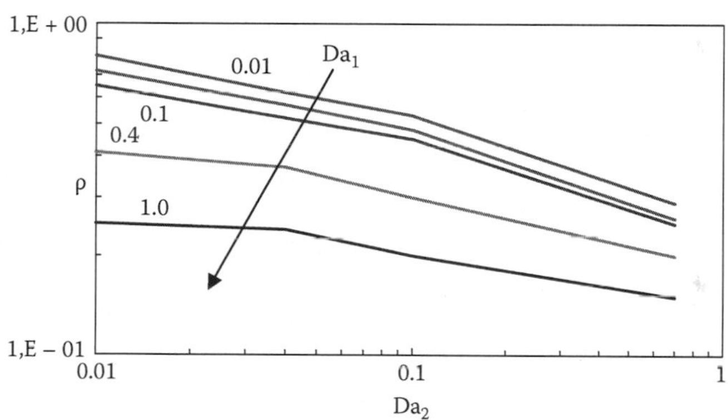

FIGURE 6.3 Representation of the effect of individual mass transfer coefficients (D_a) on dimensionless extraction efficiency (ρ).

ratios on the extraction efficiency. Assuming that concentration ratios are constant, Figure 6.2 shows the effect of individual mass transfer coefficients on extraction efficiency. Four different regions can be observed.

- Region I: This is the region where $\rho = 1$, that is, the extraction is in the chemical regime. For $p = 1.5$ this regime is achieved at $Da_2 < 0.8.10^{-2}$ and $Da_1 < 10^{-2}$.
- Region II: The extraction rate is independent of the mass transfer coefficient in the aqueous phase and decreases with decreasing values of K_X, that is, the diffusional resistance is in the organic phase only.
- Region III: The mass transfer resistance of both the aqueous and organic films affects the efficiency of extraction.

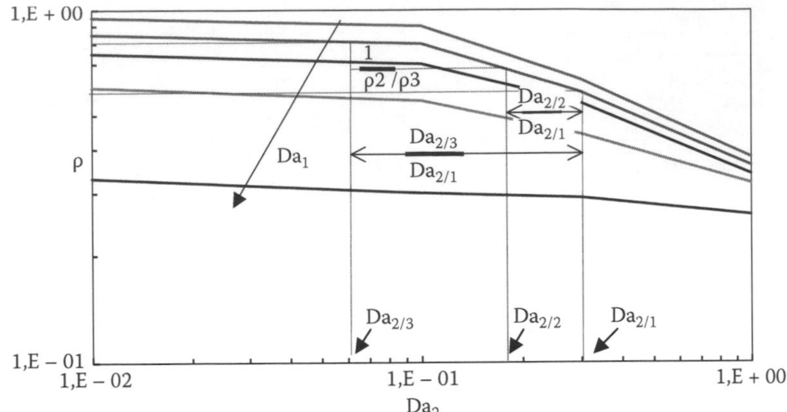

FIGURE 6.4 Representation of the effect of individual mass transfer coefficients (Da) on dimensionless extraction efficiency (ρ).

- Region IV: For $Da_2 > 0.2$ and $Da_1 > 10^{-1}$ the extraction rate is insensitive to variations of K_X but depends on the mass transfer coefficients in the aqueous phase. The diffusional control is in the aqueous phase.

Figure 6.3 shows how the value of p affects the four control regions.

This effect of the concentration ratio on the type of extraction rate control is even more obvious if changes in Da_2 are made while keeping the mass transfer coefficients constant. In this case it is possible to achieve at least three of the different control regimes.

Assuming that the stoichiometric factors and equilibrium constants are known, for a given set of concentrations Equation (6.17) contains four unknown constants, $K_{+,}$ the kinetic parameter of the forward reaction, a, b, and d, and two unknown variables, K_M and K_X. The first step is to assume values for the orders of the reaction and to obtain the log ρ versus log Da_2 graphs. For the description of the subsequent steps the assumption a = b = 1 and d = –1 leading to Figure 6.2 through Figure 6.4 will be used. Having selected the concentration ratio C_M/C_H, that is, a given value of p, two experiments must be conducted at different values of the ratio C_X/C_H, keeping the same hydrodynamic conditions and the product $C_XC_MC_H^{-1}$ constant. Under these conditions,

$$(C_X/C_H)_1/(C_X/C_H)_2 = (Da_2)_1/(Da_2)_2$$

and the ratio of the two experimental fluxes

$$R_{exp1}/R_{exp2} = \rho_1/\rho_2$$

it can easily be shown that the same is true for Equation (6.15). In the logρ versus log Da_2 graph these ratios are coordinate differences that give two sides of a triangle, as in Figure 6.4. Since neither the mass transfer coefficients nor the ratio C_M/C_H have been changed, the value of Da_1 is constant and the two values of ρ belong to the same curve. This curve is searched graphically. If a curve is found that fits the two points the model is likely to represent the system; otherwise, a different model should be tried. For the model to be checked, a third point belonging to a different region should also lie on the same curve.

Once the fitting curve has been found the values of K_+, K_M, and K_X can be calculated from Da_1 and the three points Da_2, ρ. However, the accuracy of these values depends on the sensitivity of the model to the unknown parameters in the region covered by the experimental point.

6.4 KINETIC MODELING OF MEMBRANE ASSISTED SOLVENT EXTRACTION PROCESSES

During the last decades many efforts have been directed to the research and development of a new type of contactors that solve some of the problems related to conventional contactors. Among the most investigated, hollow fiber (HF) modules occupy an important position; several advantages can be cited, namely, a very large interfacial area per unit of extractor volume without direct mixing of the aqueous and organic phases, with no need for a difference in phase densities, and with no problems due to loading and flooding.

This technology is based on the use of hollow fiber modules constituted by a set of bundle hollow fibers of cylindrical geometry inserted in a plastic or metallic carcase. One of the fluid phases circulates through the inner side of the fibers, whereas the second fluid phase circulates through the outer side of the fibers. When hydrophobic fibers are used (e.g., polypropylene HF), the pores are filled with the organic phase, and a certain differential pressure is necessary to avoid displacement of the organic phase from the pores of the membrane. Figure 6.5 shows a cross-sectional cut of a hollow fiber.

The mass transfer process of each solute from the initial phase to the receiving one is considered to take place through the following steps.

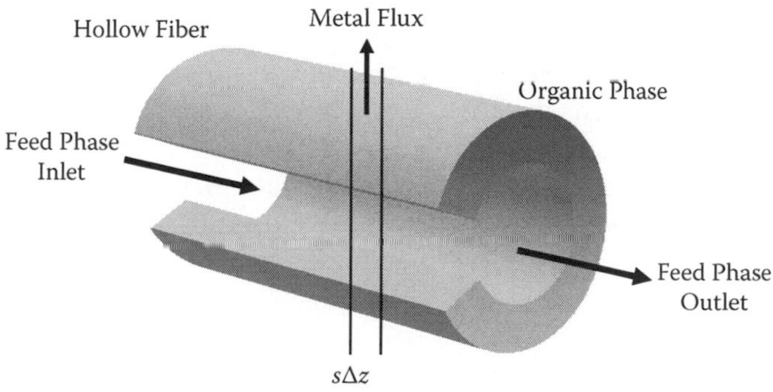

FIGURE 6.5 Cross-sectional cut of a hollow fiber (HF).

- *Step 1:* The solute in the aqueous phase (inner side of the hollow fiber) diffuses from the bulk to the aqueous–organic interface through the aqueous boundary layer.
- *Step 2:* At the aqueous–organic interface, the solute of the aqueous phase reacts with the carrier present in the organic phase in the membrane pore to form the solute-extractant complex species. The ions released to the aqueous phase boundary layer by the chemical reaction with the carrier diffuse to the bulk of the aqueous phase.
- *Step 3:* The solute-extractant complex species diffuses from the aqueous–organic interface to the outside wall of the fiber through the organic phase contained in the membrane pores; the free carrier diffuses in the opposite direction from the shell-side interface to the aqueous–membrane interface.
- *Step 4:* The solute-extractant complex species diffuses from the outside fiber wall to the bulk of the organic phase (shell side) where it flows concurrently to the aqueous phase (fiber side).

Mathematical modeling and determination of the characteristic parameters to predict the performance of membrane solvent extraction processes has been studied widely in the literature. The analysis of mass transfer in hollow fiber modules has been carried out using two different approaches. The first approach to the modeling of solvent extraction in hollow fiber modules consists of considering the velocity and concentration profiles developed along the hollow fibers by means of the mass conservation equation and the associated boundary conditions for the solute in the inner fluid. The second approach consists of considering that the mass flux of a solute can be related to a mass transfer coefficient that gathers both mass transport properties and hydrodynamic conditions of the system (fluid flow and hydrodynamic characteristics of the membrane module).

The analysis by means of the fundamental equations separates the effects of the operation variables such as hydrodynamic conditions and the geometry of the system from the mass transfer properties of the system, described by diffusion coefficients in the aqueous and organic phases and membrane permeability. When linear boundary conditions are applied at the fiber wall, it is often possible to derive analytical solutions to these equations; when it is not the case, numerical methods have to be used [14–17].

In the second approach, the total amount transferred of a given solute from the feed to the receiving phase can be assumed to be proportional to the concentration difference between both phases and to the interfacial area, defining the proportionality ratio by a mass transfer coefficient. Several types of mass transfer coefficients can be distinguished as a function of the definition of the concentration differences involved. When local concentration differences at a particular position of the membrane module are considered, the local mass transfer coefficients are obtained, which is in contrast to the average mass transfer coefficient [18].

For the kinetic analysis of the experimental system, it is sufficient to solve for the mass transport process in a single fiber to predict the performance of the HF extraction module. According to Figure 6.6, a Newtonian fluid, from which the solute is to be extracted, enters the reactive section of the hollow fiber in fully developed,

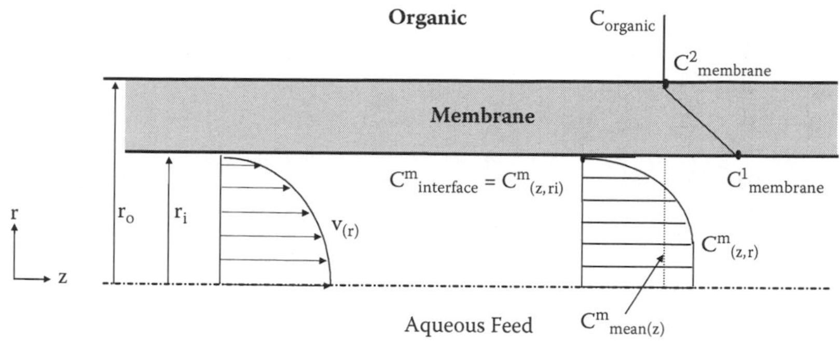

FIGURE 6.6 Schematic one-dimensional representation of the mass transport process in a single fiber of the HF extraction module.

one-dimensional, laminar flow. At z = 0, the fluid contacts the reactive membrane. At this point the concentration of the solute is uniform and equal to Co. As the fluid flows further into the reactive section of the hollow fiber, the solute diffuses through the membrane by carrier-facilitated transport and emerges into the organic solution.

To facilitate the following analysis the reversible extraction of chromate anions with Aliquat 336, a quaternary ammonium salt, will be selected as a representative example. The extraction chemical reaction can be represented by the following equation:

$$CrO_4^{2-} + 2(AQ)Cl \Leftrightarrow (AQ)_2CrO_4 + 2\,Cl^- \tag{6.17}$$

where (AQ)Cl represents the initial carrier form and $(AQ)_2CrO_4$ is the solute-carrier complex species.

For the development of the mass balance of chromate anions in the hollow fiber axial diffusion will be neglected compared with axial convection; this assumption is valid when the Peclet number is greater than 100 [19,20]. The steady-state mass conservation equation for the solute species and the associated boundary conditions will be

$$2v\left[1-\left(\frac{r}{R}\right)^2\right]\frac{\partial C}{\partial z} = D\frac{1}{r}\frac{\partial}{\partial r}\left[r\frac{\partial C}{\partial r}\right] \tag{6.18}$$

BC1

$$C = C_o,\ z = 0 \text{ for all r} \tag{6.19}$$

BC2

$$\frac{\partial C}{\partial r} = 0, \quad r = 0 \text{ for all z} \tag{6.20}$$

BC3

$$-D\frac{\partial C}{\partial r} = \bar{k}s\left(\bar{C}_i - \bar{C}\right), \quad r = R \text{ for all z.} \tag{6.21}$$

where C represents the solute concentration in the fluid, C_o the initial solute concentration, $\overline{C}_i$ the solute-carrier complex equilibrium concentration at the fluid-membrane interface, $\overline{C}$ the solute-carrier complex concentration in the organic phase, v is the average velocity, D is the diffusivity of the solute in the fluid, $\overline{K}$ is the membrane mass transfer coefficient, and s is the shape factor [21] based on the inside surface area. The third boundary condition imposes continuity of flux across the membrane-fluid interface. The left-hand side gives the species flux arriving at the wall from the lumen of the capillary. The right-hand side accounts for the complex species through the organic membrane.

The solute, chromate anions, diffuses through the aqueous phase until it reaches the fluid-membrane interface in r = R; here the Cr(VI) species reacts instantaneously to give the chromium-Aliquat complex and the Cl^- anion. The concentration of the complex species at the interface, $\overline{C}_i$, is obtained from the equilibrium expression. This solute-carrier complex diffuses through the membrane fiber to the organic phase where the concentration is $\overline{C}$. When the fluid-membrane interface becomes loaded, the complex concentration takes the value of CT, which represents the maximum loading capacity of the organic phase. At this moment, the mass transport through the membrane fiber governs the kinetic control of the extraction process until all the organic phase is completely loaded, in which case the solute flux becomes zero.

For simplicity in what follows ideal behavior of the fluid phases will be considered; thus, the equilibrium parameter will be described by the following equation according to the mass action law:

$$K_{eq} = \frac{\left[(AQ)_2(CrO_4)\right]}{\left[CrO_4^{2-}\right]} \frac{\left[Cl^-\right]^2}{\left[(AQ)Cl\right]^2} \tag{6.22}$$

Taking into account that the total carrier concentration in all forms (carrier and carrier-solute complex species) $\overline{C}_M$ has a known value, $\overline{C}_i$ can be obtained from the equilibrium equation resulting in

$$\overline{C}_i = \frac{4(C_o - C)^2 + 4K_{eq}C\overline{C}_M - \left[2(C_o - C)\right]\sqrt{4(C_o - C)^2 + 8K_{eq}C\overline{C}_M}}{8K_{eq}C} \tag{6.23}$$

By introducing the following dimensionless variables,

$$C^* = \frac{C}{C_o}, \quad r^* = \frac{r}{R}, \quad z^* = \frac{zD}{vR^2}, \quad Sh_w = \frac{R\overline{K}s}{D}$$
$$\beta = K_{eq}, \quad \alpha = \frac{C_T}{C_o}, \quad \gamma = \frac{\overline{C}}{C_T} \tag{6.24}$$

and taking into account the stoichiometric Equation (6.17) $\overline{C}_T = \overline{C}_M/2$, the system of differential equations and associated boundary conditions transforms into

$$2\left[1-r^{*2}\right]\frac{\partial C^*}{\partial z^*}=\frac{1}{r^*}\frac{\partial}{\partial r^*}\left[r^{*}\frac{\partial C^*}{\partial r^*}\right] \tag{6.25}$$

BC1

$$C^* = 1,\ z^* = 0 \text{ for all } r^* \tag{6.26}$$

BC2

$$\frac{\partial C^*}{\partial r^*}=0,\ r^* = 0 \text{ for all } z^* \tag{6.27}$$

BC3

$$-\frac{\partial C^*}{\partial r^*}=Sh_w\alpha\left[\frac{\left(1-C^*\right)^2}{2\alpha\beta C^*}+1-\left(1-C^*\right)\frac{\sqrt{4\left(1-C^*\right)^2+16\alpha\beta C^*}}{4\alpha\beta C^*}-\gamma\right] \tag{6.28}$$

r* = 1 for all z*

By solving the partial differential equations, the solution of C* as a function of r* and z* will be obtained. The mixing-cup concentration that defines the average concentration of a flowing stream can be used to obtain the outlet concentration of solute from the hollow fiber module:

$$C_M^*=\frac{\int_0^1 2v\left(1-r^*\right)^2 r^*C^*dr^*}{\int_0^1 2v\left(1-r^*\right)^2 r^*dr^*}=\sum_m 4W_mC_m^*\left(1-r_m^{*2}\right) \tag{6.29}$$

6.4.1 Simultaneous Extraction and Back-Extraction Processes

From a practical point of view, it is important not only to extract the solute but also to concentrate it [22]. For this reason, the extraction of the solute must be followed by the stripping process. Working with hollow fiber modules, different configurations of the contactors can be used for simultaneous extraction and back-extraction using two different HF modules in a series as represented in Figure 6.7. The modules are connected to each other by an organic extractant in a recirculating line. This configuration guarantees that saturation of the carrier does not occur, as it is continuously regenerated; consequently, the carrier concentration can be reduced, maintaining the mass transfer rate and therefore decreasing the associated operating costs of the process. Simultaneous membrane solvent extraction and back-extraction have been used for the removal and concentration of different solutes, such as valeric acid [23], sodium lactate [24], Cu [25], hexavalent chromium [26], Cd [27–29], and Cu(II) and Cr(VI) mixtures [30] .

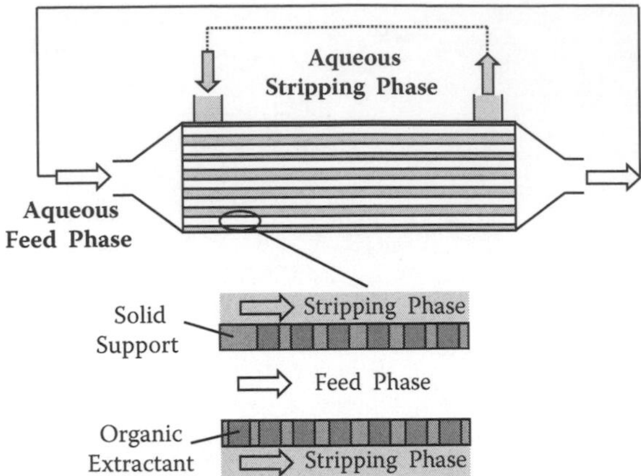

FIGURE 6.7 Schematic representation of hollow fiber contactors with simultaneous extraction and back-extraction.

Supported liquid membranes (Figure 6.8) and contained liquid membranes present a somehow different process configuration, requiring only the use of a single hollow fiber module to achieve the extraction and back-extraction processes. In supported liquid membranes, the organic solution impregnates the pores of the membranes, and the aqueous phases flow one through the lumen of the fibers and the other through the outer side of the fibers. In this way, the forward extraction occurs on the feed side, and the reverse stripping reaction occurs in the receiving phase. Different works have reported several applications of supported liquid membranes [31–33]. In general, a lack of stability is a common limitation of the supported liquid membrane, and much effort has to be devoted to improve its performance [34]. Hollow fiber contained liquid membranes comprise one set of fibers for extraction and another for back-extraction, both contained within the same shell [35,36]. The organic solution is contained in the shell side of the module, and the aqueous feed undergoing extraction flows in the lumen of one set of fibers, while the aqueous

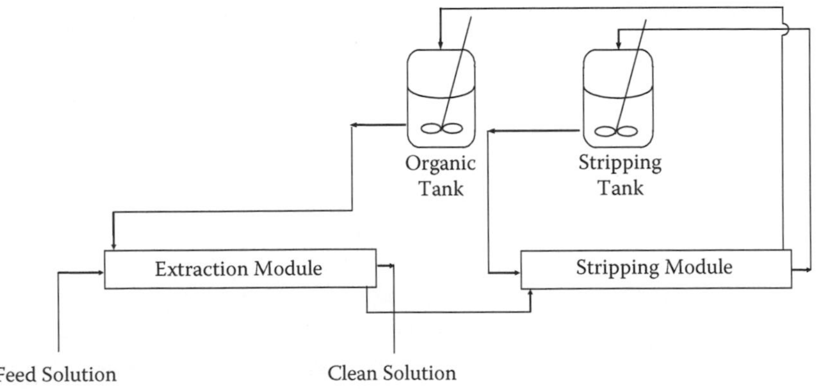

FIGURE 6.8 Representation of the experimental set-up for liquid membranes including the the extraction and back-extraction processes.

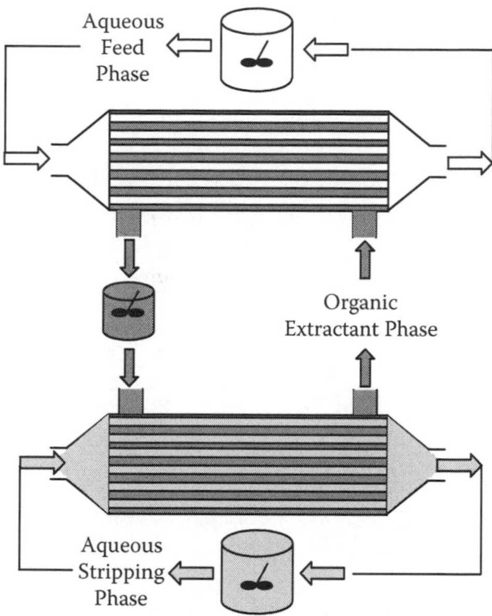

FIGURE 6.9 Experimental set-up comprising two HF modules, three stirred tanks, and accessories.

back-extraction solution passes through the lumen of the other fiber set. The main shortcoming of this configuration arises from the need of having three independent streams in motion and two different interfaces in the same module; therefore, maintaining appropriate relative pressures becomes more complicated. As a result many efforts are being devoted to the analysis, design, and optimization of the most flexible configuration: nondispersive solvent extraction.

Under the last configuration, for the simultaneous extraction (EX) and back-extraction (BEX), two hollow fiber modules in series are used flowing the organic phase from the EX module to the BEX one. In the first module, extraction of the solutes is accomplished by the organic phase that contacts the back-extraction solution in the second module; then it was mixed in the organic phase reservoir and entered again the extraction step. Figure 6.9 depicts the experimental set-up comprising two HF modules, three stirred tanks, and accessories.

In laboratory hollow fiber units, the concentration change during a single pass usually is not large enough due to the small interfacial area. To increase the change of solute concentration, the extraction and back-extraction steps are often carried out by recirculation of the aqueous and organic phases through the module and back into the aqueous and organic reservoirs, and solute concentration is measured as a function of time. However, in large-scale applications continuous operation mode where the aqueous phase changes its composition in a single pass through the fiber could be required, thus needing a larger interfacial area.

6.4.2 Kinetic Modeling of Membrane Extraction and Back-Extraction Processes

Incorporation of two HF modules connected through the organic phase and the presence of three stirred tanks in batch operation introduces a somewhat larger complexity to the kinetic modeling than in the previous case. By applying the film theory and considering that the transport mechanism in membrane extraction processes occurs through the previously described four steps—namely, diffusion through the feed aqueous layer, fast interfacial chemical reaction, diffusion through the hollow fiber membrane and diffusion process through the organic phase layer—the contribution of the individual mass transfer coefficients due to the development of liquid boundary layers and the contribution of the membrane mass transfer coefficient can be separated. According to the resistances-in-series model, overall mass transfer coefficients are a weighted average of individual mass transfer coefficients in the feed aqueous phase, across the membrane and in the organic phase. Often, one of the three individual coefficients will be much smaller than the others and hence dominates the overall mass transport coefficient. It is widely assumed that in nondispersive extraction systems, the overall mass transport resistance is dominated by the resistance in the membrane if the permeability of the membrane toward the solute is low and the fiber is thick, but when thin, high-permeability membranes are used the overall coefficient can be affected by boundary layers. The limiting step on the mass transfer process can be identified by varying independently the hydrodynamics of both aqueous and organic phases and evaluating the influence in the overall mass transfer coefficient. The membrane mass transfer coefficient does not depend on the hydrodynamics; it is only related to the membrane properties and to the diffusion coefficient of the solute in the organic phase present in the pores. Several authors have made use of the assumption that diffusion through the organic phase contained in the pores of the hollow fiber is the kinetic controlling step in the mass transport process (see, for example, [38–41]). On the other hand, other authors reported for different systems and conditions that the kinetic control of the whole proces could be shared between diffusion in the aqueous phase boundary layer and the kinetics of the chemical reaction [42–47].

Next, a kinetic model for the simultaneous EX and BEX processes will be developed. The principal assumptions underpinning the model are as follows:

1. Convective mass transfer dominates diffusive mass transfer in the fluid flowing inside the hollow fibers.
2. The resistance in the membrane dominates the overall mass transport resistance; therefore, the overall mass transfer coefficient can be set equal to the mass transfer coefficient across the membrane.
3. Chemical reactions are considered to take place instantaneously at the aqueous–membrane interfaces.

For systems working under a nonsteady state, it is also necessary to describe the change in the solute concentration with time both in the modules and in the reservoir tanks. The reservoir tanks are modeled as ideal stirred tanks.

6.4.2.1 Extraction Module

Differential mass balances of the solutes in the feed and organic solutions in the extraction module are expressed through the following equations:

6.4.2.1.1 Aqueous Solution

$$-\frac{1}{v_A}\frac{\partial C_{A_i}}{\partial t} = \frac{\partial C_{A_i}}{\partial z} + \frac{2\pi n_f r_f}{F_A} K_m (C_{OI_i} - C_{O_i}) \tag{6.30}$$

with boundary conditions

$$z = 0 \quad C_{A_i} = C_{out_i} \text{ (extraction tank)} \tag{6.31}$$

$$t = 0 \quad C_{A_i} = C_{A_i,initial} \quad i = 1, 2 \tag{6.32}$$

Suffix Ai refers to the component i in the aqueous phase, and suffix Oi refers to component i in the organic phase. C_{OIi} is the interfacial concentration of the metal in the organic phase that is related to the aqueous metal concentration through the chemical equilibrium equation

Extraction: $Me_i + n\,\overline{(HX)_m} \quad \overline{Me_iX_n(HX)_{m-n}} + n\,H^+$

With the thermodynamic equilibrium constant given by Equation (6.2)

$$K_{EX} = \frac{\left(a_{\left[\overline{Me_iX_n(HX)_{m-n}}\right]}\right)\left(a_{(H^+)}\right)^n}{\left(a_{[Me_i]}\right)\left(a_{\overline{(HX)}_m}\right)^n}$$

that can be expressed as a function of the concentrations parameter by

$$K_{EX} = \frac{\left[\overline{Me_iX_n(HX)_{m-n}}\right]\left[(H^+)\right]^n}{[Me_i]\left[\overline{(HX)}_m\right]^n} \times \frac{\left[\gamma_{\left[\overline{Me_iX_n(HX)_{m-n}}\right]}\right]\left[\gamma_{(H^+)}\right]^n}{[\gamma_{Me_i}]\left[\gamma_{\overline{(HX)}_m}\right]^m}$$

For simplicity, it will be assumed that the ratio of the activity coefficients of the species in the organic phase keeps constant, whereas in the aqueous phases the values of the activity coefficients of the metal ions and hydrogen ions will be calculated using the Debye-Hückel equation. Therefore, Equations (6.2) and (6.3) can be expressed as follows:

$$K'_{EX} = K_{EX}\frac{\left[\gamma_{\overline{(HX)}_m}\right]^m}{\left[\gamma_{\left[\overline{Me_iX_n(HX)_{m-n}}\right]}\right]} = \frac{\left[\overline{Me_iX_n(HX)_{m-n}}\right]\left[(H^+)\right]^n}{[Me_i]\left[\overline{(HX)}_m\right]^n} \times \frac{\left[\gamma_{(H^+)}\right]^n}{[\gamma_{Me_i}]} \tag{6.33}$$

6.4.2.1.2 Organic Solution

$$\frac{1}{v_O}\frac{\partial C_{O_i}}{\partial t} = -\frac{\partial C_{O_i}}{\partial z} + \frac{2\pi n_f r_f}{F_O} K_m (C_{OI_i} - C_{O_i}) \tag{6.34}$$

Boundary conditions of the organic phase are

$$z = 0 \quad C_{O_i} = C_{out_i} \text{ (organic tank)} \tag{6.35}$$

$$t = 0 \quad C_{O_i} = C_{O_i,\text{ initial}} \tag{6.36}$$

6.4.2.2 Back-Extraction Module

The extraction and back-extraction steps take place consecutively, being connected by the concentration of the metal-extractant complex species in the organic phase.

Similar equations to those used in the description of the extraction step are used to describe the back-extraction process, considering in this case the equilibrium of the interfacial reaction between the organic complex species and the back-extraction agent.

6.4.2.2.1 BEX Aqueous Solution

$$\frac{1}{v_S}\frac{\partial C_{S_i}}{\partial t} = -\frac{\partial C_{S_i}}{\partial z} + \frac{2\pi n_f r_f}{F_S} K_m (C_{O_i} - C_{OI_i}) \tag{6.37}$$

where the suffix Si refers to component i in the stripping or back-extraction phase.

Boundary conditions for the stripping aqueous phase are

$$z = 0 \quad C_{S_i} = C_{out_i} \text{ (stripping tank)} \tag{6.38}$$

$$t = 0 \quad C_{S_i} = C_{S_i,\text{ initial}} \quad i = 1, 2 \tag{6.39}$$

with the equilibrium back-extraction reactions

$$\overline{Me_i X_n (HX)_{m-n}} + n\,H^+ \qquad Me_i + n\,\overline{(HX)_m} \tag{6.40}$$

and

$$K_{BEX} = \frac{\left(a_{[Me_i]}\right)\left(a_{\left[\overline{(HX)}_m\right]}\right)^n}{\left(a_{\left[\overline{Me_i X_n (HX)_{m-n}}\right]}\right)\left(a_{[(H^+)]}\right)^n} \tag{6.41}$$

and

$$K_{BEX} = \frac{[Me_i]\left[\left(\overline{HX}\right)_m\right]^n}{\left[\,\overline{Me_iX_n(HX)_{m-n}}\,\right]\left[(H^+)\right]^n} \times \frac{[\gamma_{Me_i}]\left[\gamma_{\left(\overline{HX}\right)_m}\right]^n}{\left[\gamma_{\left[\,\overline{Me_iX_n(HX)_{m-n}}\,\right]}\right]\left[\gamma_{(H^+)}\right]^n} \tag{6.42}$$

which with similar assumptions to those made in the extraction reactions would transform into

$$K'_{BEX} = K_{BEX}\frac{\left[\gamma_{\left[\,\overline{Me_iX_n(HX)_{m-n}}\,\right]}\right]}{\left[\gamma_{\left(\overline{HX}\right)_m}\right]^n} = \frac{[Me_i]\left[\left(\overline{HX}\right)_m\right]^n}{\left[\,\overline{Me_iX_n(HX)_{m-n}}\,\right]\left[(H^+)\right]^n} \times \frac{[\gamma_{Me_i}]}{\left[\gamma_{(H^+)}\right]^n} \tag{6.43}$$

6.4.2.2.2 BEX Organic Solution

$$-\frac{1}{v_O}\frac{\partial C_{O_i}}{\partial t} = \frac{\partial C_{O_i}}{\partial z} + \frac{2\pi n_f r_f}{F_O} K_m\,(C_{O_i} - C_{OI_i}) \tag{6.44}$$

Boundary conditions of the organic phase are

$$z = 0 \quad C_{O_i} = C_{O_i}\ \text{(outlet extraction module)} \tag{6.45}$$

$$t = 0 \quad C_{O_i} = C_{O_i,\,\text{initial}} \tag{6.46}$$

6.4.2.3 Stirred Tanks

The dynamic response of the system is determined by simultaneously solving the differential Equation (6.30) through Equation (6.46) together with the modeling equations for the three vessels considered as ideal stirred tanks. It is assumed that the solute concentrations at the reservoir and at the module inlet for both phases are identical.

$$V\frac{dC_{out_i}}{dt} = F\,(C_{in_i} - C_{out_i}) \tag{6.47}$$

$$t = 0 \quad C_{out_i}\ \text{(organic tank)} = C_{O_i,\,\text{initial}} \tag{6.48}$$

$$C_{out_i}\ \text{(extraction tank)} = C_{A_i,\,\text{initial}} \tag{6.49}$$

$$C_{out_i}\ \text{(stripping tank)} = C_{S_i,\,\text{initial}} \tag{6.50}$$

Consequently, it has been considered the most basic model to describe the separation of metallic components using hollow fiber membranes, which consists of a set of coupled differential equations corresponding to the mass balances of the metallic solutes in the extraction and back-extraction modules together with the algebraic equations corresponding to the description of the chemical equilibrium reactions and six total differential equations describing the mass balances in the stirred tanks. The use of the mathematical model would require of the knowledge of the design parameters—that is, the membrane mass transport coefficient and the parameters of the interfacial chemical equilibria. Extension of the mathematical model to multicomponent systems could be done by duplicating the mass balances equations including the corresponding equilibrium expressions for all the components analyzed [48].

Furthermore, if necessary the model could include concentration gradients in the radial direction due to the development of velocity profiles as shown in Equation (6.18). In this case diffusivity of metallic components in the aqueous phase would be needed, and more computational efforts would be required to solve the system of differential plus algebraic equations.

6.5 OPTIMIZATION OF MEMBRANE ASSISTED SOLVENT EXTRACTION PROCESSES

The potential of membrane assisted solvent extraction processes is high as it can be deduced from the large number of applications of this technology that has been mentioned in the literature, but there is very little information on the analysis and optimisation of these processes.

Many decisions regarding the selection of the operating mode, batch or semicontinuous module configurations, size of equipment, and operating conditions need to be made to promote its industrial application.

In what follows the methodology for the selection of the operating conditions of a nondispersive solvent extraction process will be developed. As an example the removal and recovery of Cr(VI) from an industrial effluent of a surface treatment plant will be considered. The kinetic modeling including the extraction reactions, Equation (6.17) and Equation (6.22), and the mass balances of chromium compounds to the three fluid phases and considering the hollow fiber modules and the homogeneization stirred tanks, Equation (6.30) through Equation (6.50) were described in Sections 6.3 and 6.4.

Surface treatment industries employ huge amounts of rinse waters that become contaminated because of metals solubilization. Thus, the treatment technology should reduce continuously the metallic content in the generated volumes of waste effluents. Figure 6.9 shows a schematic flow diagram of a nondispersive solvent extraction process where the aqueous effluent flows in a continuous mode, whereas the organic and stripping solutions are recycled to the homogeneization tanks.

The aim of the process is to remove Cr(VI) from wastewater and to concentrate it in the stripping solution for reuse, with the objective function to determine the maximum wastewater flowrate Fe that the industrial plant with a known membrane area is able to treat. In this case the maximum wastewater Cr(VI) concentration (Ce,outlet) for disposal is fixed according to environmental regulations, and the minimum Cr(VI) concentration of the stripping stream required for reuse is also established, Csmin. Both values are posed as inequality endpoint constraints w(tf) in the definition of the optimisation problem. When the stripping phase reaches the desired concentration level, Csmin, it is replaced by a new batch with fresh solution; in this moment the concentration of the metal in the organic phase must be the same as that at the beginning of the batch to maintain process conditions constant with time. The same organic stream is used in different batches. Thus, to keep the initial Cr(VI) composition in the organic phase (Co,initial) constant, an extra equality constraint on this control variable is assumed, imposing that the initial u(0) and final u(tf) Cr(VI) composition in the organic phase should be the same.

For a semicontinuous operation mode with the fluid phases flowing cocurrently and for a given value of the initial chromium concentration, the operating conditions of the NDSX plant are selected solving the following optimization problem [48,49]:

$$
\begin{aligned}
&\max_{u(t),v,t_f} \quad F_e(u(t), v, t_f) \\
&f(x(t), \dot{x}(t), u(t), v) = O \quad t \in [O, t_f] \\
&I(x(O), u(O), v) = O \\
&u(t_o) = u(t_f) \\
&t_f^{\min} \le t \le t_f^{\max} \\
&v^{\min} \le v \le v^{\max} \\
&u^{\min} \le u(t) \le u^{\max} \\
&w^{\min} \le w(t_f) \le w^{\max}
\end{aligned}
\tag{6.51}
$$

The objective function of maximizing the wastewater flowrate F_e is a function of the time horizon t_f, the time invariant parameters v, and the control variables $u(t)$. The process model is represented by the set of differential and algebraic equations f and initial conditions I, shown in the previous sections Equation (6.30) through Equation (6.50), where $\dot{x}(t)$ are the time derivatives of the variables $x(t)$. There are lower and upper bounds on the end point constraints $w(t_f)$, time invariant parameters, control variables, and time horizon.

Using a proper software code (for example, code *gOPT*, the gPROMS optimization tool), the solution of the optimization problem will give the values of the operating variables or time invariant parameters: volume of the organic and stripping tanks, organic and stripping flow rates, as well as the initial Cr(VI) composition in the organic phase. Thus, the application of optimization tools allows selection of the operating conditions of nondispersive solvent extraction processes for wastewater treatment and metal recovery. This is an important step in promoting the industrial application of this new technology, due to the influence of the operating conditions on the cost and the economical viability of the project.

REFERENCES

1. M. Cerná, V. Bízek, J. St'astova, and V. Rod, Extraction of nitric acid with quaternary ammonium bases, *Chem Eng Sci* 48:99–103 (1993).
2. L. Calvarin, B. Roche, and H. Renon, Anion exchange and aggregation of dicyanocobalamin with quaternary ammonium salts in a polar environment, *Ind Eng Chem Res* 31:1705–1709 (1992).
3. B. Galán, A. M. Urtiaga, A. I. Alonso, J. A. Irabien, and I. Ortiz, Extraction of anions with Aliquat 336: Chemical equilibrium modelling, *Ind Eng Chem Res* 33:1765–1770 (1994).
4. A. I. Alonso, B. Galán, A. Irabien, and I. Ortiz, Separation of Cr(VI) with Aliquat 336: Chemical Equilibrium Modelling, *Sep Sci Technol* 32(9):1543–1555 (1997).
5. P. R. Danesi, The relative importance of diffusion and chemical reactions in liquid-liquid extraction kinetics, *Solvent Extrac Ion Exch* 2:29–44 (1984).
6. D. S. Flett, D. N. Okuhara, and D. R. Spink, Solvent extraction of copper by hydroxyoximes, *J Inorg Nucl Chem* 35:2471–2487 (1973).
7. R. J. Whewell, M. A. Hughes, and C. Hanson, The kinetics of the solvent extraction of copper (II) with LIX reagents, *I. J Inorg Nucl Chem* 37:2303–2307 (1975).
8. L. A. Ajawin, E. S. Pérez de Ortiz, and H. Sawistowski, Extraction of zinc by di(2-ethylhexyl)phosphoric acid, *Chem Eng Res Dev* 61:62–66 (1983).
9. I. Komasawa and T. Otake, Extraction of copper with 2-hydroxy-5-nonylbenzophenone oxime and the catalytic role of bis (2-ethylhexyl) phosphoric acid, *Ind Eng Chem Fundam* 22:122–126 (1983).
10. K. Yoshizuka, K. Kondo, and F. Nakashio, The kinetics of copper extraction with N-8-quinolyl-p-dodecylbenzene-sulfonamide, *J Chem Eng Jpn* 18:163–168 (1985).
11. E. S. Pérez de Ortiz, M. Cox, and D. Flett, The effect of hydrodynamic conditions on the kinetics of copper extraction by LIX 65N, *CIM 21* (special volume):198–202 (1979).
12. A. Irabien, I. Ortiz, and E. S. Pérez de Ortiz, Kinetics of metal extraction. Model discrimination and parameter estimation, *Chem Eng Process* 27:13–18 (1990).
13. M. A. Hughes and V. Rod, On the use of the constant interface cell for kinetic studies, *Hydrometallurgy* 12:267–273 (1984).
14. A. M. Urtiaga and A. Irabien, Internal mass transfer in hollow fiber supported liquid membranes, *AIChE J* 39:521–525 (1993).
15. Y. Qin and M. S. Cabral, Lumen mass transfer in hollow-fiber membrane processes with constant external resistances, *AIChE J* 43:1975–1988 (1997).
16. Y. Qin and M. S. Cabral, Theoretical analysis on the design of hollow fiber modules cascades for the separation of diluted species, *J Membrane Sci* 143:197–205 (1998).
17. A. Alonso, A. M. Urtiaga, A. Irabien, and M. I. Ortiz, Extraction of Cr(VI) with Aliquat 336 in hollow fiber contactors: Mass transfer analysis and modelling, *Chem Eng Sci* 49:901–909 (1994).
18. E. L. Cussler, *Diffusion mass transfer in fluid systems*, 2nd ed., Cambridge, UK: Cambridge University Press (1997).
19. P. J. Schneider, Effect of axial fluid conduction on heat transfer in the entrance regions of parallel plates and tubes, *Trans ASME* 79:765–773 (1957).
20. E. Papoutsakis, D. Y. Ramkrishna, and H. C. Lim, The extended Graetz problem with Dirichlet wall boundary conditions, *Appl Sci Res* 36:13–40 (1980).
21. R. D. Noble, Shaped factors in facilitated transport through membranes, *Ind Eng Chem Fundam* 22:139–144 (1983).
22. R. Basu, P. Prasad, and K. K. Sirkar, Nondispersive membrane solvent back extraction of phenol, *AIChE J* 36(3):450–460 (1990).

23. J. Rodriguez, R. M. C. Viegas, S. Luque, I. M. Coelhoso, J. P. S. G. Crespo, and J. R. Alvarez, Removal of valeric acid from wastewaters by membrane contractors, *J Membrane Sci* 137:45–53 (1997).
24. I. M. Coelhoso, P. Silcivestre, R. M. C. Viegas, J. P. S. G. Crespo, and M. J. T. Carrondo, Membrane-based solvent extraction and stripping of lactate in hollow-fiber contactors, *J Membrane Sci* 134:19–32 (1997).
25. S.-Y. Hu and J. M. Wiencek, Emulsion-liquid-membrane extraction of copper using a hollow-fiber contactor, *AIChE J* 44(3):570–581 (1998).
26. A. Alonso, B. Galán, M. González, and M. I. Ortiz, Experimental and theoretical analysis of a NDSX pilot plant for the removal of Cr(VI) from a galvanic process waste waters, *Ind Eng Chem Res* 38:1666–1675 (1999).
27. I. Ortiz, A. I. Alonso, A. M. Urtiaga, M. Demircioglu, N. Kocacik, and N. Kabay, Integrated process for the removal of cadmium and uranium from wet phosphoric acid, *Ind Eng Chem Res* 38:2450–2459 (1999).
28. A. M. Urtiaga, S. Zamacona, and M. I. Ortiz, Analysis of a NDSX process for the selective removal of Cd from phosphoric acid, *Sep Sci Technol* 34:3279–3296 (1999).
29. B. Galán, F. San Román, A. Irabien, and I. Ortiz, Viability of the separation of Cd from highly concentrated Ni-Cd mixtures by non-dispersive solvent extraction, *Chem Eng J* 70:237–243 (1998).
30. Z.-F. Yang, A. K. Guha, and K. K. Sirkar, Simultaneous and synergistic extraction of cationic and anionic heavy metallic species by a mixed solvent extraction system and a novel contained liquid membrane device, *Ind Eng Chem Res* 35:4214–4220 (1996).
31. P. R. Danesi, Separation of metal species by supported liquid. Membranes, *Sep Sci Tech* 19:857–894 (1984).
32. R. D. Noble, C. A. Koval, and J. J. Pellegrino, Facilitated transport membrane systems, *Chem Eng Prog* 58:26 (1989).
33. M. Rovira and A. M. Sastre, Modelling of mass transfer in facilitated supported liquid-membrane transport of palladium(II) using di-(2-ethylhexyl) thiophosphoric acid, *J Membrane Sci* 149:241–250 (1998).
34. M. F. Paugam and J. Buffle, Comparison of carrier-facilitated copper(II) ion transport mechanisms in a supported liquid membrane and in a plasticized cellulose triacetate membrane, *J Membrane Sci* 147:207–215 (1998).
35. D. K. Mandal, A. K. Guha, and K. K. Sirkar, Isomer separation by a hollow fiber contained liquid membrane permeator, *J Membrane Sci* 144:13–24 (1998).
36. Z.-F. Yang, A. K. Guha, and K. K. Sirkar, Novel membrane-based synergistic metal extraction and recovery processes, *Ind Eng Chem Res* 35:4214–4220 (1996).
37. R. Basu and K. K. Sirkar, Citric acid extraction with microporous hollow fibers, *Ion Exch* 10(1):229–243 (1992).
38. R. S. Juang, Modelling of the competitive permeation of cobalt and nickel in a di(2-ethylhexyl)phosphoric acid supported liquid membrane process, *J Membrane Sci* 85:157–166 (1993).
39. R.-S. Juang and H.-C. Huang, Non-dispersive extraction separation of metals using hydrophilic microporous and cation exchange membranes, *J Membrane Sci* 156:179–186 (1999).
40. H. Escalante, A. I. Alonso, I. Ortiz, and A. Irabien, Separation of L-phenyl alanine by non-dispersive extraction and back-extraction. Equilibrium and kinetic parameters, *Sep Sci Tech* 33:119–139 (1997).
41. K. Yoshizuka, Y. Sakamoto, Y. Baba, K. Inoue, and F. Nakashio, Solvent extraction of holmium and yttrium with bis(2-ethylhexyl)phosphoric acid, *Ind Eng Chem Res* 31:1372–1378 (1992).

42. K. Kondo, Y. Yamamoto, and M. Matsumoto, Separation of indium(III) and gallium(III) by a supported liquid membrane containing diisostearylphosphoric acid as a carrier, *J Membrane Sci* 137:9–15 (1997).
43. F. Kubota, M. Goto, F. Nakashio, and T. Hano, Extraction kinetics of rare earth metals with 2-ethylhexyl phosphonic acid mono-2-ethylhexyl ester using a hollow fiber membrane extractor, *Sep Sci Tech* 30:777–792 (1995).
44. C. H. Yun, R. Prasad, A. K. Guha, and K. K. Sirkar, Hollow fiber solvent extraction removal of toxic heavy metals from aqueous waste streams, *Ind Eng Chem Res* 32:1186–1195 (1993).
45. S. A. Mohammed, K. Inoue, and K. Yoshizuka, Kinetics of palladium extraction with bis(2-ethylhexyl)monothiophosphoric acid in a hollow fiber membrane extractor, *Ind Eng Chem Res* 35:3899–3906 (1996).
46. Y.-S. Hsu and H.-M. Yeh, Microporous membrane solvent-extraction in multiple-fiber passes and one-shell pass hollow-fiber modules, *Sep Sci Tech* 33:757–765 (1998).
47. I. Ortiz, B. Galán, F. San Román, and R. Ibáñez, Kinetics of the separation of multicomponent mixtures by non-dispersive solvent extraction, *AIChE J* 47(4):895–904 (2001).
48. A. M. Eliceche, S. M. Corvalán, A. I. Alonso, and I. Ortiz, Proceedings of the 10th European Symposium on Computer Aided Process Engineering, Florence, 877–882 (2000).
49. A. M. Eliceche, A. I. Alonso, and I. Ortiz, Optimal operation of selective membrane separation processes for wastewater treatment, *Comp Chem Eng* 24(9–10):2115–2123 (2000).

7 New Materials in Solvent Extraction

Lawrence L. Tavlarides, Jun S. Lee, and Sergio Gomez-Salazar

CONTENTS

7.1 INTRODUCTION

Traditionally solvent extraction involves contacting an aqueous phase and an organic phase to extract solutes such as metal ions or biochemical products and separate them from mixtures or concentrate them. The organic "solvent" may contain a chelation ligand or an ion exchanger with a modifier dissolved in a diluent. Along with the contactor, process equipment requires fluid phase separators to separate the loaded organic extractant from the aqueous feed. Subsequently, a stripping step is required to recover the metal ion or biochemical and, perhaps, to regenerate the solvent. Such liquid extraction processes have wide applicability for high-volume hydrometallurgical, radionuclear, and biochemical separations.

An alternative approach is to immobilize the chelating or ion exchange ligands to a polymeric or inorganic solid support. The former support consists of a solid polymer network within which the exchange ligands are embedded. Another form is an aqueous phase soluble polymer to which the ligand is attached. The class of inorganic supports discussed here are silica-based materials. These can be classified into two categories: (1) preformed silica gels onto which ligands are placed; or (2) organo-ceramic polymers with the ligand embedded within the ceramic matrix. This discussion is limited to metal ion separations for which these materials find application in hydrometallurgical processing for finishing steps, electrochemical metal refining, electroplating operations, and remediation of acid mine drainage solutions. The lecture covers topics of selection of ligands, synthesis of solid supported extractant materials, characterization, adsorption isotherms, kinetics of adsorption, mass transfer characteristics, and applications.

7.2 SYNTHESIS OF SOLID SUPPORTED EXTRACTION MATERIALS

The classification of solid supported extraction materials can be categorized by the nature of the support matrix. The materials can also be categorized according to whether they are ion exchangers or chelating agents as shown in Figure 7.1 [1]. The capability of metal ion separation of a solid extractant is mainly dependent on the acid-base characteristics of the chelating ligand and metal ions, polarizability, molecular structure of the functional ligand, coordination number of the metal ions, and solid matrix geometry.

For selective metal ion separations, coordinating chelating extractants are normally used. The chelating ligands act as bases in metal separations. The metal binding characteristics of a ligand are determined by which atom is involved in the coordination with the metal atoms such as those shown in Table 7.1 [2].

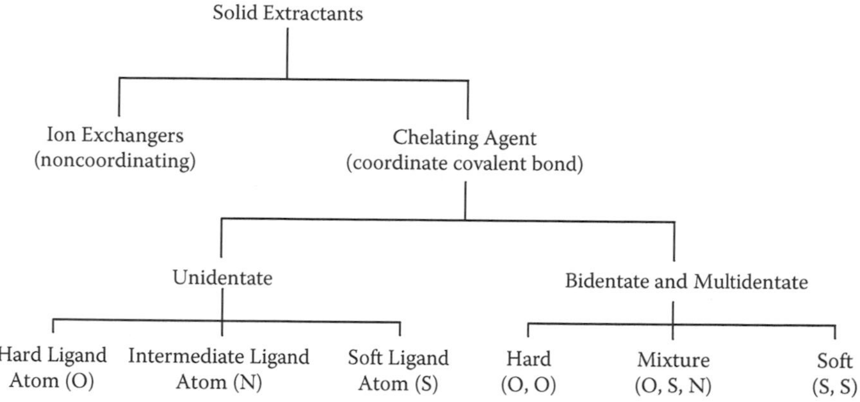

FIGURE 7.1 The classification of solid extractants based on functional ligand or donor atoms. (From C. Kantipuly, S. Katragadda, A. Chaw, D. H. Gesser. *Talanta* 37:491–517, 1990. With permission.)

TABLE 7.1
Classification of Lewis Acids

Hard	Soft	Borderline
H^+, Li^+, Na^+, K^+	Cu^+, Ag^+, Au^+, Tl^+, Hg^+, Cs^+	Fe^{2+}, Co^{2+}, Ni^{2+}, Cu^{2+}
Be^{2+}, Mg^{2+}, Ca^{2+}, Sr^{2+}, Sn^{2+}	Pd^{2+}, Cd^{2+}, Pt^{2+}, Hg^{2+}, CH_3Hg^+	Zn^{2+}, Pb^{2+}
Al^{3+}, Sc^{3+}, Ga^{3+}, In^{3+}, La^{3+}	Tl^{3+}, I^+, Br^+	
Cr^{3+}, Co^{3+}, Fe^{3+}, As^{3+}, Ir^{3+}	M_0 (metal atoms)	
UO_2^{2+}		

Source: From R. G. Peterson. *J. Am. Chem. Soc.* 85:3533–3539, 1963. (With permission.)

Hard metal ions such as Fe^{3+}, Co^{3+}, and UO_2^{2+} easily chelate with molecules that have oxygen atoms, whereas soft metal ions such as noble metal ions, Cd^{2+}, Hg^+, and Hg^{2+} can easily form complexes with ligands containing sulfur. However, the functional ligands in solid extractants usually contain polyfunctional groups instead of one functional atom. Hence, the multifunctional character of a functional ligand can form very specific complexes with a specific metal ion. For example, in the soft acid soft base case, 2,5-bis(butylthiomethyl)ethylenethiourea and platinum(II) form complexes as shown in Figure 7.2 [3]. However, the soft acid, platinum chloride, does not form complexes with the multifunctional (intermediate-hard base) dipicolinic acid ligand while the soft acid, gold chloride and palladium chloride, form complexes with it over the pH range from 0 to 7.0 such as shown in Figure 7.3 [4].

Realization of the aqueous phase metal speciation is important to match a ligand to a specific metal ion. Heavy metal ions, for instance, can exist in various complex forms depending on the conditions of the solutions to be treated such as pH, counter ion concentrations, and ionic strength. For example, copper (I) exists in three different complex forms in cyanide solutions, such as $Cu(CN)_2^-$, $Cu(CN)_3^{2-}$, and

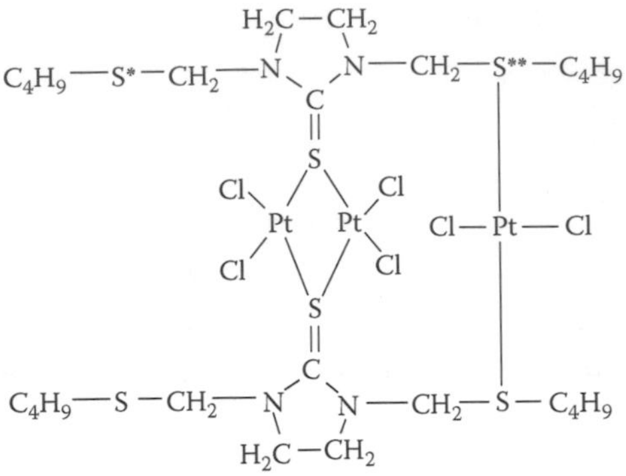

Only one of the sulfur atoms of the sulfide group (S** or S*, with equal probability) takes part in the coordination with the platinum atom, which can be explained by steric hindrances.

FIGURE 7.2 Complexes of platinum(II) chloride with thiourea moiety. (From A. R. Khisamutdinov, G. N. Afzaletdinova, I. Yu. Murinov, and V. E. Vasil'eva. *Russian J. Inorg. Chem.*, 42:1560–1566, 1997. With permission.)

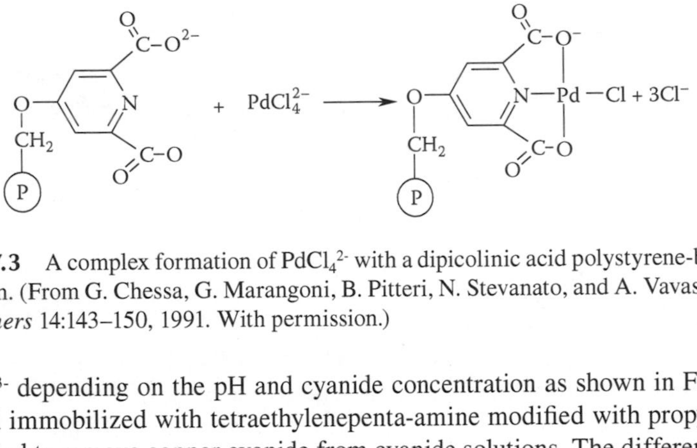

FIGURE 7.3 A complex formation of $PdCl_4^{2-}$ with a dipicolinic acid polystyrene-based chelating resin. (From G. Chessa, G. Marangoni, B. Pitteri, N. Stevanato, and A. Vavasori. *Reactive Polymers* 14:143–150, 1991. With permission.)

$Cu(CN)_4^{3-}$ depending on the pH and cyanide concentration as shown in Figure 7.4. Silica gel immobilized with tetraethylenepenta-amine modified with propyl groups was applied to remove copper cyanide from cyanide solutions. The different species of copper cyanide complexes were adsorbed at different pH values as shown in Figure 7.5 [5]. The figure shows that the total amount of adsorbed copper cyanide is the sum of the two complexes adsorbed on the surface. Therefore, a chelating agent must be selected that can chelate with the metal ion at the expected conditions of extraction. Alternatively, the composition of the solution (e.g., pH, counter ion concentrations) must be altered to have favorable binding conditions.

7.2.1 Polymeric Network Systems

A variety of polymeric systems have been developed and applied for metal separations from aqueous solutions. Organo-polymeric synthesis techniques are well established, and polymeric metal extractants can be tailored and adopted for specific applications, such as noble metal separations, nuclear waste treatments, and

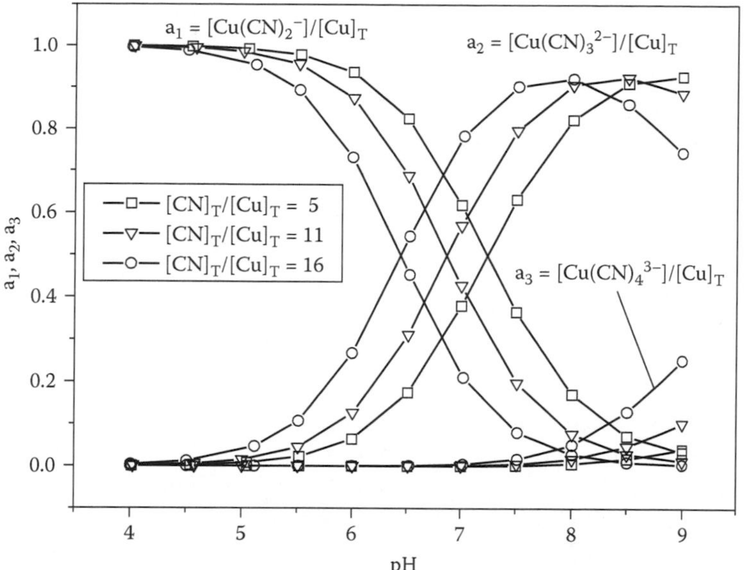

FIGURE 7.4 Computed speciation of copper cyanide complexes in cyanide solutions with respect to the solution pH.

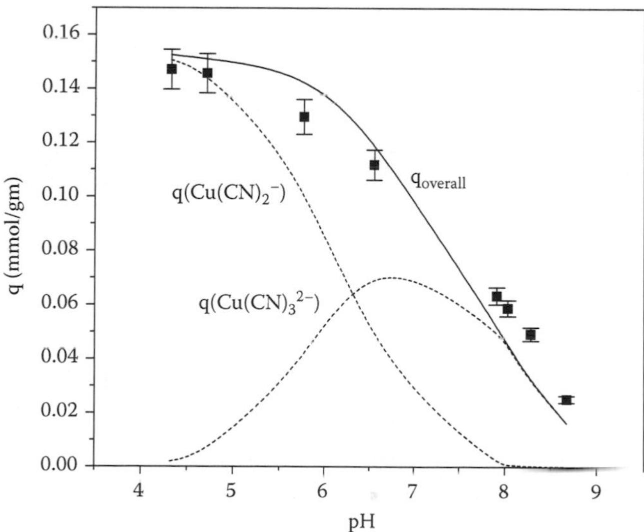

FIGURE 7.5 pH isotherm of copper cyanide: $[Cu(I)]_{ini}$ = 200 mg/L; $[CN]_T/[Cu]_T$ = 16; wt. of extractant = 0.5 gm; Contact time: 1.0 hr; ■: experimental data, ——: computed overall copper(I) uptake; – – – – : computed individual copper(I) complex uptake. (From J. S. Lee. M.S. Thesis, Syracuse University, 1997. With permission.)

electroplating wastes clean-up. Polymeric extractants can be classified as (1) solid resin systems and (2) water-soluble polymer systems. The former system has been studied widely and applied for many applications for more than fifty years. Research works and reviews for these solid resins in metal separations can easily be found in the literature [1,6,7]. The latter system consists of metal extraction with water-soluble polymers and the separation of the polymers from the aqueous solution by ultrafiltration. An attractive feature of this system includes the elimination of the diffusive mass transfer resistance that exists in solid supports. Hence, the extraction rate is faster than solid resin systems [8,9]. Extensive literature is available, so these systems are not discussed in this chapter.

7.2.2 Inorganic Network Systems

Solid extractants made of functional ligands and inorganic supports have attracted much attention because of their mechanical strength, thermal stability, wide range of particle size, and well-defined pore structure of these materials. In particular, the well-defined pore structure provides a good environment for diffusion of metal ions in the solid matrix.

A variety of synthesis methods using different functional ligands and inorganic supports have been employed. However, it is useful to classify the inorganic supported extractants into three categories depending on the methods of synthesis. The immobilization techniques using inorganic network systems that are discussed here are (1) solvent deposition, (2) covalent bonding, and (3) formation of organo-ceramic polymers.

7.2.2.1 Solvent Deposition on Functionalized Silica

As shown in Figure 7.6, chelating agents are immobilized on functional groups bonded to the surface. The carbon chain of the extractant is retained on the bonded functional group through van der Walls forces. To maximize diffusion of metal ions into the pores, the degree of functionalization (chain length of functional group, polymeric functionalization) and properties of support (pore size and pore volume) must be controlled to prevent steric hindrance.

Surface silanol groups (Si-OH) are common to all inorganic supports and provide reactive sites for surface modification. The density of the SiOH group is constant for a fully hydroxylated surface (for silica gel 8 μmol OH/m^2). Silica gel can be functionalized with desired functional groups (e.g., methyl, ethyl) by conducting silinization reactions on the hydroxilated surface. Various silylating agents can be used. For example, we have functionalized silica gel with methyl groups by using dimethyl-dichloro-silane.

$$2 \equiv Si\text{-}OH + Cl_2 Si(CH_3)_2 \rightarrow (Si\text{-}O)_2\text{-}Si\text{-}(CH_3)_2 \ldots$$

The chain length of functional groups is selected by considering pore size and pore volume of the silica gel. Titanium coatings can also be added to the surface through the silanol group to prevent dissolution of the surface in highly caustic solutions.

Chelating agents are deposited on the functionalized silica gel. To prepare inorganic chemically active adsorbents (ICAAs) with high stability, capacity, and kinetic

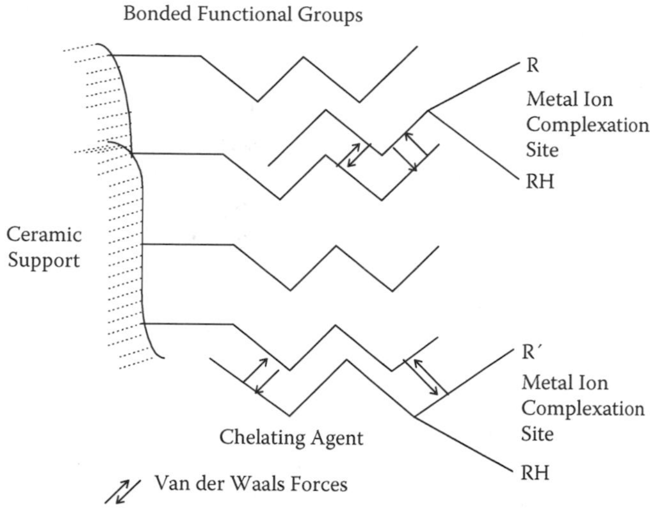

FIGURE 7.6 Inorganic chemically active adsorbent prepared by solvent deposition of chelating agents on the functionalized surface. (From N. V. Deorkar and L. L. Tavlarides. *Emerging Separation Technologies for Metals II,* The Minerals, Metals & Materials Society, Warrendale, PA, 1996. With permission.)

rates equal or greater than pore diffusion rates, the properties that the chelating agents should possess include the following:

1. Very low solubility in water.
2. Hydrocarbon chains away from the complexing moiety to retain hydrophilicity at the complexing end and to prevent steric hindrance to the formation of chelate rings.
3. Sufficient thermal stability so that the extractant is not destroyed or altered during immobilization since the mixture is heated to remove excess solvent.
4. Sterically compact geometry comparable with the pore size and pore volume of functionalized support so that the extractant can penetrate into the pores and interact with the bonded functional groups.
5. Sufficient chemical stability so that it retains its activity during operation.

Some properties of adsorbents prepared by this method [10] are shown in Table 7.2.

7.2.2.2 Covalent Attachment to the Support

Tailor-making specific ligands by covalently attaching them to inorganic supports is a very elegant approach. This method can produce ICAAs with greater stability, selectivity, and adsorption rates. One can form covalent bonds between an organic moiety and a substrate through an intermediary coupling agent. Selection of the functional group to be immobilized depends on intended applications. The groups taking part in the formation of chelate rings usually include nitrogen, oxygen, and

TABLE 7.2
Surface Modified and Solvent Deposited Inorganic Adsorbents

Support	Chelating Agent	Metal Ions	Applications/Comments
Silica gel Alumina	Kelex-100®[a]	Pb(II), Cu(II), Ni(II), Cd(II)	Remove Pb(II) to less than 1 ppm and stable for 20 adsorption/stripping cycles
Silica gel	Cyanex-302®[b]	Cd(II), Zn(II), Pb(II)	Simultaneous removal and separation by selective stripping
Silica gel	Alamine®[a]	Cr(VI), and $Cu(CN)_2$, $Cu(N)_3^{2-}$	Recover and recycle copper cyanide from cyanide waste solutions

[a] Henkel Co.
[b] American Cyanamide Co.
Source: From N. V. Deorkar and L. L. Tavlarides. *Emerging Separation Technologies for Metals II,* The Minerals, Metals & Materials Society, Warrendale, PA, 1996. (With permission.)

sulfur atoms. The attachment of specific complexing groups into inorganic matrices makes them capable of reacting with metal ions, owing to the coordinate/covalent/ionic bond. The interaction between metal ion and functional groups depends on the properties of the metal (charge, size, coordination number), adsorption condition (solution pH, ionic strength), functional group, and physical nature of the matrix (steric factors). Chelate rings can be formed with the participation of donor atoms situated in one unit of matrix or at the matrix chain. Accordingly, highly selective ICAAs can be prepared by careful planning and execution of synthesis schemes to introduce desired donor atoms in a preferred geometry.

Of great importance today is the availability of silica gels with variable size and microporous structure that provides better permeability into the matrix, facilitating an introduction of active groups. A variety of organic-functional silanes and their derivatives can be used to introduce reactive groups, such as chlorides or aminos, into a silica matrix. These reactive groups can take part in coupling of desired chelating groups. We prepared ICAAs for different metal ions by covalently attaching to the silica surface the selected chelating agents or derivatives of functional groups capable of complexing the desired metal ions (Table 7.3) [10]. The functional groups were attached by using commercially available silane coupling agents. First, suitable derivatives of functional groups and silane coupling agents are prepared to facilitate covalent bonding. The covalent attachment of a functional group to the silica surface through the use of silane coupling agents is achieved by two general methods and is described in the two following examples.

Method A—where the coupling agent is first attached to the silica surface followed by an attachment of the functional group precursor to the lattice, is outlined in Figure 7.7.

Method B—where a functionalized silane coupling agent is formed by reacting the silane-coupling agent with functional groups. Then, the functionalized silane coupling agent is attached to the silica surface. This procedure is outlined in Figure 7.8.

TABLE 7.3
ICAAs Prepared by Covalent Bonding

Support	Functional Group	Method/Coupling Agent	Metal Ions
Silica gel	5-methyl-8-hydroxy-quinoline	Organic functional silane derivatives in solutions and surface	Pb(II), Cu(II), Ni(II), Cd(II)
Silica gel	ThioSulfide acid	Organic functional silane in solution	Cd(II), Hg(II), Zn(II), Pb(II)
Silica gel	Primary, secondary, and tertiary amines, and diazole	Organic functional silane on surface	Cu(II), Ni(II), anionic cyanide complexes, $Cr_2O_7^{2-}$, CrO_4^{2-}
Silica gel	Pyrogallol	Derivatization on surface modified with organo-functional silane	Antimony(III) Al(III), Cu(II)

Source: From N. V. Deorkar and L. L. Tavlarides. *Emerging Separation Technologies for Metals II,* The Minerals, Metals & Materials Society, Warrendale, PA, 1996. (With permission.)

In general, it was observed that Method B produces high-capacity materials due to high-coverage density of functional groups and homogeneous attachment. However, in some instances, when functionalization of silane coupling agent is not possible, Method A was used. Moreover, it may not be possible for bulky functionalized silanes, prepared by Method B, to penetrate into the pores to achieve high-coverage density. Figure 7.9 shows functional groups covalently bonded to the surface using the silane coupling agent. The details of the synthesis procedure are presented elsewhere [11–14].

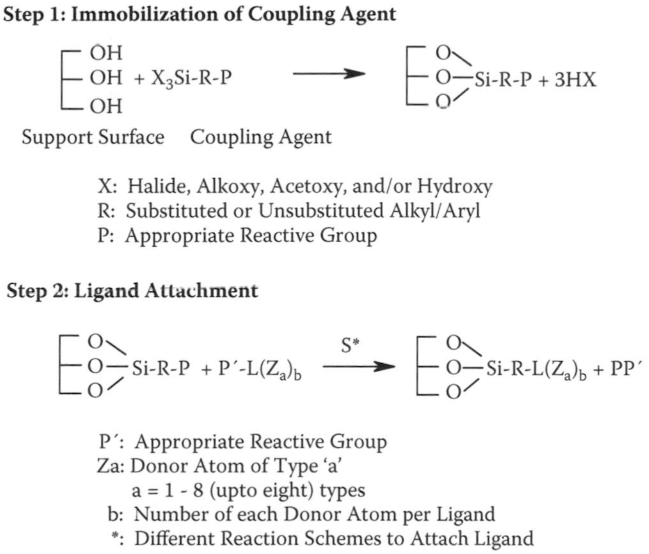

Step 1: Immobilization of Coupling Agent

$$\text{[Support Surface]}(OH)_3 + X_3Si\text{-}R\text{-}P \longrightarrow \text{[Surface]}(O)_3Si\text{-}R\text{-}P + 3HX$$

Support Surface Coupling Agent

X: Halide, Alkoxy, Acetoxy, and/or Hydroxy
R: Substituted or Unsubstituted Alkyl/Aryl
P: Appropriate Reactive Group

Step 2: Ligand Attachment

$$\text{[Surface]}(O)_3Si\text{-}R\text{-}P + P'\text{-}L(Z_a)_b \xrightarrow{S^*} \text{[Surface]}(O)_3Si\text{-}R\text{-}L(Z_a)_b + PP'$$

P′: Appropriate Reactive Group
Za: Donor Atom of Type 'a'
a = 1 - 8 (upto eight) types
b: Number of each Donor Atom per Ligand
*: Different Reaction Schemes to Attach Ligand

FIGURE 7.7 Covalent attachment to support by method A.

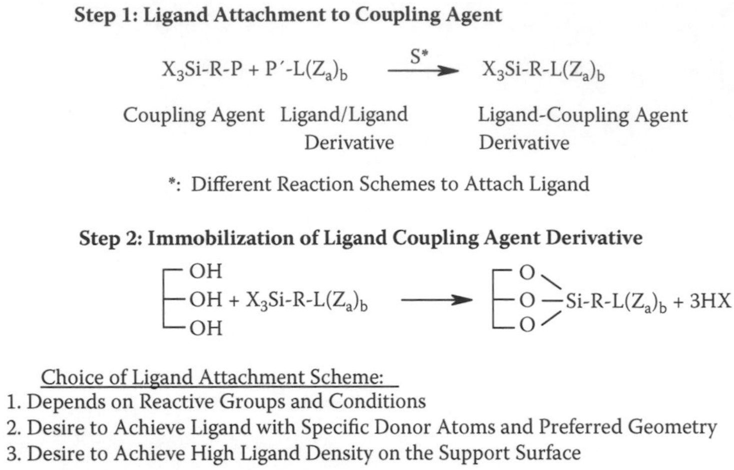

Step 1: Ligand Attachment to Coupling Agent

$$X_3Si\text{-}R\text{-}P + P'\text{-}L(Z_a)_b \xrightarrow{S^*} X_3Si\text{-}R\text{-}L(Z_a)_b$$

Coupling Agent — Ligand/Ligand Derivative — Ligand-Coupling Agent Derivative

*: Different Reaction Schemes to Attach Ligand

Step 2: Immobilization of Ligand Coupling Agent Derivative

$$\left[\begin{matrix}\text{OH}\\ \text{OH}\\ \text{OH}\end{matrix}\right. + X_3Si\text{-}R\text{-}L(Z_a)_b \longrightarrow \left[\begin{matrix}\text{O}\\ \text{O}\\ \text{O}\end{matrix}\right. \!\!> Si\text{-}R\text{-}L(Z_a)_b + 3HX$$

Choice of Ligand Attachment Scheme:

1. Depends on Reactive Groups and Conditions
2. Desire to Achieve Ligand with Specific Donor Atoms and Preferred Geometry
3. Desire to Achieve High Ligand Density on the Support Surface

FIGURE 7.8 Covalent attachment to support by method B.

Similar approaches were employed by using specially designed meso-porous silica materials by Feng and colleagues [15] and by Mercier and Pinnavaia [16]. Mesoporous silica materials are attractive supports because they have high surface area (up to 1,500 m^2/gm) and well-defined uniform pore size. In the synthesis of these extractant materials, a series of silanizations with a silane containing a functional moiety was performed to increase the ligand density on the meso-porous silica. With this technique, up to 3.2 mmol/gm of mercury uptake capacity was reported by Mattigod and colleagues [17].

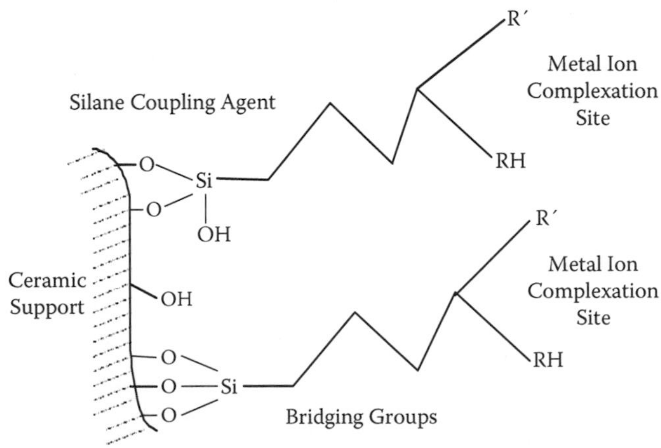

FIGURE 7.9 Inorganic chemically active adsorbent prepared by covalent bonding using silane-coupling agent. (From N. V. Deorkar and L. L. Tavlarides. *Emerging Separation Technologies for Metals II,* The Minerals, Metals & Materials Society, Warrendale, PA, 1996. With permission.)

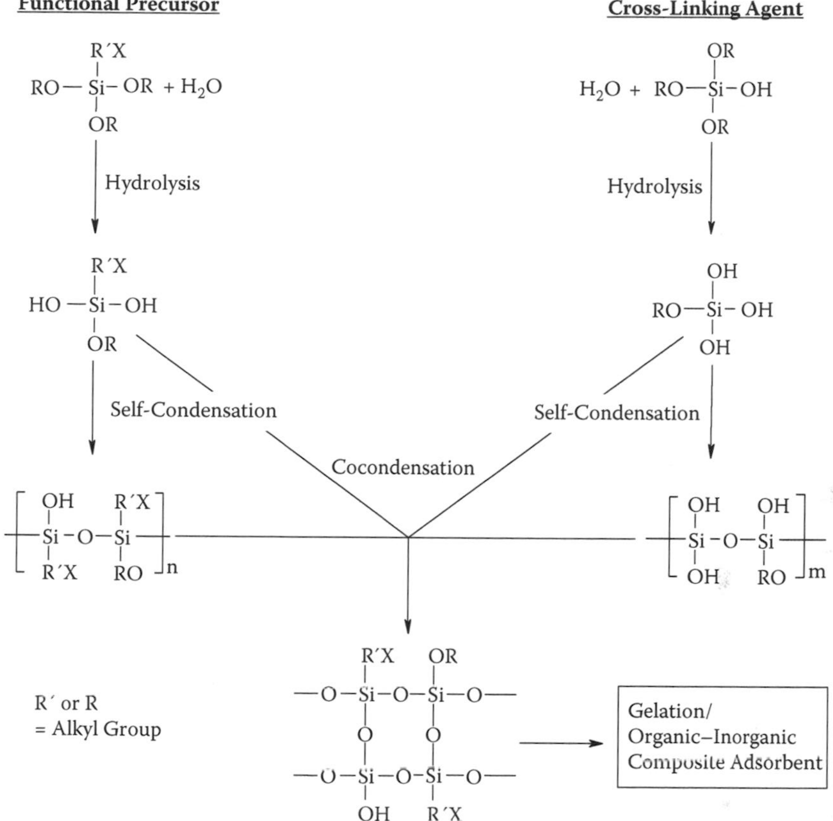

FIGURE 7.10 Sol-Gel synthesis of an adsorbent : R` and R are alkyl groups and X is the chelating molecule. (From J. S. Lee, S. Gomez-Salazar, and L. L. Tavlarides. *Reactive and Functional Polymers* 49:159–172, 2001. With permission.)

7.2.2.3 Organo-Ceramic Polymers

The use of sol-gel chemistry provides a fundamentally different approach to synthesizing organo-ceramic composite adsorbents. Sol-gel synthesis has been widely studied in glass making, optics, and coating processes [18,19]. Recently the incorporation of specific organic molecules in the silica network has been intensively studied [20,21]. The incorporation of organic molecules with chelation/ion exchange properties is possible by cocondensing a hydrolyzed functional precursor with a hydrolyzed cross-linking agent. Advantages of these materials are high ligand densities, homogeneous distributions of the functional moiety throughout the matrix, and controlled pore characteristics. The concept is illustrated in Figure 7.10 [22].

A series of adsorbents called SOL-AD were developed by this sol-gel processing [23]. For the synthesis of these adsorbents, the functional precursor containing an active group, such as 3-mercapto-propyltrimethoxysilane ($(CH_3O)_3$-Si-$(CH_2)_3SH$), is hydrolyzed and undergoes self-condensation. Likewise, a cross-linking agent, such as tetraethoxysilane ($(CH_3CH_2O)_4$-Si), is hydrolyzed and undergoes self-condensation. The two reaction systems can be combined at a certain time to maximize the density of ligand in the matrix and the properties of the matrix. The two choices for combinations are (1) combine the hydrolyzed, partially self-condensed functional precursor with the hydrolyzed, partially self-condensed cross-linking agent, or (2) combine the hydrolyzed, partially self-condensed functional precursor with the hydrolyzed cross-linking agent. The extent to which the individual reactions are executed before combining depends on the chemistry requirements of the individual reactions, the desired structure of the matrix, and the desired mechanical characteristics of the resulting gels. Parameters that greatly influence the properties of the resulting gels are molar ratio of precursor to cross-linking agent, extent of hydrolysis and homo-condensation of the precursor, and aging and drying methods. Figure 7.11 [22,24] shows the evolution of the hydrolysis–condensation reaction of the sol-precursor, 3-mercaptopropyltrimethoxysilane (MPS).

The reaction progress is monitored with ^{29}Si-nuclear magnetic resonance. Hydrolysis and condensation of MPS produces hydrolyzed monomers, dimers, and oligomers. As the condensation proceeds, the formation of cyclic compounds (T_Δ) and completely self-cross-linked species T_3 is observed. These species are not preferable compounds as they are either hindered in the silica network or they do not combine with the cross-linking agent. Oligomers composed mostly with T_2 silicons are the most favorable for the formation of functional clusters.

The synthesis route of the SOL-AD series extractants is optimized for the homo-condensation of MPS and the molar ratio of MPS/tetraethyoxysilane (TEOS). The homo-condensation kinetics is followed using ^{29}Si-NMR. Surface area and average pore diameter are measured using Brunauer-Emmett-Teller (BET) analysis, and cadmium uptake capacity using atomic absorption spectroscopy characterize the sol-gel. The results are shown in Figures 7.12, 7.13, and 7.14. It is seen from Figures 7.12 and 7.13 that the optimum uptake capacity occurs when the concentration of T_2 approaches a maximum. Higher concentration of T_2 produces higher capacity adsorbents. Figure 7.14 shows that the cross-linking silane (CS)/functional precursor silane (FPS) molar ratio of 2.0 optimizes the uptake capacity as it provides a sufficiently high ratio of functional groups and a suitable pore structure and surface area.

7.3 CHARACTERIZATION OF EXTRACTION MATERIALS

The synthesized materials are usually characterized by a variety of chemical and physical properties. Chemical properties of the materials include pH dependency of metal uptake, maximum metal uptake capacity, acid-base characteristics, metal selectivity, and stability. Physical properties of the materials include mechanical strength and pore-network characteristics. A number of techniques discussed in the following section are employed to determine these properties.

7.3.1 Characterization of Chemical Properties of Solid Extractants

The mechanism of metal extraction by using solid extractants can be categorized as cation extraction and anion extraction. These extraction mechanisms are dependent

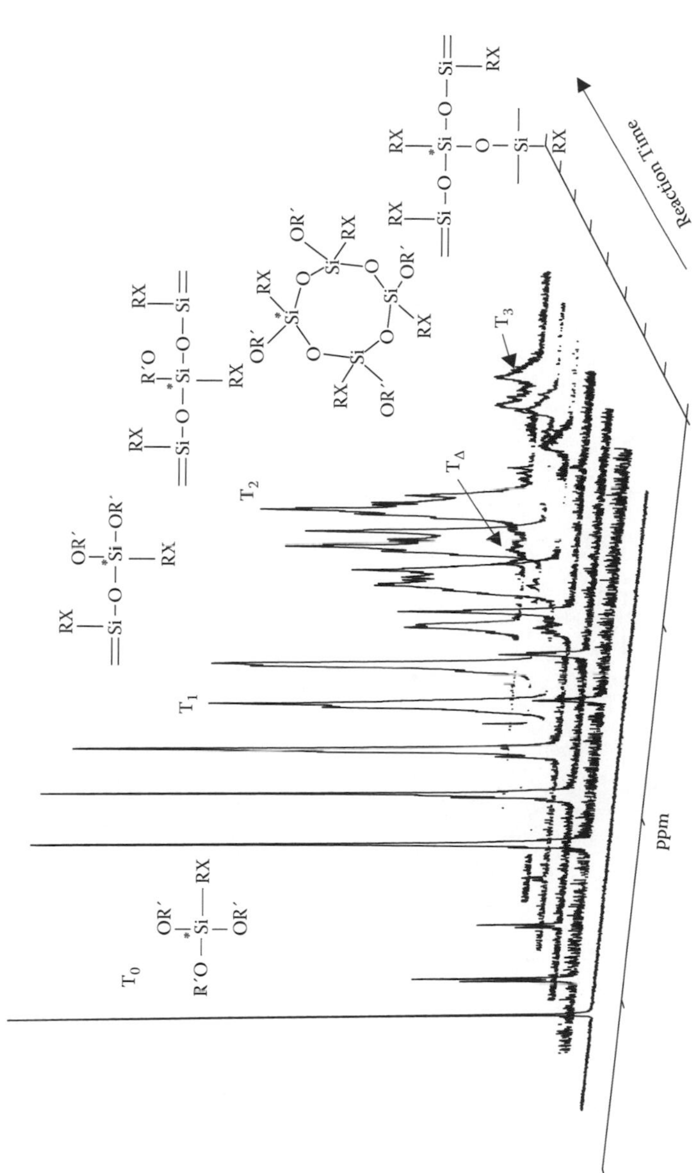

FIGURE 7.11 ^{29}Si-NMR spectra taken under low temperature (–50°C) condition to reduce reaction rate during the NMR analysis (for NMR analysis details elsewhere see [24]). (From J. S. Lee, S. Gomez-Salazar, and L. L. Tavlarides. *Reactive and Functional Polymers* 49:159–172, 2001. With permission.)

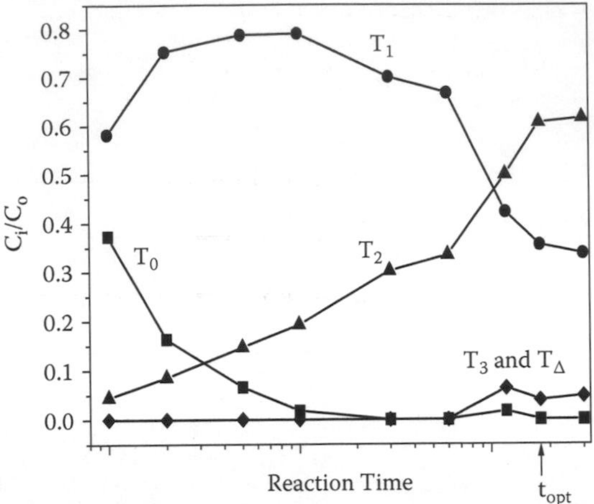

FIGURE 7.12 The results of NMR analysis for the hydrolysis/condensation reaction of MPS in methanol solution. (From J. S. Lee, S. Gomez-Salazar, and L. L. Tavlarides. *Reactive and Functional Polymers* 49:159–172, 2001. With permission.)

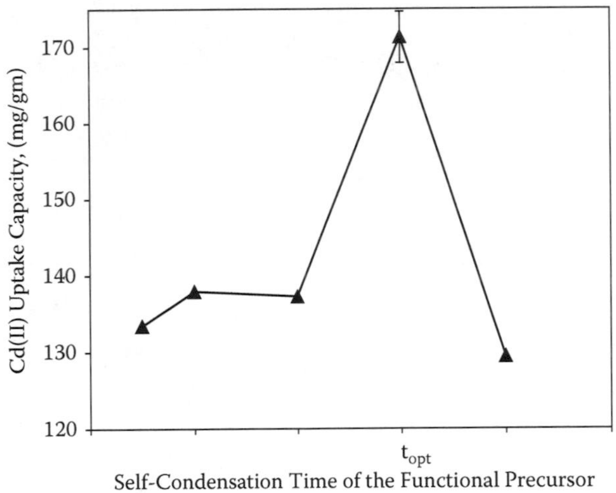

FIGURE 7.13 A trend of cadmium uptake capacity change of the adsorbent, SOL-AD-I as the degree of self-condensation of the functional precursor increases. (From J. S. Lee, N. V. Deorkar, and L. L. Tavlarides. Proceedings of a symposium sponsored by the Engineering Foundation Conference and National Science Foundation, The Minerals, Metals & Materials Society, Hawaii, 1999, pp. 43–52. With permission.)

on the acid-base characteristics of the ligands and the speciation of metal ions in the aqueous solutions.

Metal cations can react with a neutrally charged ligand having hydrogen bonding by replacing the hydrogen ion, or the metal cations can react with negatively charged ligands as shown in the following equations.

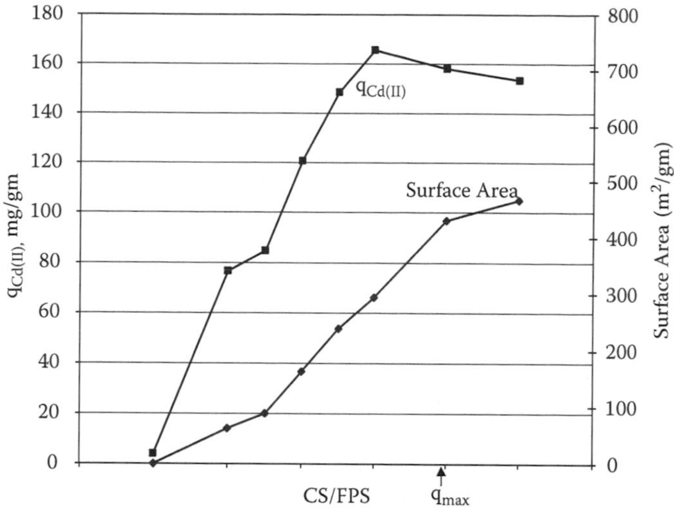

FIGURE 7.14 Effects of molar ratio of CS and FPS on cadmium uptake capacity and surface area of SOL-AD-I. CS: cross-linking silane; FPS: functional precursor silane. (From J. S. Lee, S. Gomez-Salazar, and L. L. Tavlarides. *Reactive and Functional Polymers* 49:159–172, 2001. With permission.)

Cation Extraction

$$M_{aq}^{z+} + n\overline{RH} \leftrightarrow \overline{R_nM^{(z-n)+}} + nH^+, \tag{7.1}$$

or

$$\overline{RH} \leftrightarrow \overline{R^-} + H^+, \tag{7.2}$$

$$M_{aq}^{z+} + n\overline{R^-} \leftrightarrow \overline{R_nM^{(z-n)+}} \tag{7.3}$$

Here M_{aq}^{z+} is a metal cation, $\overline{RH}$ is an unreacted neutral ligand on the solid surface, $\overline{R_nM^{(z-n)+}}$ is a metal ion – ligand complex on the solid surface, H^+ is a hydrogen ion, and $\overline{R^-}$ is a negatively charged ligand on the solid surface. As an example of Equation (7.1), thiol ligands (-SH) react with cadmium ions and release hydrogen ions. It is known that when the ligand has hydrogen bonding the extraction kinetic rate is slow because of the hydrogen bonding strength. As an example of Equations (7.2) through (7.3), carboxylic groups (COOH) can easily be deprotonated (COO⁻) at pH higher than 6 and can form complexes with cobalt.

Anionic metal ion species can be extracted when the functional ligands are protonated, as shown in the following equations.

Anion Extraction

$$\overline{RN} + H^+ \leftrightarrow \overline{RNH^+}, \tag{7.4}$$

$$M(A)_n^{z-} + m\overline{RNH^+} \leftrightarrow \overline{[(RNH)_m M(A)_n]^{(z-m)^-}} \tag{7.5}$$

$\overline{RN}$ is an anionic extractant ligand on the solid surface, $\overline{RNH^+}$ is a protonated ligand on the solid surface, and $\overline{[(RNH)_m M(A)_n]^{(z-m)-}}$ is a metal ion–ligand complex on the solid surface.

In anionic metal ion extraction, the available functional ligand for the metal extraction is strongly dependent on the proton uptake of the extractant at a given solution pH. Hence, the metal uptake capacity of an anionic extractant is determined by both the proton uptake capacity of the extractant and the metal anionic speciation in the aqueous solution at a given pH.

7.3.1.1 pH Dependency of Extraction

The pH dependency of extraction can be studied by conducting pH isotherm experiments. The pH isotherm provides useful information as it relates the extent of metal ion extraction with the solution pH for a given solution composition. This information can be employed to determine conditions to extract and strip metal ions from the solid extractant. Figure 7.15 shows a typical cation extraction pH isotherm: Negligible metal uptake occurs at a lower pH, and high metal uptake occurs at a higher pH. Thus, one can extract cadmium ion effectively at a pH higher than 4 and strip

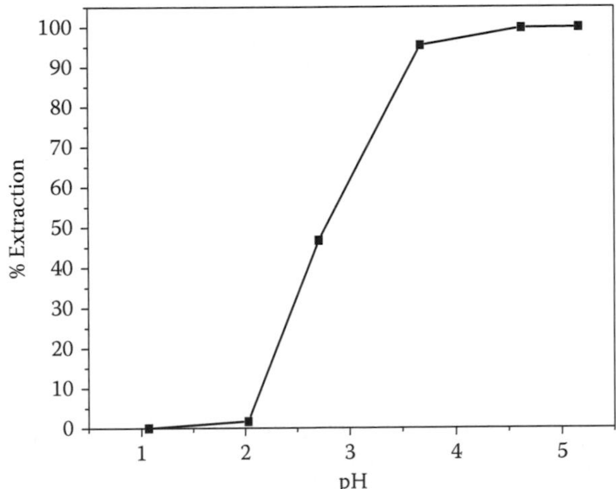

FIGURE 7.15 Effect of the pH of aqueous solutions on the adsorption of cadmium ion on SOL-AD-IV. (From J. S. Lee, N. V. Deorkar, and L. L. Tavlarides. Proceedings of a symposium sponsored by the Engineering Foundation Conference and National Science Foundation, The Minerals, Metals & Materials Society, Hawaii, 1999, pp. 43–52. With permission.)

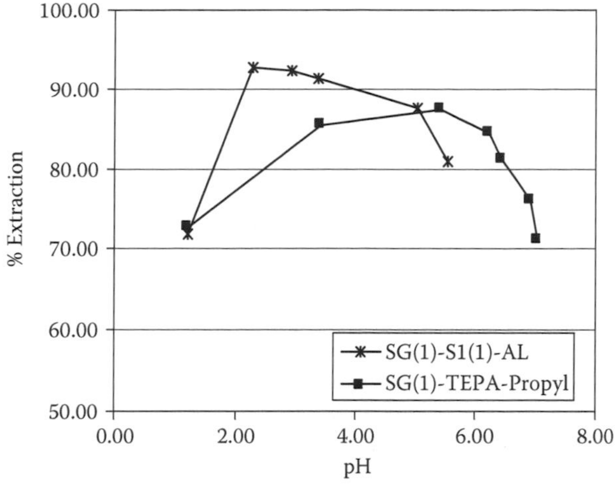

FIGURE 7.16 pH isotherms of Cr(VI) adsorption on SG(1)-TEPA-Propyl, and SG(1)-S1-AL.

the loaded cadmium from the extractant by using a moderately concentrated mineral acid such as 0.5 M HCl or 0.5 M H_2SO_4.

As an example of the effect of solution pH on anion adsorption, chromium solutions at various pH levels are contacted with two amine functionalized silica gels. The results show that the extraction efficiency increases as the pH increases up to a maximum value for each extractant and then starts to decrease as shown in Figure 7.16. This behavior could be a result of (1) the protonation of amine groups on the multidonor atom ligand (see Figure 7.22 in Section 7.4), (2) preferential adsorption of multicharged anionic species, and (3) steric hinderance factors between the anionic species and the adjacent protonated amine donor atoms. However, sufficient information is not yet available at the present to analyze this behavior conclusively.

7.3.1.2 Capacity

Theoretical uptake capacity of a solid extractant for a specific metal ion is the stoichiometric amount of the functional ligand on the solid surface. The theoretical uptake capacity can be evaluated by determining the number of moles of ligands per unit weight of the solids and is referred to as the ligand density. This measurement can be conducted by elemental analysis or acid-base titration. For example, Zhmud and Sonnefeld [26] synthesized organo-ceramic extractants by using triethoxy-amino-propylsilane and tetraethoxysilane. They performed three different methods to determine the ligand densities of the materials synthesized: elemental analysis, thermal analysis, and acid-base titration. The results from three different methods are compatible as shown in Table 7.4. The details of the analysis can be found in their documents [26,27]. Actual maximum uptake capacities of the metal by the solid extractants can be determined either by batch experiments or by breakthrough experiments. In a batch system, a specified amount of the extractant is equilibrated with a concentrated metal solution for which complete saturation of the extractant is expected. Moreover, it is possible to determine the maximum metal uptake capacity

TABLE 7.4
The Amino Group Loadings (mmol g^{-1})

Method of Determination	Matrix APS-1	APS-2	MAPS-1	MAPS-2
Elemental Analysis	3.1 ± 0.2	1.5 ± 0.1	1.5 ± 0.1	2.1 ± 0.2
Titration[a]	3.3 ± 0.3	1.0 ± 0.1	–	2.2 ± 0.2
Thermal Analysis	3.5 ± 0.3	2.4 ± 0.2[b]	–	

[a] MAPS-1 proved to be hydrophobic, so it was not titrated.

[b] When there are two or more different groups being destroyed at or near the same temperature, the thermal analysis only allows for the determination of the summed loading of all the groups present and not the distinguishing of the individual contribution of each of them.

Source: From B. V. Zhmud, J. Sonnefeld. *J. Non-Cryst. Solids* 195:16–27, 1996. (With permission.)

by conducting a breakthrough curve experiment. This method is more practical than the batch method because the maximum metal uptake capacity can be determined at operational conditions. For example, Figure 7.17 showns the results of a mercury loading breakthrough experiment. The mercury uptake capacity of SOL-AD-IV determined by this experiment conducted with a feed concentration of 250 mg/L is 884 mg/gm at pH 4, whereas the maximum loading in batch is 1280 mg/gm when it is contacted with a 600 mg/L mercury solution at pH 5.

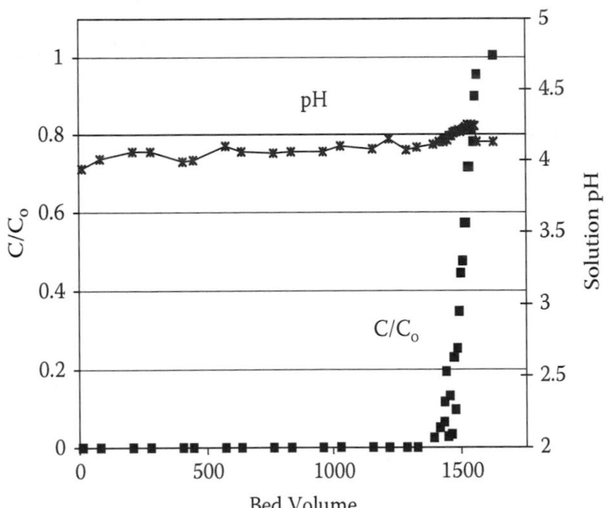

FIGURE 7.17 Breakthrough curve of mercury on SOL-AD-IV bed: 1 bed vol. = 8.65 cm^3; C_0 = 266 mg/l; pH_{feed} = 4.0; breakthrough capacity = 824 mg/g (Caq < 0.1 ppm); total adsorbed amount = 884 mg/g. (From J. S. Lee, N. V. Deorkar, and L. L. Tavlarides. Proceedings of a symposium sponsored by the Engineering Foundation Conference and National Science Foundation, The Minerals, Metals & Materials Society, Hawaii, 1999, pp. 43–52. With permission.)

TABLE 7.5
Metal Selectivity of SOL-AD-IV

Metal Ion	% Extraction
Cu(II)	98.9
Cd(II)	92.7
Pb(II)	57.5
Zn(II)	2.0
Ni(II)	0.0

Source: From J. S. Lee, N. V. Deorkar, and L. L. Tavlarides. Proceedings of a symposium sponsored by the Engineering Foundation Conference and National Science Foundation, The Minerals, Metals & Materials Society, Hawaii, 1999, pp. 43–52. (With permission.)

7.3.1.3 Selectivity

The metal selectivity of a specific solid extractant depends on the physicochemical properties of the functional ligand incorporated in the solid matrix, the oxidation state of metal ions, and the geometry of the surface. As it was mentioned earlier, a functional ligand that contains oxygen generally prefers to chelate with hard metal ions, and ligands containing sulfur can easily form complexes with soft metal ions. However, chelation between metal ions and the functional ligands on the solid surface may occur in a different way compared to the liquid–liquid system due to the complexity of the solid matrix and static hinderance effects. One can determine the selectivity of a solid extractant by equilibrating a unit mass of the solid extractant with a solution containing equimolar metal ions. For example, the results of a selectivity experiment of SOL-AD-IV are shown in Table 7.5. It is observed that the thiol groups in this adsorbent tend to react with metal ions in the order of Cu(II) > Cd(II) > Pb(II) > Zn(II) > Ni(II) when the adsorbent is contacted with an aqueous solution containing equimolar concentrations of these metal ions. The selectivity of a solid extractant for metal ions may also be changed as the solution condition changes. Hence, one can study the dependency of the selectivity of the extractant on the solution pH by performing a pH isotherm experiment. In addition, one can simply compare the pH isotherms independently obtained for various metal ions to estimate the selectivity of the extractant. As an example, Figure 7.18 shows the percent extraction of various metal ions with respect to the solution pH.

7.3.1.4 Stability

Stability of solid extractants is directly related to the life-cycle costs. Effects such as ligand oxidation, irreversible metal adsorption on ligands, and cleavage of the ligand from the solid network reduce the metal uptake capacity of a solid extractant while the extractant is used repeatedly. One can determine the stability of an extractant by repeating metal loading and stripping cycles at extreme conditions such as using concentrated acid or base for metal ion stripping. For example, the stability

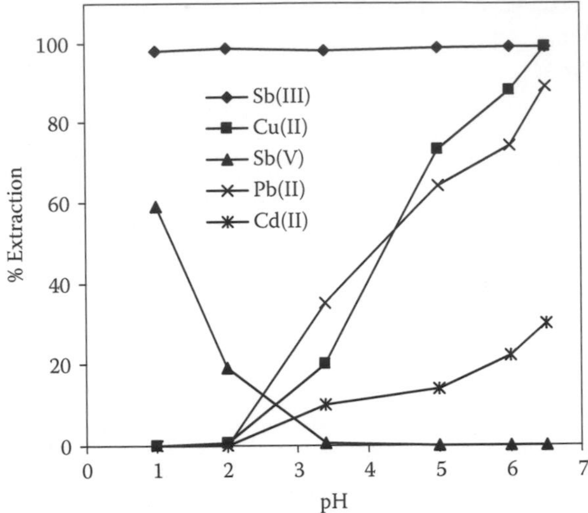

FIGURE 7.18 Effect of the pH of aqueous solutions on the adsorption of metal ions on ICAA-PPG. Equilibrium time = 1 h; weight of ICAA-PPG = 1 g; volume of aqueous solution = 25 ml; Co = 200 mg/l. (From N. V. Deorkar and L. L. Tavlarides. *Environ. Progress* 17:120–125, 1998. With permission.)

of SOL-AD-IV is tested by this method. Figure 7.19 shows the results of subsequent loading and stripping cycles. From this figure, a 30% loss of capacity after 26 loading and stripping cycles is observed.

7.3.2 Characterization of Physical Properties of Solid Extractants

In addition to having high ligand density, the solid extractant should have accessible pores and good mechanical strength. The mechanical strength of the solid extractants can be easily measured by standard methods usually used for polymer resins. Hence, only pore characteristics of the solid extractants are discussed in this section.

7.3.2.1 Pore Characteristics of the Solid Extractants

Relatively uniform pore size distribution and moderate pore diameters of the solid extractants (10–100 A°) are important to permit pore accessibility and ease diffusion of metal ions. The matrix characteristics, especially pore diameter, surface area, and pore size distribution, can be measured by BET analysis with nitrogen adsorption. The cadmium uptake capacities of the series of SOL-AD-I are correlated with BET surface areas to show the relationship between metal uptake capacity and the surface area as shown in Figure 7.20. In this figure, the theoretical metal uptake capacity is computed based on the number of moles of functional precursor silane added to the reaction mixture and the resulting weight of the extractant after the synthesis. It is observed from this figure that the actual metal uptake capacity increases as CS/FPS increases (decrease of ligand density) until CS/FPS reaches 2 even though the theoretical capacity consistently decreases. On other hand, the surface area of

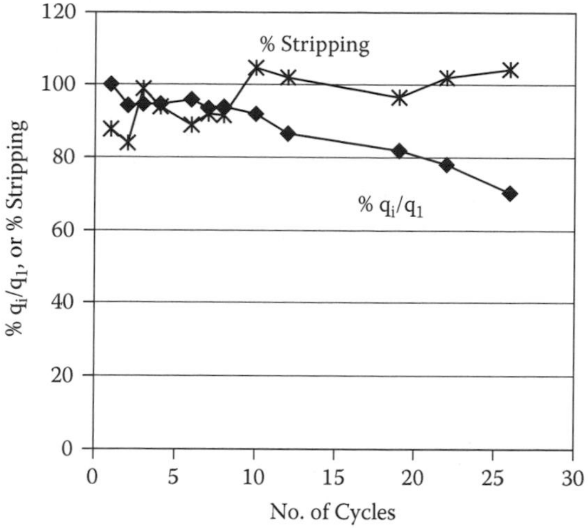

FIGURE 7.19 Stability Test of SOL-AD-IV material at extreme condition. (From J. S. Lee, S. Gomez-Salazar and L. L. Tavlarides. *Reactive and Functional Polymers* 49:159–172, 2001. With permission.)

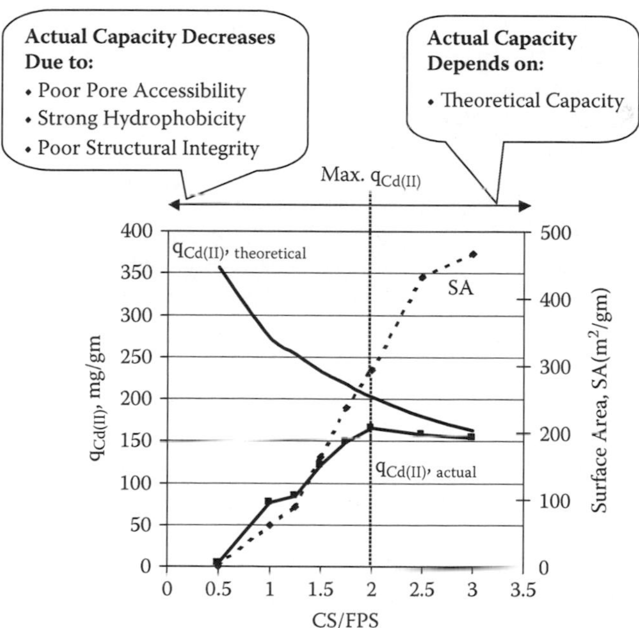

FIGURE 7.20 The effect of CS/FPS ratio on cadmium uptake capacity ($q_{Cd(II)}$) and surface area (SA) of SOL-AD-I; FPS: functional precursor silane; CS: cross-linking silane. (From J. S. Lee, S. Gomez-Salazar, and L. L. Tavlarides. *Reactive and Functional Polymers* 49:159–172, 2001. With permission.)

the material consistently increases as CS/FPS increases. Thus, it can be concluded that the metal uptake capacity of SOL-AD-I is controlled by the pore characteristics on the left-hand side of CS/FPS = 2 and by the ligand density on the right-hand side of CS/FPS = 2.

7.4 METAL ION ADSORPTION ON SOLID SUPPORTED EXTRACTANTS

The adsorption of metal cations and inorganic metal anion species (either in free or complex forms) from aqueous solutions onto chelating ligands or ionic charged ligands on the support involves a surface reaction. Although the extent of adsorption can be described by conventional isotherm expressions such as the Langmuir or Fruendlich equations, it is more appropriate to consider ion adsorption with models based on chemical reactions or complexation models. As this topic is too broad to cover in general, for our purpose we show one comprehensive example with the adsorbent SG(1)-(Tetraethylenepenta-amine)-propyl (TEPA) prepared in our laboratories via the covalent attachment to silica gels. For a broader treatment refer to Tien [29].

7.4.1 The SG(1)-TEPA-propyl/Copper Cyanide Adsorption Process

The adsorbent SG(1)-TEPA-propyl is useful in adsorbing anionic copper cyanide complexes from electroplating solutions. Modeling this anionic adsorption phenomena requires the description of copper cyanide ion speciation in the aqueous solution, the acid-base characteristics of the adsorbent, and the complexation reactions between the copper cyanide anions and the protonated ligand sites. Details can be found elsewhere [5,30].

7.4.1.1 Copper Cyanide Ion Speciation

Copper cyanide complexes are usually present as monovalent ($Cu(CN)_2^-$), divalent ($Cu(CN)_3^{2-}$), and trivalent ($Cu(CN)_4^{3-}$) species in aqueous solutions, as shown in Figure 7.4. Computations for copper cyanide speciation show that for ratios of total copper cyanide to total copper greater than 16, speciation at any given pH does not depend on the $[CN]_T/[Cu]_T$ ratio. The solution pH, however, greatly affects the speciation of complexes (see Figure 7.4). The fraction of monovalent species decreases whereas the fraction of the divalent species increases as pH increases. Concentrations of trivalent species are negligible below pH 7.5. These results show the need to consider pH dependency of speciation in the adsorption isotherm expression.

7.4.1.2 Characterization of SG(1)-TEPA-propyl

The molecular structure of SG(1)-TEPA-propyl is shown in Figure 7.21. There are five functional nitrogen atoms on a branch available for the anionic species after protonation. The mechanism of protonation of a ligand can be described if we assume each numbered site in Figure 7.21 has a distinctive pKa value:

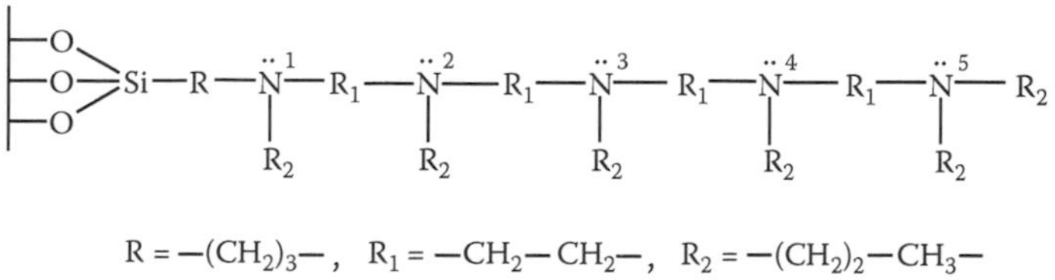

R = —(CH$_2$)$_3$—, R$_1$ = —CH$_2$—CH$_2$—, R$_2$ = —(CH$_2$)$_2$—CH$_3$—

FIGURE 7.21 Molecular structure of SG(1)-TEPA-propyl. (Reproduced from J. S. Lee, N. V. Deokar, and L. L. Tavlarides. *Ind. Eng. Chem. Res.* 37:2812–2820, 1998. With permission.)

$$\overline{RNH^+}_i \leftrightarrow \overline{RN}_i + H^+, \quad Ka_i, \quad i = 1,\ldots,5 \tag{7.6}$$

where i refers to the ith site on a single ligand chain, $\overline{RN}$ is the unprotonated site, and $\overline{RNH^+}$ is the protonated site.

The sum of the fraction of total sites that are protonated can be expressed as

$$\frac{[\overline{RNH^+}]_T}{[\overline{RN}]_{0,T}} = \frac{1}{5}\sum_{i=1}^{5}\frac{1}{10^{pH-pKai}+1} \tag{7.7}$$

where $[\overline{RN}]_{0,T}$ is total number of amine groups per unit mass of adsorbent and equals $5[\overline{RN}]_{0,T}$ and $[\overline{RNH^+}]_T$ is the total number of protonated amine groups per unit mass of adsorbent.

A nonlinear parameter estimation was used to obtain the five pKa values of $pKa_{,1} = 3.26$, $pKa_{,2} = 4.46$, $pKa_{,3} = 5.61$, $pKa_{,4} = 6.59$ and $pKa_{,5} = 7.98$, and $[RN_0]_T = 1.25$ mmol/g using titration curve data.

7.4.1.3 Adsorption Equilibria in a Batch System

The adsorption mechanism of copper cyanide complexes can be described as follows:

$$RN + H^+ \leftrightarrow RNH^+ \qquad K_a \tag{7.8}$$

$$A^- + RNH^+ \leftrightarrow RNHA \qquad K_1 \tag{7.9}$$

$$B^{2-} + RNH^+ \leftrightarrow RNHB^- \qquad K_2 \tag{7.10}$$

$$C^{3-} + RNH^+ \leftrightarrow RNHC^{2-} \qquad K_3 \tag{7.11}$$

where A^- is $Cu(CN)_2^-$, B^{2-} is $Cu(CN)_3^{2-}$, and C^{3-} is $Cu(CN)_4^{3-}$.

To model the adsorption of copper cyanide complexes, the multicomponent Langmuir isotherm can be modified to correlate the aqueous phase speciation and the basicity of the ICAA. The available protonated sites on the ligands attached to the surface also can be assumed to be a function of pH.

The adsorption of trivalent copper cyanide from solution can be neglected when the pH is lower than 8 because it exists in very small amounts. If we assume that all the protonated sites can adsorb copper cyanide species, and the effect of ionic strength can be neglected in the adsorption processes, the modified multicomponent Langmuir isotherm can be written for total amount adsorbed q as follows:

$$q = [RNHA] + [RNHB^{-}] \tag{7.12}$$

$$q = f\frac{K_1C_A + K_2C_B}{1 + K_1C_A + K_2C_B}[RNH^+]_T \tag{7.13}$$

Here, $[RNH^+]_T$ is the total number of protonated sites and depends on pH as seen in Equation (7.7), C_A is $[Cu(CN)_2^-]$, C_B is $[Cu(CN)_3^{2-}]$, and $C_{aq} = C_A + C_B$. It is observed that copper cyanide complexes cannot be adsorbed to the theoretical maximum due to geometrical constraints; thus, a correlation factor f in the equation is incorporated.

Adsorption isotherms were generated for the copper cyanide complex adsorption at three pH values, and the Langmuir constants and geometrical factor were found to have values of K_1 = 2.43, L/mmol, K_2 = 3.95 L/mmol and f = 0.40. A comparison of the experimental values and those calculated with the model and obtained parameters is given in Figure 7.22.

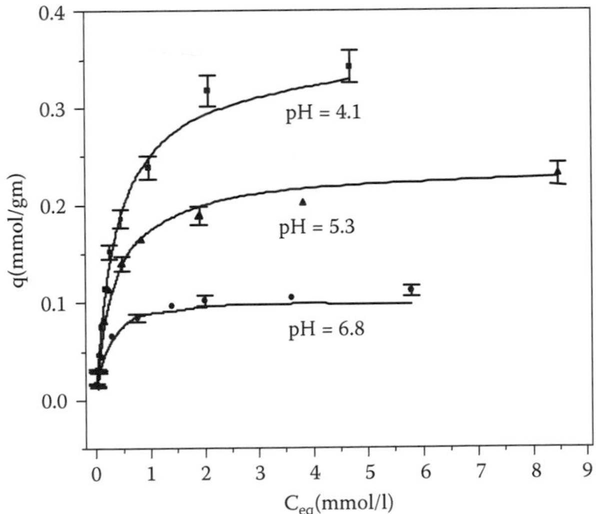

FIGURE 7.22 Adsorption isotherms of copper cyanide complexes at three different pH values. Symbols represent experimental data for the different pH conditions and solid lines represent the the modified multicomponent Langmuir Equation (7.13). (Reproduced from J. S. Lee, N. V. Deokar, and L. L. Tavlarides. *Ind. Eng. Chem. Res.* 37:2812–2820, 1998. With permission.)

7.5 CHEMICAL KINETICS OF METAL ION ADSORPTION ON SOLID EXTRACTANTS

There are cases of metal ion adsorption on solid extractant surfaces where the rate of chemical chelation or ion exchange is the rate determining process. Under these conditions, it is useful to develop a kinetic model to describe this rate process for use in design equations. The kinetic processes are similar to chemical catalytic reactions and also may include electrostatic effects [31].

In the example that follows, a possible mechanism is presented to illustrate the procedure followed to obtain an acceptable kinetic model and estimates of the kinetic parameters.

For the adsorption of cadmium ions by the solid extractant SOL-AD-IV synthesized by the technique described in Section 7.2.2.3, the reaction can be assumed to be a cationic exchange with the thiol proton. The general reversible reaction can be assumed as

$$mCd^{+2} + n\overline{RH} = \overline{R_nCd_m} + nH^+ \tag{7.14}$$

One postulated reaction mechanism consists of the surface adsorption of one cadmium ion ($m = 1$) on one surface ligand site ($n = 1$) followed by a surface reaction with a nearby site to give

$$Cd^{+2} + \overline{RH} = \overline{RCd} + H^+ \tag{7.15}$$

$$\overline{RCd} + \overline{RH} = \overline{R_2Cd} + H^+ \tag{7.16}$$

Several possibilities can be followed to develop a rate equation. It can be assumed that both steps are of similar reaction rate to obtain a rate expression. Otherwise, it can be assumed that one reaction is at equilibrium while the other is rate determining. Taking the later approach, assuming that reaction (Equation 7.15) is at equilibrium, no electrostatic effects exist, and the activity coefficients are unity, one obtains

$$K = \frac{[\overline{RCd}][H^+]}{[Cd^{2+}][\overline{RH}]} \tag{7.17}$$

$$r_{surf} = \frac{d[\overline{R_2Cd}]}{dt} = k_1' [\overline{RCd}][\overline{RH}] - k_{-1}'[\overline{R_2Cd}][H^+] \tag{7.18}$$

where k_1, k_{-1} and k'_1, k'_{-1} represent the forward and reverse reaction constants for reactions (Equations 7.15 and 7.16), respectively, K represents the equilibrium constant for reaction (Equation 7.15), and $[i]$ denotes the concentration of i.

To obtain expression (Equation 7.18) in terms of the observable bulk solution concentration $[Cd^{2+}]_o$, we employ a mass balance on surface sites and on total cadmium adsorbed on the solid extractant surface:

$$[\overline{RH}]_T = [\overline{RH}] + [\overline{RCd}] + 2[\overline{R_2Cd}] \tag{7.19}$$

$$q_T = [\overline{RCd}] + [\overline{R_2Cd}] \tag{7.20}$$

Using these expressions we obtain expressions for $[\overline{RH}]$, $[\overline{RCd}]$, $[\overline{R_2Cd}]$ as functions of $[Cd^{2+}]_o$ in the bulk solution.

7.5.1 Design Equation

To evaluate the appropriate form of Equation (7.18) we must employ the design equation used to obtain the kinetic data. We studied the kinetics of adsorption using a batch reactor with recycle operation in the differential mode. The reactor consists of a packed column with the adsorbent between two layers of glass beads. Pore diffusion and mass transfer resistances were minimized by using small particle sizes (180 to 120 μm) and high flow rates. The design equation written for the aforementioned metal cation extraction is

$$r_{Cd^{2+}} = \left(\frac{V_r + V_T}{V_r}\right)\frac{d[Cd^{2+}]}{dt} \tag{7.21}$$

where V_r and V_T are the reactor and reservoir volumes, respectively.

Combining Equations (7.17) through (7.21) and the relation,

$$q_T = \frac{V_T}{W}\left(\left[Cd^{2+}\right]_0 - \left[Cd^{2+}\right]\right)$$

where $[Cd^{2+}]_o$ is the initial concentration of cadmium, W is the weight of adsorbent, provides the differential equation which can be employed to optimize the parameters of the kinetic model using techniques such as successive quadratic programming (SQP) [32]. The equation is

$$r_{surf} = \frac{d\left[\overline{R_2Cd}\right]}{dt} = k_1' \frac{K}{\left[H^+\right]}\left[Cd^{2+}\right]\left\{\frac{\left[\overline{RH}\right]_T - 2q_T}{1 - \frac{K\left[Cd^{2+}\right]}{\left[H^+\right]}}\right\}^2 - k_{-1}'\left[H^+\right]\left\{q_T - \frac{K\left[Cd^{2+}\right]}{\left[H^+\right]}\left[\frac{\left[\overline{RH}\right]_T - 2q_T}{1 - \frac{K\left[Cd^{2+}\right]}{\left[H^+\right]}}\right]\right\} \tag{7.22}$$

Evaluation of the standard deviation of the measurements from the fit of the model can be used to compare various mechanisms to determine the best representative model.

7.5.2 Kinetic Measurements

The batch recycle differential reactor is used to obtain adsorption rate data to evaluate various kinetic models for cadmium ion adsorption on Sol-AD-IV. Approximately 0.25 g of the solid extractant is placed between glass beads inserted in a glass tube of 1.0 cm internal diameter. Solutions of various cadmium ion concentrations (25 to 300 mg/l) are contacted in a batch mode over 720 minutes, with intermediate sample taken for atomic absorption spectroscopic measurements of the cadmium ions. The external mass transfer resistances are minimized by operating at 16 ml/min. Results of the experiments are shown in Figure 7.23.

7.6 METAL ION EXTRACTION IN FIXED BEDS

The solid supported extractants can be employed in fixed bed contactors to extract metal ions from solutions. Other geometries include slurry extractors and moving bed adsorbers. We consider a fixed bed geometry. In this case the following mass transfer processes may be present: (1) interpellet mass transfer, which refers to the diffusion and mixing of metal ion in fluid occupying the spaces between pellets; (2) interphase mass transfer, which is the transfer of metal ion across the fluid pellet interface; and (3) intraparticle mass transfer, which is the diffusion of metal ions in

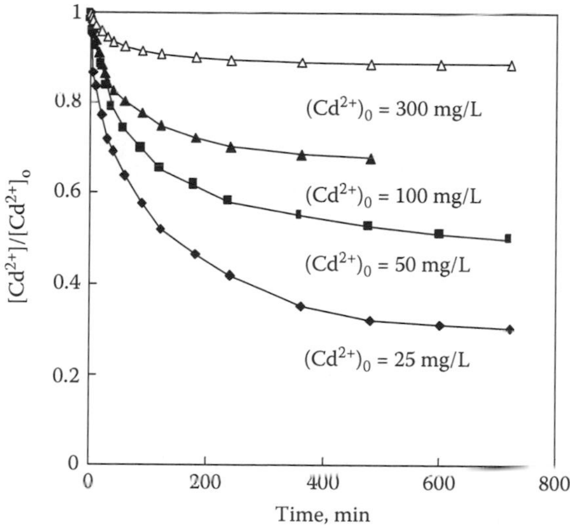

FIGURE 7.23 Kinetic data of Cd^{2+} adsorption on Sol-AD-IV with batch recycle differential reactor. (From J. S. Lee, S. Gomez-Salazar, and L. L. Tavlarides. *Reactive and Functional Polymers* 49:159–172, 2001. With permission.)

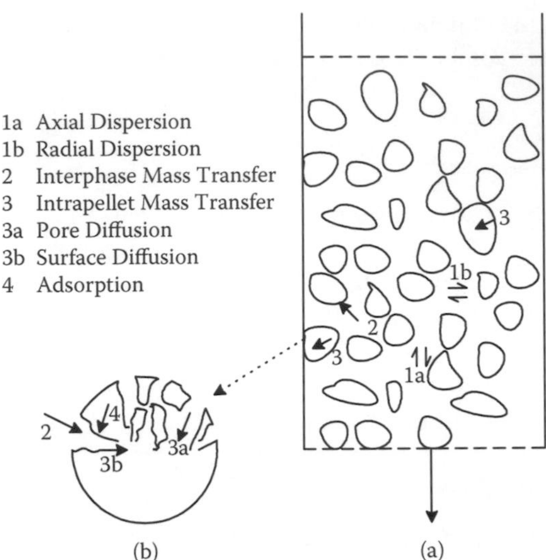

FIGURE 7.24 Mass transfer in adsorption processes: (a) fixed beds and (b) intrapellet mass transfer. (Reproduced from C. Tien. Butterworth-Heinemann Series in Chemical Engineering. Butterworth-Heinemann, Boston, 1994. With permission.)

the pellet (either in the pore solution or adsorbed on the pore surface). A schematic is shown in Figure 7.24.

7.6.1 Interpellet Mass Transfer

In fixed-bed adsorption, diffusion and mixing of metal ions in fluids occur due to adsorbate concentration gradients and non-uniformity of fluid flow. This gives rise to axial dispersion in the main direction of fluid flow and radial dispersion in the direction transverse to the main flow. The former is usually undesirable since it reduces separation efficiency, whereas the latter is desirable as it equalizes concentrations at the same axial position and reduces axial dispersion. For the simple case of single-phase flow in a packed bed of cylindrical configuration, the conservation equation for a metal ion in the solution is

$$\frac{\partial C_i}{\partial t} = \frac{E_r}{r}\frac{\partial}{\partial r}\left[r\frac{\partial C_i}{\partial r}\right] + E_z\frac{\partial^2 C_i}{\partial z^2} - \left(u_s/\varepsilon\right)\frac{\partial C_i}{\partial z} \tag{7.23}$$

A great deal of effort has been expended to relate E_r, E_z to the operation and physical parameters and is not discussed here.

7.6.2 Interphase Mass Transfer

The interphase mass transport rate of adsorbable metal ion from the bulk of the fluid phase to the external surface of the solid extractant can be written for spherical pellets as

$$\frac{\partial \overline{q_i}}{\partial t} = k_f a \, (C_{i,b} - c_{i,s}) \tag{7.24}$$

Correlations are available in the literature for k_f and are not discussed here.

7.6.3 Intraparticle Mass Transfer

The particle is assumed to have randomly distributed uniform pores throughout. In the absence of potential gradients and assuming Fick's law applies, the metal ion species equation for adsorption can be written as

$$\varepsilon_p \frac{\partial c_i}{\partial t} + \rho_p \frac{\partial q_i}{\partial t} = \frac{\partial}{\partial r}\left(D_p r^2 \frac{\partial c_i}{\partial r} \right), \quad 0 \le r \le a_p \tag{7.25}$$

When the reactions are rapid relative to the aforementioned mass transfer processes, then there will be an equilibrium between the concentration of the metal ion species in the pore fluid and at the pore surface at any point in the particle. Under these conditions we need an equilibrium relationship for the overall adsorption reactions. This relationship can be similar to a Langmuir or Freundlich isotherm, and in general

$$q_i = f(c_i) \tag{7.26}$$

Equation (7.25) is subjected to the boundary conditions

$$c_i = 0 \qquad \text{at } t = 0 \tag{7.27a}$$

$$\frac{\partial c_i}{\partial r} = 0 \quad \text{at } r = 0 \tag{7.27b}$$

$$k_f a (C_{i,b} - c_{i,s}) = D_{p,i} \frac{\partial c_i}{\partial r} \quad \text{at } r = a_p \tag{7.27c}$$

These equations can be solved for special cases using appropriate assumptions to describe the behavior of a fixed bed solid extractant column. Numerous examples can be found in the literature [29,31]. One example is given here.

The extraction of copper cyanide species using SG(1)-TEPA-propyl discussed in Section 7.4 occurs in a fixed bed column filled with 75 to 270 μm particles and is modeled by the previous equations.

7.6.4 Model

It was observed that the extraction rate of copper-cyanide complexes on SG(1)-TEPA-propyl is too rapid to measure due to fast kinetics and pore diffusion (small particles). Accordingly, the adsorption behavior can be easily modeled using the equilibrium model of Section 7.4.

$$q = f\frac{K_1C_A + K_2C_B}{1 + K_1C_A + K_2C_B}[RNH^+]_T \tag{7.28}$$

Equation (7.25) becomes

$$\varepsilon_p\frac{\partial c_i}{\partial t} + \rho_p\frac{\partial q_i}{\partial t} = 0, \quad 0 \le r \le a_p \tag{7.29}$$

Under the assumptions of no radial and axial dispersion ($E_r = E_z = 0$) and since $C_i = c_i$ when no mass transfer resistances exist in the pellet

$$\rho_b\frac{\partial q_i}{\partial t} = (u_s/\varepsilon)\frac{\partial c_i}{\partial z} \tag{7.30}$$

where $\rho_b = \rho_p/\varepsilon_p$.

The first derivative in Equation (7.30) can be evaluated using Equation (7.31). Accordingly, the interpellet mass transfer equation can be written in nondimensional form as

$$\frac{\partial C^*}{\partial z^*} + \frac{K}{(1 + KC^*)^2}\frac{\partial C^*}{\partial \theta^*} = 0 \tag{7.31}$$

$$\begin{aligned} C^*(0, \theta^*) &= 1 \\ C^*(z, 0) &= 0 \end{aligned} \tag{7.32}$$

where $K = (K_1\alpha_1 + K_2(1 - \alpha_1))C_0$,

$C^* = C_{aq}/C_0$,

$z^* = z/L$,

$\theta^* = \dfrac{\theta \cdot u_s}{L}\dfrac{C_0}{\rho_b f[RNH^+]_T}$,

$= t - z \cdot \varepsilon/u_s$,

C_0 is the feed concentration of total copper, C_{aq} is the total copper concentration in the solution, L is the height of the packed column and α_1 is $[Cu(CN)_2^-]/C_{aq}$. The differential equation with the given boundary conditions is solved numerically using a finite difference scheme [5].

The results of the breakthrough curve experiments and the model computation are shown in Figure 7.25. The experimental result shows that the adsorption of copper cyanide complexes on SG(1)-TEPA-propyl is controlled by equilibrium because the slope of the curve at the breakthrough point is a sharp, step-like response. This response suggests that there are no film resistances at the particle surface, fast diffusion of copper cyanide complexes through the pores, and fast adsorption reaction on the functional sites. Hence, the developed equilibrium model well represents the experimental data since it is developed based on the assumption of equilibrium

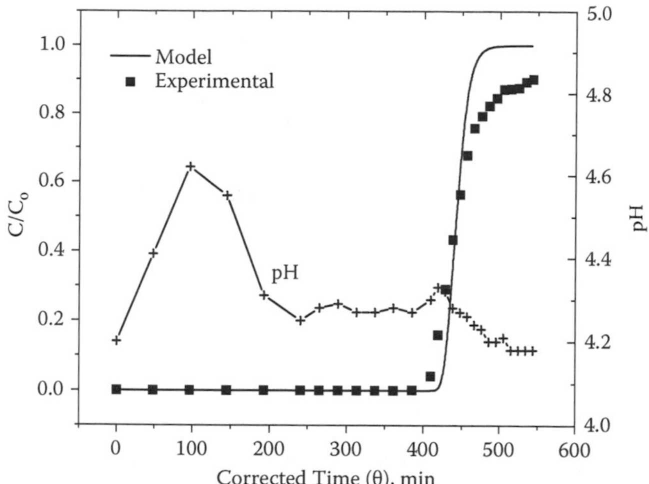

FIGURE 7.25 Breakthrough curve of copper cyanide complexes on SG(1)-TEPA-propyl at pH = 4.1. (Reproduced from J. S. Lee, N. V. Deorkar, and L. L. Tavlarides. *Ind. Eng. Chem. Res.* 37:2812–2820, 1998. With permissio.)

between the adsorbed copper cyanide complexes and the free copper cyanide complexes in the bulk solution.

7.7 APPLICATIONS

7.7.1 Antimony Separation from Copper Electrorefining Solutions

Antimony and its compounds are listed among the most toxic elements of priority pollutants by the U.S. Environmental Protection Agency. Antimony exists in several industrial and mining wastes, such as chemical and allied products, glass products, electrical and electronic equipment, lead acid storage batteries, and copper electrorefining solutions.

The solid extractant immobilized with pyrogallol moiety (ICAA-PPG) shown in Figure 7.26 has been applied for the separation of antimony from copper electrorefining solutions. The ICAA-PPG material has 43.1 mg/gm of antimony saturation capacity at pH 6. This extractant can selectively remove antimony from the aqueous solution containing other metal ions at pH less than 2 as shown in Figure 7.18. Moreover, antimony can be selectively separated after loading with other metal ions on the extractant by subsequent stripping with different stripping solutions as shown in Figure 7.27. Further details of this application can be found elsewhere [33].

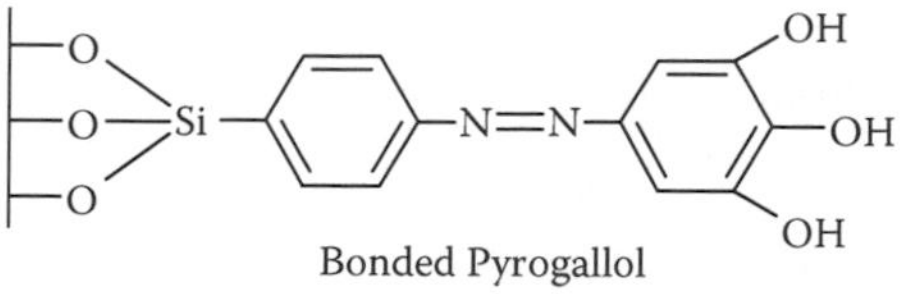

O
O—Si—N=N—OH
O
OH
OH
Bonded Pyrogallol

FIGURE 7.26 Molecular structure of bonded pyrogallol(ICAA-PPG).

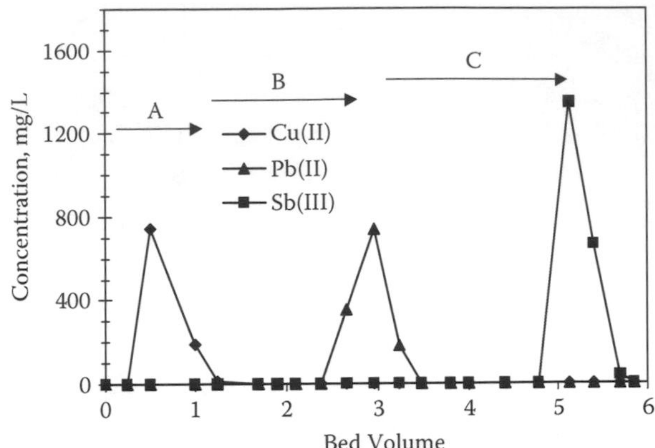

FIGURE 7.27 Separation of copper, lead and antimony(III) by selective stripping from ICAA-PPG bed: stripping solution A = 0.5 M sulfuric acid; stripping solution B = 0.5 M hydrochloric acid; stripping solution C = 4 M hydrochloric acid with 0.05 M potassium hydrogen tartrate. (From N. V. Deorkar and L. L. Tavlarides. *Hydrometallurgy* 46:121–135, 1997. With permission.)

7.7.2 Zinc, Cadmium, and Lead Separation

For the removal of zinc, cadmium, and lead from aqueous waste streams, an inorganic solid-phase extractant (ISPE) was synthesized in our laboratory by the method of ligand deposition on hydrophobized silica-gels. The functional ligands deposited on the silica are bis(2,4,4-trimethylpentyl)monothiophosphinic acid (Cyanex-302, American Cyanamide Co.). The synthesized material showed a selectivity series of $Cd^{2+} > Pb^{2+} > Zn^{2+}$. A breakthrough curve experiment for the solution containing Cd^{2+}, Pb^{2+}, and Zn^{2+} ions was performed to show the selective extraction of metal ions as shown in Figure 7.28. The breakthrough of zinc ion occurred at 61 bed volumes although both lead and cadmium continued to be removed to less than 8.0×10^{-6} mol/dm^3. After 70 bed volumes the concentration of zinc in the effluent is greater than the feed concentration, whereas lead and cadmium are being adsorbed. The increase of the zinc concentration in the effluent is due to displacement of zinc by the Cd^{2+} and Pb^{2+} ions, which have higher affinities for the active sites than zinc. Similar behavior for displacement of adsorbed lead after 104 bed volumes shows that cadmium has the greatest affinity for ISPE-302.

The adsorbed metals can be recovered by selective stripping similar to the previous example for antimony (III) separation. Here, adsorbed lead, zinc, and cadmium are shown to be selectively stripped using 0.1 M and 1.0 M nitric acid and 2.0 M hydrochloric acid, respectively. Further details on application can be found elsewhere [34].

7.7.3 Berkeley Pit Problem

The Berkeley Pit is a vast open-pit mine located in an ore-rich section of southwestern Montana, which was closed in the 1980s and has been targeted by the U.S. Department of Energy as a "demonstration site" for environmental reclamation. The Berkeley Pit is filled with some 17 billion gallons of water which contains various types

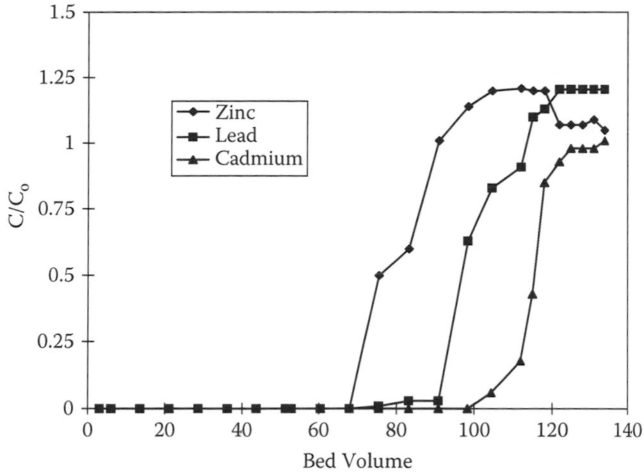

FIGURE 7.28 Simultaneous removal of zinc, lead and cadmium with ICAA-302; flow rate = 1.0 - 1.2 cm^3/min; pH = 6.3; [Zn^{2+}] = 55 mg/L; [Cd^{2+}] = 48 mg/L; [Pb^{2+}] = 44 mg/L. (Reproduced from J. S. Lee, N. V. Deokar, and L. L. Tavlarides. *Ind. Eng. Chem. Res.* 37, 2812–2820, 1998. With permission.)

of metal ions, including iron, copper, zinc, manganese, magnesium, and aluminum at considerably high concentrations. A typical composition of the Berkeley Pit water is shown in Table 7.6 [35]. Metal separation from the solution is necessary from an environmental aspect, and also one may recover useful metals economically. A process has been proposed to separate these metals by using various solid extractants prepared in our laboratory.

TABLE 7.6
Target Metals and Other Constituents in Berkeley Pit Water

Ions	Concentration mg/L	Other Constituents	Value/Concentration
Aluminum	2.60	pH	2.85
Cadmium	2.14	[Fe^{3+}]/[Fe]	0.15
Calcium	456		
Copper	172		
Iron	1068		
Lead	0.031		
Magnesium	409		
Manganese	185		
Sodium	76.5		
Zinc	550		
Nitrate	<1.0 (as N)		
Sulfate	7,600		

Source: From Resource Recovery Project Technology Demonstrations, Series IV, A60488, MSE Technology Applications, Inc., Butte, MT, pp. 1C–7C, March 1, 1996. (With permission.)

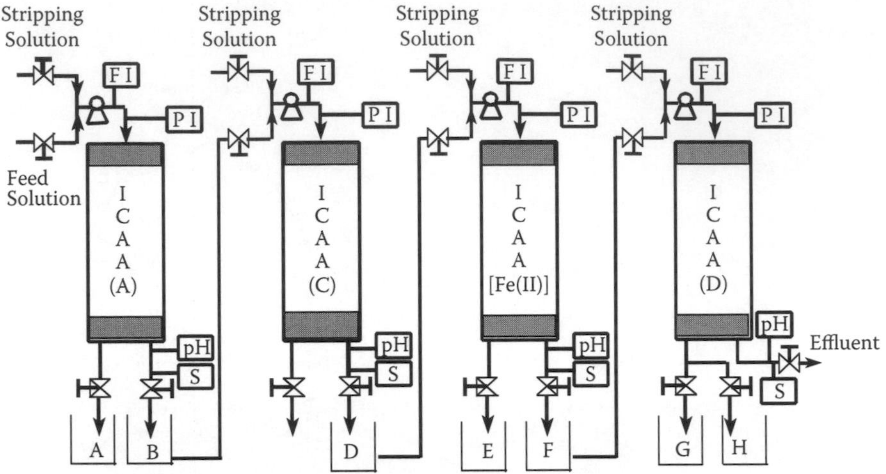

FI: Flow Indicator; PI: Pressure Indicator; pH: pH Probe; S: Sample Port

⊢⊠: Solonoid Valve

Tank A: Concentrated Fe(III); B: Bed 1 Effluent; C: Concentrated Cu(II); D: Bed 2 Effluent; E: Concentrated Fe(II); F: Bed 3 Effluent; G: Concentrated Zn/Cd; H: Concentrated Lead

FIGURE 7.29 Integrated adsorption process for recovery of iron, copper, zinc, lead and cadmium from Berkeley pit waters. (Reproduced from N. V. Deorkar, L. L. Tavlarides. *Environ. Progress* 17:120–125, 1998. With permission.)

One proposed scheme is to use three different adsorbents shown in Table 7.7 for a train of four columns for the separation of the major components. A schematic of the process is shown in Figure 7.29. Here column A selectively removes Fe(III) as shown in Table 7.7. The second column with ICAA-C selectively removes Cu(II)

TABLE 7.7
Adsorption of Metal Ions from Simulated Berkeley Pit Water

ICAAs Material	pH	% Extraction					
		Fe(III)	Cu(II)	Zn(II)	Cd(II)	Pb(II)	Fe(II)
ICAA-A	2.5–2.7	98.7[a]	6.4[a]	ND[a]	ND[a]	ND[a]	5.1[c]
		NA[b]	NA[b]	5.6[b]	5.6[b]	34.0[b]	NA[b]
ICAA-C	2.5–2.7	43.7[a]	93.1[a]	ND[a]	ND[a]	ND[a]	11.7[c]
		NA[b]	NA[b]	ND[b]	ND[b]	ND[b]	NA[b]
ICAA-D	2.5–2.7	NA[b]	NA[b]	99.6[b]	99.8[b]	99.7[b]	98.7[c]

Notes: ND: not detectable; NA: not applicable.

[a]Feed solution composition: [Fe(III)] = 200 ppm; [Cu(II)] = 200 ppm; [Zn(II)] = 200 ppm; [Cd(II)] = 3.8 ppm; [Pb(II)] = 4.1 ppm; [SO42–] = 8000 ppm; pH = 2.5.

[b]Feed solution composition: [Zn(II)] = 200 ppm; [Cd(II)] = 3.8 ppm; [Pb(II)] = 4.1 ppm; [SO42–] = 8000 ppm; pH = 2.5.

[c][Fe(II)] = 200 ppm; [SO42–] = 8000 ppm; pH = 2.5.

Source: From N. V. Deorkar and L. L. Tavlarides. *Environ. Progress* 17:120–125, 1998. (With permission.)

and some Fe(II). In order to separate Fe(II) and Cd(II), Pb(II) and Zn(II), the third column with ICAA-D will remove Fe(II) preferentially over the other three metals. Other adsorbents for the removal of Fe(II) have been suggested; however, such developments have not been conducted as of yet. The final column, also with ICAA-D can remove Cd(II), Pb(II), and Zn(II). A major additional benefit with ICAA-D is that the pH of the solution is increased to approximately 8–10, providing a nonacidic water for release to the environment.

GLOSSARY OF SYMBOLS

a Specific surface area (area per unit mass of solid extractant)
a_p Particle radius
C_i Concentration of metal ion species i
$C_{i,b}$, $c_{i,s}$ Metal ion concentration of species, i in the bulk of the fluid and that at the fluid-pellet interface
c_i, q_i Metal ion concentration in the pore fluid and adsorbed phase, respectively
D_M Molecular diffusivity in the bulk
D_p Diffusivity of the metal ion in the pore $[D_p = \varepsilon_p D_M/\tau]$
E_r, E_z Radial and axial dispersion coefficients
k_f Interphase (or external) mass transfer coefficient.
$\bar{q}$ Average adsorbed-phase concentration (on a mass basis)
t, r, z Independent variables of time, radial distance, and axial distance, respectively
u_s Superficial velocity
ε Fixed-bed porosity
ε_p Particle porosity
ρ_p Particle density
τ Tortuosity factor

ACKNOWLEDGMENTS

The able assistance of our research member, Kwan H. Nam, in preparation of this manuscript is acknowledged and gratefully appreciated. The financial support of the National Science Foundation through Grant CTS-9805118 is gratefully acknowledged.

REFERENCES

1. C. Kantipuly, S. Katragadda, A. Chaw, D. H. Gesser. *Talanta,* 37:491–517, 1990.
2. G. R. Pearson. *J. Am. Chem. Soc.,* 85:3533–3539, 1963.
3. A. R. Khisamutdinov, G. N. Afzaletdinova, I. Yu. Murinov, V. E. Vasil'eva. *Russian J. Inorg. Chem.,* 42:1560–1566, 1997.
4. G. Chessa, G. Marangoni, B. Pitteri, N. Stevanato, A. Vavasori. *Reactive Polymers,* 14:143–150, 1991.
5. J. S. Lee. Adsorption of Copper Cyanide on Inorganic Chemically Active Adsorbents, M.S. Thesis, Syracuse University, 1997.
6. S. J. Al-Bazi, A. Chow. *Talanta*, 31:815–836, 1984.
7. S. D. Alexandratos, A. W. Trochimczuk, D. W. Crick, E. P. Horwitz, R. C. Gatrone, R. Chiarizia. *Macromolecules,* 29:1021–1026, 1996.

8. F. B. Smith, W. T. Robison, D. G. Jarvinen. Metal-Ion Separation and Pre-concentration-Progress and Opportunities, ACS Symposium Series 716, 294–330, 1999.
9. D. G. Jarvinen, F. B. Smith, W. T. Robinson. Metal Separation Technologies Beyond 2000: Integrating Novel Chemistry with Processing, Proceedings of a symposium sponsored by the Engineering Foundation Conference and National Science Foundation, The Minerals, Metals & Materials Society, Hawaii, June 13–18, 1999, pp. 131–138.
10. N. V. Deorkar, L. L. Tavlarides. *Emerging Separation Technologies for Metals II,* The Minerals, Metals & Materials Society, Warrendale, PA, 1996.
11. L. L. Tavlarides, N. V. Deorkar. Ceramic Compositions with a Hydroxyquinoline Moiety, U.S. Patent No. 5,668,079, 1997.
12. L. L. Tavlarides, N. V. Deorkar. Ceramic Compositions with Thio and Amine Moiety, U.S. Patent No. 5,616,533, 1997.
13. L. L. Tavlarides, N. V. Deorkar. Ceramic Compositions with a Pyrogallol Moiety, U.S. Patent No. 5,624,881, 1997.
14. L. L. Tavlarides, N. V. Deorkar. Ceramic Compositions with a Phospho-acid Moiety, U.S. Patent No. 5,612,175, 1997.
15. X. Feng, G. E. Fryxell, L.-Q. Wang, A. Y. Kim, J. Liu, K. M. Kemner. *Science*, 276:923–926, 1997.
16. L. Mercier, J. T. Pinnavaia. *Environ. Sci. Technol.,* 32:2749–2754, 1998.
17. S. Mattigod, G. E. Fryxell, X. Feng, J. Liu. Metal Separation Technologies Beyond 2000: Integrating Novel Chemistry with Processing, Proceedings of a symposium sponsored by the Engineering Foundation Conference and National Science Foundation, The Minerals, Metals & Materials Society, Hawaii, 1999, pp. 71–79.
18. E. M. Rabinovich. *J. Mater. Sci.,* 20:4259–4297, 1985.
19. J. Boilot, F. Chaput, J. Galaup, A. Veret-Lemarinier, D. Riehl, Y. Levy. *AIChE Journal,* 43:2820–2826, 1997.
20. B. M. De Witte, D. Commers, J. B. Uytterhoeven. *J. Non-Cryst. Solids,* 202:35–41, 1996.
21. S. Parbakar, R. A. Assink. *J. Non-Cryst. Solids,* 211:39–48, 1997.
22. J. S. Lee, S. Gomez-Salazar, L. L. Tavlarides. *Reactive and Functional Polymers,* submitted (November, 2000).
23. R. Deshpande, D. Hua, M. D. Smith, C. J. Brinker. *J. Non-Cryst. Solids,* 144:32–44, 1992.
24. A. H. Boonstra, T. N. M. Bernards. *J. Non-Cryst. Solids,* 108:249–259, 1989.
25. J. S. Lee, N. V. Deorkar, L. L. Tavlarides. Metal Separation Technologies Beyond 2000: Integrating Novel Chemistry with Processing, Proceedings of a symposium sponsored by the Engineering Foundation Conference and National Science Foundation, The Minerals, Metals & Materials Society, Hawaii, 1999, pp. 43–52.
26. B. V. Zhmud, J. Sonnefeld. *J. Non-Cryst. Solids,* 195:16–27, 1996.
27. H. Opfermann, W. Lugwig, G. Wilke, Beitr. FSU Jena, Thermische Analysen Verfahren in Industrie und Forschung, 159, 1983.
28. N. V. Deorkar, L. L. Tavlarides. *Environmental Progress,* 17:120–125, 1998.
29. C. Tien. *Adsorption Calculations and Modeling,* Butterworth-Heinemann Series in Chemical Engineering, Butterworth–Heinemann, Boston, 1994.
30. J. S. Lee, N. V. Deorkar, L. L. Tavlarides. *Ind. Eng. Chem. Res.,* 37:2812–2820, 1998.
31. S. Yacuomi, C. Tien. *Kinetics of Metal Ion Adsorption from Aqueous Solutions: Models, Algorithms, and Applications,* Kluwer Academic Publishers, Boston, 1995.
32. L. T. Biegler, J. E. Cuthrell. *Comput. Chem. Engng.,* 9:257–267, 1985.
33. N. V. Deorkar, L. L. Tavlarides. *Hydrometallurgy,* 46:121–135, 1997.
34. N. V. Deorkar, L. L. Tavlarides. *Ind. Eng. Chem. Res.,* 36:399–406, 1997.
35. Resource Recovery Project Technology Demonstrations, Series IV, A60488, MSE Technology Applications, Inc., Butte, MT, pp.1C–7, March 1, 1996.

8 Solid Polymeric Extractants (TVEX)

Synthesis, Extraction Characterization, and Applications for Metal Extraction Processes

Vadim Korovin, Yuri Shestak, Yuri Pogorelov, and José Luis Cortina

CONTENTS

8.1 INTRODUCTION

Solvent extraction and ion exchange are widely used for metal recovery and separation. However, increasing demands to environmental protection as well as the need to optimize a wide range of industrial processes facilitate the development of advanced separation techniques. For this reason impregnation of selective extractants onto polymer beads was proposed to "fill" the gap between the aforementioned techniques. These materials are produced either by impregnation of the porous carrier by liquid extractants (solvent impregnated resins) [1] or by polymerization of styrene and divinylbenzene in the presence of the extractant. In this second case these materials have been commercialized by Bayer AG as levextrel-type resins [2,3] and as solid extractants (TVEX, the first letters from Russian words that represent *solid extractants*) in the former USSR [4,5]. TVEX combine the advantages of solvent extraction and ion exchange in the following ways:

- They differ from impregnated resins by optimal well-developed meso- and macroporous structure formed during synthesis, which provides high diffusion rate of metal ions into granules.
- They differ from ion-exchange resins by production simplicity, low cost defined mostly by extractant cost, frost resistance, and absence of special procedures for storage and transportation; they have relatively low swelling coefficient as compared with ion-exchange resins, which increases TVEX sorption capacity and improves technological parameters; their low density allows effective TVEX separation to be performed from pulp and recovery processes to be carried out in countercurrent pulsating continuous technology.
- They differ from solvent extraction in their fire safety due to absence of solvent, in the possibility to extract aimed components directly from complicated pulps, in the absence of organic phase emulsification, in the reduction of extractant losses, and in the simplicity of equipment.

This chapter provides a description of material synthesis, characterization of the metal extraction reactions involved, and the applications of TVEX materials for metal recovery.

8.2 SYNTHESIS OF SOLID POLYMERIC EXTRACTANTS

TVEX are macroporous styrene/divinylbenzene copolymers containing a metal selective extractant that has been added directly to the mixture of the monomers during the bead polymerization process. The amount of extractant and degree of cross-linking in the final bead product may be varied at will. The conditions of polymerization can be varied, thus compensating for any differences in the properties of the individual extractants. Generally, by this method optimum conditions can

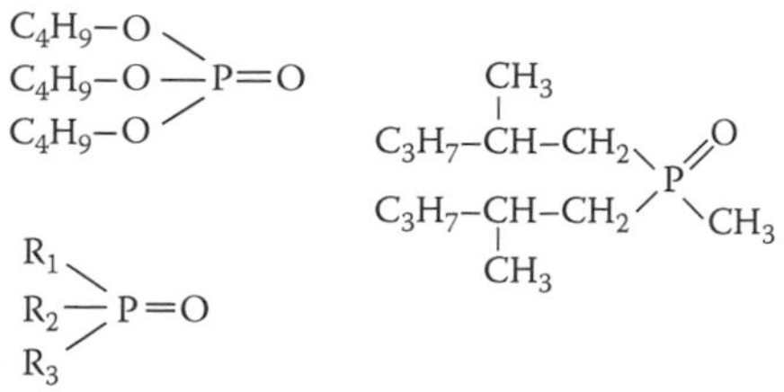

$C_4H_9{-}O$
$C_4H_9{-}O{-}P{=}O$
$C_4H_9{-}O$

CH_3
$C_3H_7{-}CH{-}CH_2$ P(=O)
$C_3H_7{-}CH{-}CH_2$ CH_3
CH_3

R_1
$R_2{-}P{=}O$
R_3

FIGURE 8.1 (a) TBP: tri-butylphosphate, (b) DIOMP: di-isooctyl methyl phosphonate, and (c) POR: phosphine oxide with different radicals; R1–R3 are alkyl radicals, 85 to 90% C8H17, 10 to 15% from C5H11 to C8H17.

be achieved to incorporate the extractant concerned, which is retained in the resin structure rather than by chemical bonding.

TVEX synthesis is carried out in an agitated water-jacket reactor [4,5]. Once the reactor is loaded with a 0.5% starch solution as an emulsion stabilizer, the reaction mixture containing styrene, divinylbenzene (cross-linking agent), the extractant, and benzoyl peroxide as initiator of radical copolymerization is fed into the reactor at 50^0C. After 8 hours of polymerization, TVEX granules are obtained and then are washed with water and size classified.

Although different extractants can be incorporated as active components during the resin synthesis process, TVEX produced are mainly under the basis of neutral and acidic organophosphorus compounds, their synergic mixtures, quaternary ammonium salts, and crown-etchers.

TVEX industrial batches are nowadays produced based on the following extractants: tri-butylphosphate (TBP), diisooctyl methyl phosphonate (DIOMP), and phosphine oxide (POR) (with different radicals $R_1(R_2)(R_3)P = O$; $R_1 - R_3$ are alkyl radicals, 85 to 90% C_8H_{17}, 10 to 15 % $C_5H_{11} - C_8H_{17}$) (Figure 8.1). These extractants act as a base due to its phosphoryl group (P = O). The alkoxy oxygenes do not play any direct role on the metal extraction process.

8.3 PHYSICOCHEMICAL PROPERTIES OF SOLID EXTRACTANTS

The ideal TVEX material should have a polymer skeleton rigidity to maintain the resin macroreticular character and constant pore volume in different external media. On the macroporous level the resin must demonstrate the combination of high pore volume and high specific surface area. This means that the resin has to contain only very narrow pores. This is significant because leakage of the extractant from narrower pores could be lower than from wider pores. In addition, a higher amount of liquid extractant will be in immediate contact with the resin pore walls to promote the effective extractant transfer between the pore space and the polymeric gel. However, any increase of the surface area of the macroreticular resin could be achieved only by an increase in the rigidity of the polymer skeleton, thus leading to a decrease in the ability of the polymer mass to swell and to accommodate extractant molecules. Selection of a perfect relationship between flexibility and rigidity of the polymeric support will depend on the type of the liquid extractant used in the TVEX preparation.

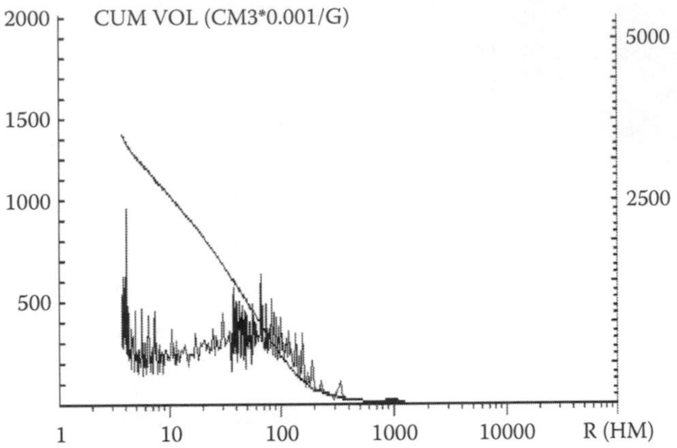

FIGURE 8.2 Integral and differential curves of pore volume distribution on radiuses for the matrix of TVEX-50%TBP·25%DVB.

The morphological characterization of TVEX containing neutral organophosphorus extractants (Figure 8.2 and Figure 8.3) indicates the presence of a mesoporous structure [6,7]. The presence of a mesoporous structure will provide better mass transfer properties. Porous matrix formation of TVEX containing tri-butylphosphate (TVEX-TBP) depends on the extractant content and cross-linking degree [6]. The pore specific surface area (S), total pore volume (V) and average pore radius (r) increase monotonously at constant divinylbenzene (DVB) content (10 and 30%) except for dependence of specific surface at 30% DVB content that has maximum value at TBP content of 40 to 60% (Figure 8.4a,b,c). Formation of the porous structure (appearance of pores with effective radius more than 5 nm) occurs at DVB content more than 10% and TBP content less than 30% in initial mixture of monomers (Figures 8.4a,b,c). At constant extractant content, DVB forms cross-links between

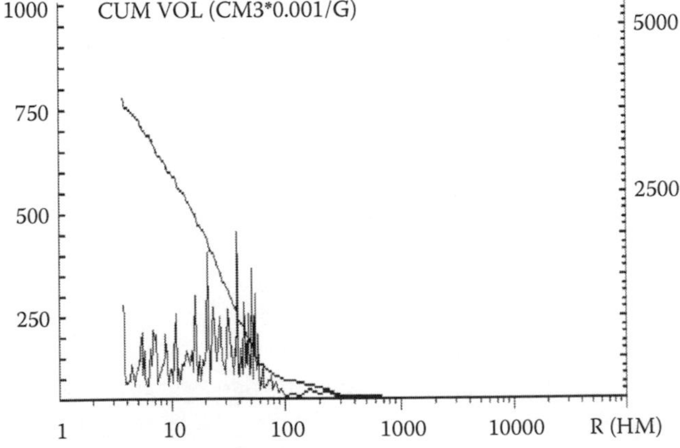

FIGURE 8.3 Integral and differential curves of pore volume distribution on radiuses for the matrix of TVEX-50%DIOMP·25%DVB.

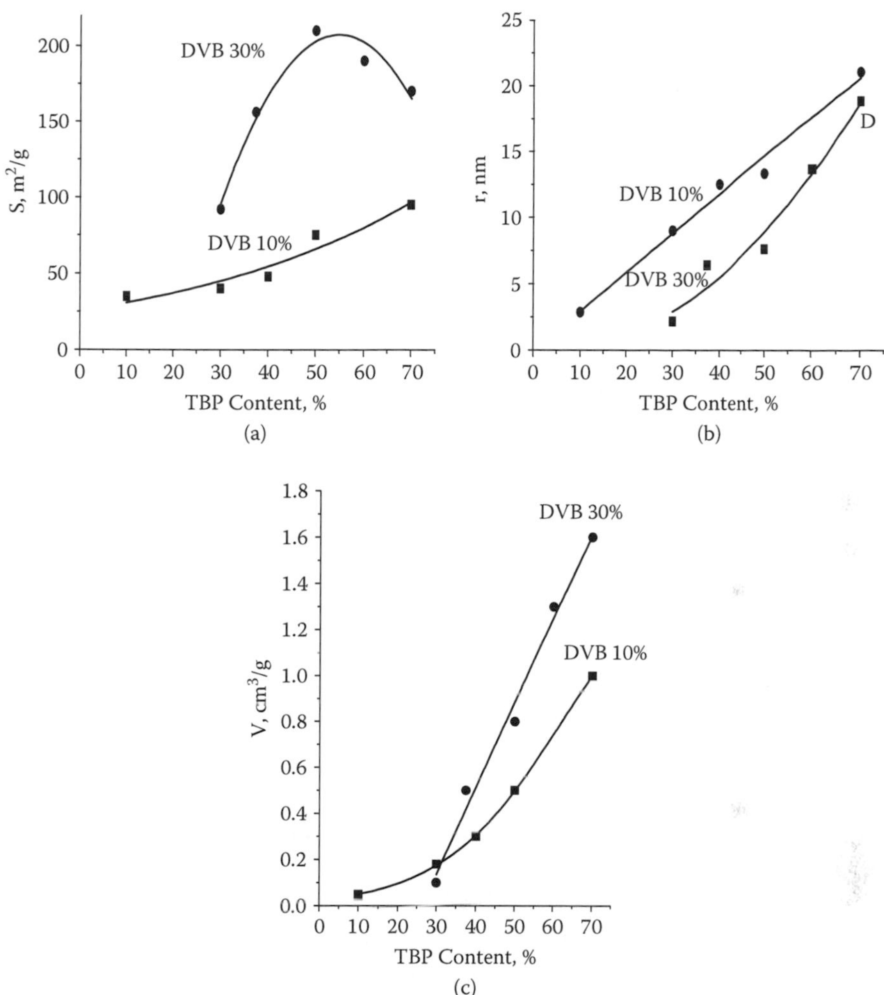

FIGURE 8.4 Specific surface area (a), total pore volume (b), and average effective pore radius (c) of TVEX-TBP matrix depending on extractant content.

styrene linear chains, increasing specific surface area and total pore volume and decreasing the average effective pore radius (Figure 8.5).

The pore specific surface area an the average radius values could be used to determined the fractal properties of TVEX resins following the mathematical models describing the disorder and polydispersity of materials. The fractal dimension D_f, defines self-similarity of porous materials [8]. It was calculated based on differential function of pore volume distribution of radius [9]:

$$D_f = 2 - \frac{d[\ln S(r)]}{d\ln(r)} \tag{8.1}$$

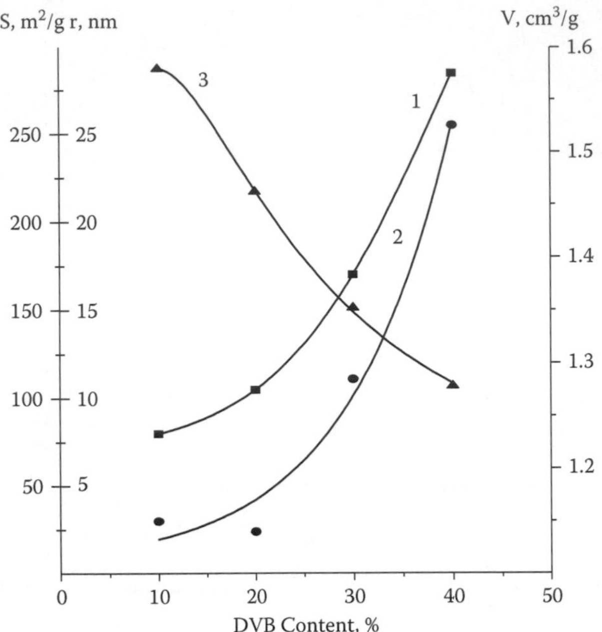

FIGURE 8.5 Specific surface area (1), total pore volume (2), and average effective pore radius (3) of TVEX-TBP matrix depending on divinylbenzene concentration.

where ln $S(r)$ is natural logarithm of specific pore area for TVEX-TBP matrix and ln (r) is a natural logarithm of pore radius.

Fractal dimension obtained by linear regression of the function ln $S(r)$ versus ln (r) (Figure 8.6) leads to a D_f value of 2.9 ± 0.1, typical for polycondensated systems with corpuscular structure [8]. Linearity is kept within the range 3.7 to 40 nm

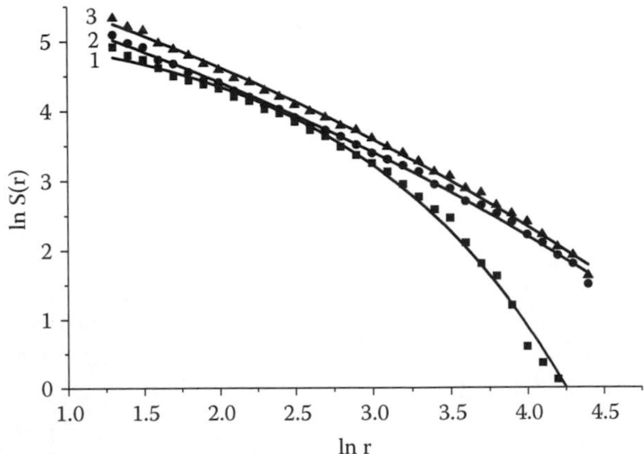

FIGURE 8.6 Dependence of pore specific surface area on pore radius for TVEX with following TBP/DVB content, %: 1–50/10; 2–50/25; 3–50/45.

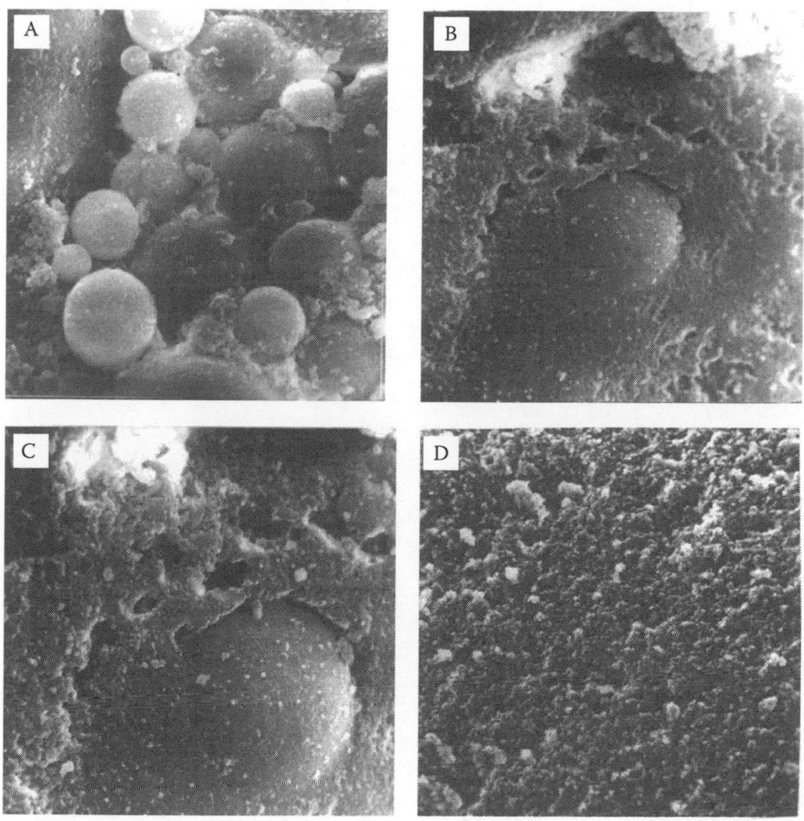

FIGURE 8.7 Scanning Electron Microscopy (SEM) microphotography of surface of TVX-TBP granules at different magnification (a) × 375, (b) × 750, (c) × 1250, and (d) × 2750.

for TVEX-TBP(50)DVB(45) and TVEX-TBP(50)DVB(25) and within the range 3.7 to 16 nm for TVEX-TBP(50)DVB(10), indicating fractality of the studied objects within these intervals. Thus, TVEX-TBP macrostructure is formed by aggregation of large number of individual clusters. This type of structure formation leads to approximately constant density of space packing by corpuscles or their aggregates corresponding to $D_f = 3$. The space between corpuscles or their aggregates are TVEX matrix pores filled by the extractant.

Scanning electron microscopy analysis of solid extractants (Figure 8.7a,b,c,d) shown TVEX-TBP granules consist of a large amount of regular spherical granules sizing from 2.5 to 100 μm. Granules are interconnected through an amorphous material and granule, in turn, is composed of small copolymer, interconnected globules. The amorphous material (Figure 8.7d) is composed by fine copolymer globules that did not form larger spherical granules. The space between globules is filled by tri-butylphosphate.

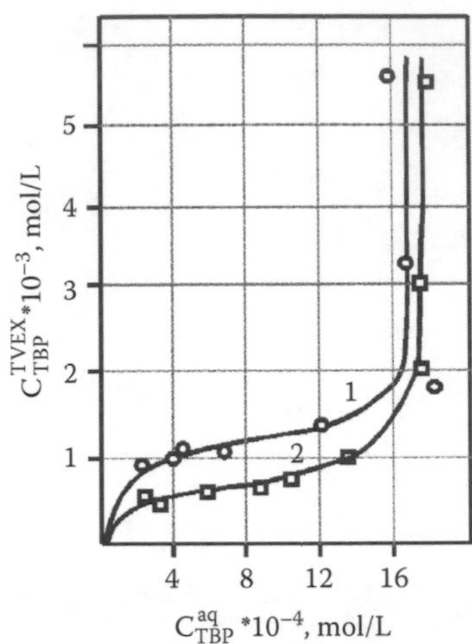

FIGURE 8.8 Equilibrium TBP distribution between water and TVEX–50%TBP (1) and TVEX-65%TBP (2).

8.4 EXTRACTANT STATE AND ACTIVITY ON SOLID POLYMERIC ADSORBENTS

Activity of the extractant molecules on impregnated materials is critical to predict and describe the interactions occurring at solid–liquid interfaces due to their technological implications. Although the fixation of metal cations on impregnated resins has been extensively investigated, little has been done to characterize the physicochemical interactions between the extractant and the solid support. The characterization of the extractant activity could be achieved by using chemical methods or by using spectroscopic techniques as nuclear magnetic resonance (NMR) spectroscopy [6,7].

Chemical methods based on the study of extractant distribution (TBP) on TVEX-TBP resins were used to demonstrate that TBP is adsorbed by the TVEX porous matrix by physical adsorption. Extractant wash-out from the TVEX matrix is an important characteristic for metal extraction applications of solid adsorbents as TVEX since it may influence the physical-chemical properties of the extraction process. Infrared (IR) spectroscopy analysis of TVEX-TPB resins show the absence of shift for stretching frequency of the P = O group in comparison to the stretching frequency assignment for liquid TBP as an indication that TBP is hold by matrix surface due to physical adsorption as was confirmed from the by enthalpy values of TBP distribution from the TVEX matrix of 45.0 ± 0.5 kJ/mol [10].

Equilibrium washout of tri-butylphosphate between the aqueous phase and the TVEX matrix (Figure 8.8) shows S-shaped curves typical for physical adsorption with formation of an extractant polymolecular adsorbed layer. The increase of temperature leads to a decrease of equilibrium distribution (Figure 8.9) due to the decrease of TBP solubility in water with temperature increase. Thus, TBP losses from TVEX matrix are small and are caused by tri-butylphosphate solubility in the aqueous phase, depending on solution acidity, ionic strength, and temperature. For example, during the pilot tests, TBP leakage from TVEX-TBP resins after 100

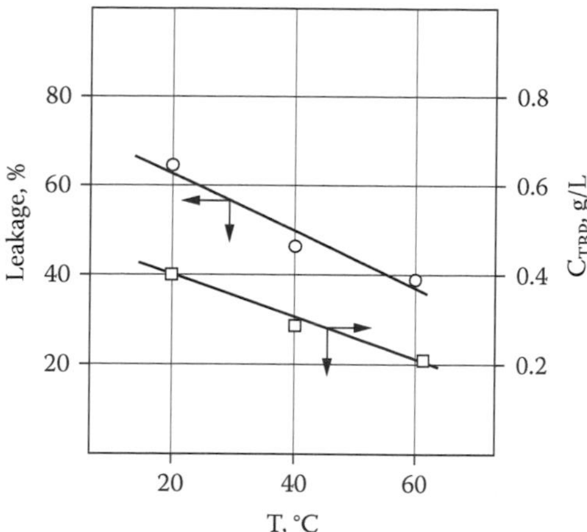

FIGURE 8.9 Influence of temperature on TBP leakage and concentration in water.

operation cycles of Sc extraction was less than 10% [7]. Extractant distribution rates are compared with rates measured for metal extraction reactions for TBP containing TVEX resins, and the diffusion coefficients ranged from 1.01×10^{-12} $m^2.s^{-1}$ at 298 K, 1.01×10^{-12} $m^2.s^{-1}$ at 298 K and 1.01×10^{-12} $m^2.s^{-1}$ at 298 K.

^{31}P NMR spectroscopy was used as a nondestructive method to determine TBP activity on TVEX-TBP materials by comparison with activity of TBP in pure samples. The signal width of TBP in TVEX matrix is only 1.5 to 2 times broader as compared with signal width for liquid tri-butylphosphate, indicating the drop-liquid state of extractant (TBP) in matrix pores. TBP signal width increases at extractant partial washout by acetone due to the increase of portion of extractant adsorbed by TVEX matrix; the total washout leads to the extractant signal disappearance. However, according to chemical analysis, TVEX matrix still contains 2 to 5% TBP after complete extractant washout. This fact indicates that tri-butylphosphate is dissolved partially in TVEX matrix and acts as plasticizer of styrene-divinylbenzene copolymer.

The extractant state in TVEX matrix could be determined by measuring longitudinal relaxation of ^{13}C {^{1}H} nuclei of the extractant molecules by using ^{13}C and ^{1}H NMR spectroscopy [11]. Unlike liquid TBP, high-resolved ^{1}H NMR spectra of TVEX-50%TBP contain two broad, partially overlapped signals. Their position and integral intensity ratio (2:7) allow attributing the less intensive signal to CH_2-1 group, another more intensive—to CH_2-2, CH_2-3 and CH_3-4 groups. Calculated proton longitudinal relaxation times increase with temperature increase with a maximum value at 359 K for liquid TBP (Figure 8.10). In these conditions the activation energy of dipole–dipole interaction for H atoms of liquid TBP and TVEX-50%TBP was estimated to range between 9590 and 10650 J/mol, respectively, as is given in Table 8.1 [12].

^{13}C {^{1}H} NMR spectrum of liquid TBP contains four resolved signals assigned to four nonequivalent carbon atoms in TBP molecule. ^{13}C {^{1}H} NMR spectrum of both TBP in pure TPB and TVEX matrix also contain four resolved broadened signals. The spectrum differs from ^{13}C {^{1}H} spectrum of 100% TBP only by signal width caused probably by nonuniformity of the magnetic field in TVEX.

TABLE 8.1
Activation Energy of Dipole–Dipole ($E_{c(dd)}$) and Spin-Rotational ($E_{c(sr)}$) H Atom Interaction for Liquid TBP and Supported on TVEX (50%TBP)

H Atom Group	TVEX-50%TBP $E_{c(dd)}$, J/mol	TBP $E_{c(dd)}$ (J/mol)	TBP $E_{c(sr)}$ (J/mol)
CH_2-1	10,625	9,592	−11,661
CH_2-2, 3		8,002	−1,448
CH_3-4	10,672	7,029	−18,309

The temperature dependence of longitudinal relaxation time ^1H, ^{13}C atoms in TBP molecule shows (Figures 8.10 and 8.11) that longitudinal relaxation time is lower for TBP molecule in TVEX matrix as compared with liquid TBP. This phenomenon is typical for a heterogeneous system and is caused by the partial adsorption of TBP molecules by the TVEX matrix [13]. Dependence of longitudinal relaxation time of ^{13}C nuclei for liquid TBP with temperature has a typical shape (Figure 8.11) for a liquid state where dipole–dipole and spin-rotational relaxation mechanisms prevail. Unlike liquid TBP, dependence of spin-rotational relaxation time with temperature for carbon atoms C-1, C-2, C-3 of TBP molecule in the TVEX matrix linearly increases with temperature increase with the exception of the C-4 carbon atom. Longitudinal relaxation time of carbon atoms of TBP are collected in Table 8.2 for both liquid TBP and TVEX-TBP resins. Based on the longitudinal relaxation times

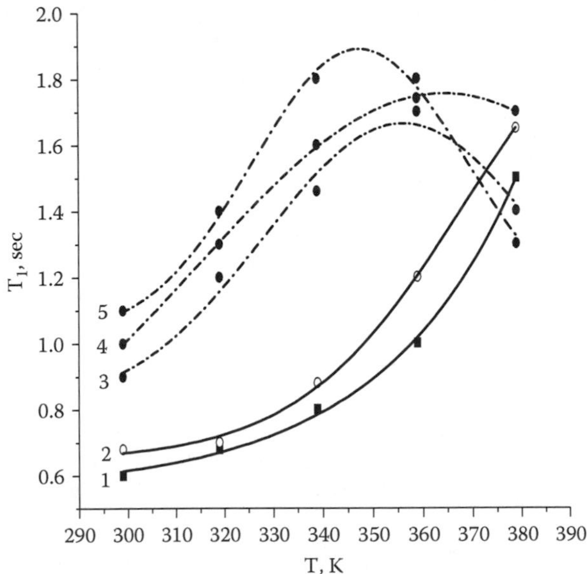

FIGURE 8.10 Temperature dependence of H atom longitudinal relaxation time for TVEX-TBP (1–groups CH2 -1, 2, 3; 2–groups CH3-4), and TBP (3–group CH2-1; 4–groups CH2-2, 3; 5–group CH3-4).

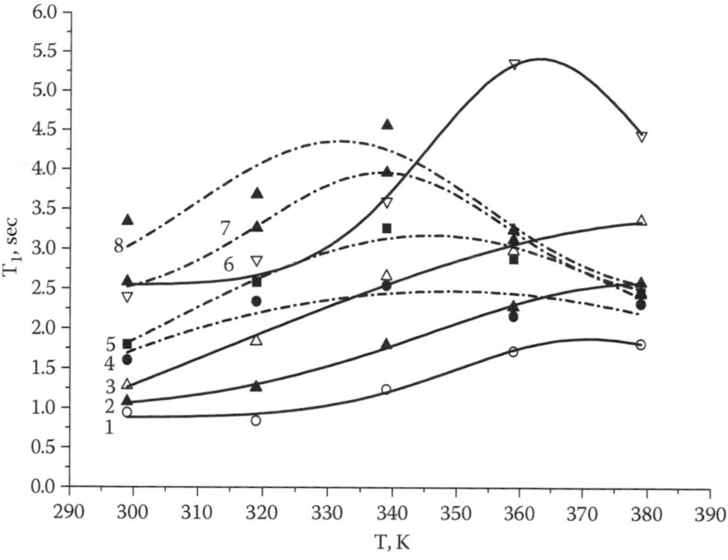

FIGURE 8.11 Temperature dependence of longitudinal relaxation time of carbon atoms for TVEX-TBP (1–C-1; 2–C-2; 3–C-3; 6–C-4) and TBP (4–C-1; 5–C-2; 7–C-3; 8–C-4).

TABLE 8.2
Longitudinal Relaxation Time of Carbon Atoms for Tri-Butylphosphate (TBP) in Liquid and Supported on TVEX (50%TBP)

	Longitudinal Relaxation Time (sec) Carbon Atom			
System	C-1	C-2	C-3	C-4
100% TBP	1.59	1.75	2.57	3.30
TVEX-TBP	0.98	1.12	1.31	2.38
	Relative Value of Longitudinal Relaxation Time TVEX-TBP/(100%TBP)			
	60%	64%	51%	72%

the lower ratio C-1 and C-3 atoms as compared with C-2 and C-4 atoms indicates that adsorption of tri-butylphosphate molecules occurs at the expense of odd carbon atoms of butyl radicals.

The activation energy of dipole–dipole $E_{c(dd)}$ and spin-rotational $E_{c(sr)}$ interaction between carbon atoms for both liquid TBP and supported TVEX resins are collected in Table 8.3. The lower difference on the activation energy of spin-rotational interaction $E_{c(sr)}$ for the C-4 atom of TBP in TVEX matrix compared with C-4 atom of 100% TBP is connected with effect of polymer matrix on mobility of extractant butyl chains. Increase of activation energy of dipole–dipole interaction for carbon C-3 and C-4 atoms of tri-butylphosphate in TVEX (as compared with liquid TBP) is caused by change of TBP conformation composition due to influence of TVEX matrix.

TABLE 8.3
Activation Energy of Dipole–Dipole $E_{c(dd)}$ and Spin-Rotational $E_{c(sr)}$ Interaction between Carbon Atoms in Extractant Molecules for Liquid TBP and TVEX-TBP

	TVEX-50% TBP		100% TBP	
Carbon Atom	$E_{c(dd)}$, J/mol	$E_{c(sr)}$, J/mol	$E_{c(dd)}$, J/mol	$E_{c(sr)}$, J/mol
C-1	9,410	—	10,409	−3,053
C-2	10,873	—	13,178	−7,513
C-3	11,977	—	9,186	−13,343
C-4	11,512	−5,170	6,536	−18,091

The data of ^{13}C {^1H} spectra (Table 8.4) show that C-1 carbon atom is most sensitive to change of environment since it has the biggest shift to weak field (C-1: 67.2 ppm for TVEX-50%TBP and 68.6 ppm for TVEX- 8.8%TBP). This signal was interpreted as a sum of high- and low-frequency ones attributed correspondingly to adsorbed and free TBP in the TVEX grain. It was found that relative integral intensity of high-frequency signal increases at decrease of TBP content in TVEX. This increase is caused probably by increase of the ratio (adsorbed TBP)/(total TBP content).

The conformation peculiarities of liquid tri-butylphosphate and TBP supported on TVEX matrix were determined by molecular mechanics and quantum chemistry [14,15]. Analysis of conformer cards allowed separating the regions of angles (rotation angles of butyl radicals around P-O bonds, α, β, γ) within which conformation energy E is minimum. A minimum value of conformation energy E = 99.9 kJ/mol corresponds to the angles $\beta = 90^0$, $\gamma = 180^0$, $\alpha = 55^0$ (sc, ap, ac). Eight conformers with the minimum energy values lower than 4.2 kJ/mol were found as indication of the good conformation flexibility of TBP molecule.

Five conformers (A, B, C, D, F) with the lowest conformation energy were selected to analyze possible locations of TBP molecule on the inner surface of the TVEX matrix. These conformers differ by mutual orientation of P = O and O-C bonds as is shown for the x–y plane projections in Figure 8.12 and Figure 8.13. The

TABLE 8.4
Signal Chemical Shifts in ^{13}C {^1H} NMR Spectra for TVEX Resins with Different TBP Content

	Chemical Shift (ppm)			
TBP Content in TVEX (%)	C-1	C-2	C-3	C-4
100	59.4	29.2	16.9	12.2
50	67.6	32.8	19.0	13.8
21.5	67.4	33.0	19.0	13.8
8.8	68.6	34.1	20.0	14.5

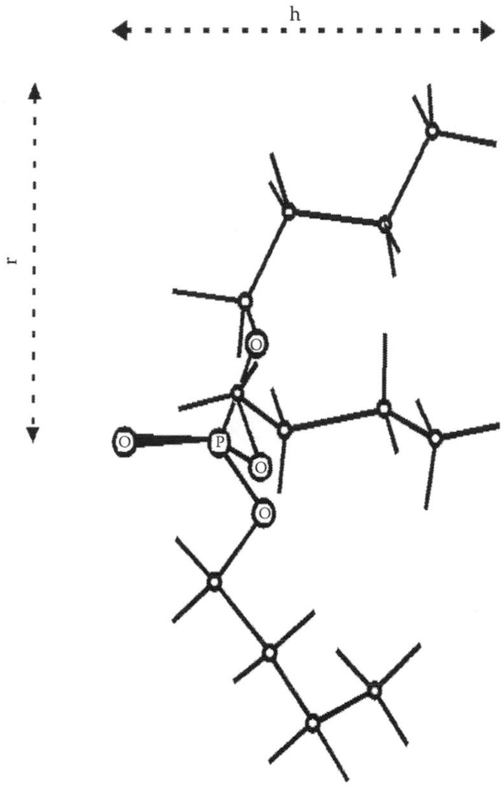

FIGURE 8.12 Projection of conformer A on x–y plane.

molecule description is based on the height *h* and radius *r* of the circle in which it is possible to inscribe alkyl chains of the extractant. The limits of these parameters for the five considered conformers are h = 0.47…0.64 mn and r = 0.65…0.85 nm. Conformers B and F have the lower value of h (0.48 nm and 0.47 nm, respectively) as compared with conformer A (0.51 nm). Conformers C and D have the highest h values (0.63 and 0.65 nm, respectively).

Accordingly, extractant adsorption takes place through the Van der Waals interaction between alkyl radicals with a matrix surface (physical adsorption). Apparently, such an interaction will be stronger at the increase of contact of butyl radicals with the surface. An increase of TBP interaction with matrix corresponds to a decrease of *h* and increase of *r*. Since conformers F and B have the lowest *h* (0.47 and 0.48 nm) and the highest *r* (0.82 and 0.78 nm), it may expected that their highest bonding strength is with the matrix surface. Conformer A has the lowest radius *r* = 0.65 nm and, hence, is adsorbed worst by matrix.

Dipole moments of A, B, C, D, F conformers were determined by quantum chemical calculations as 1.69, 4.08, 4.31, 4.20, and 4.13 D, respectively. These conformers may be classified attending to the dipole moment as low polar, as is the case of conformer A, or as high polar, as is the case of B, C, D, and F. The measure dipole moment for TBP is 3.07 D [16], close to the calculation taking into account

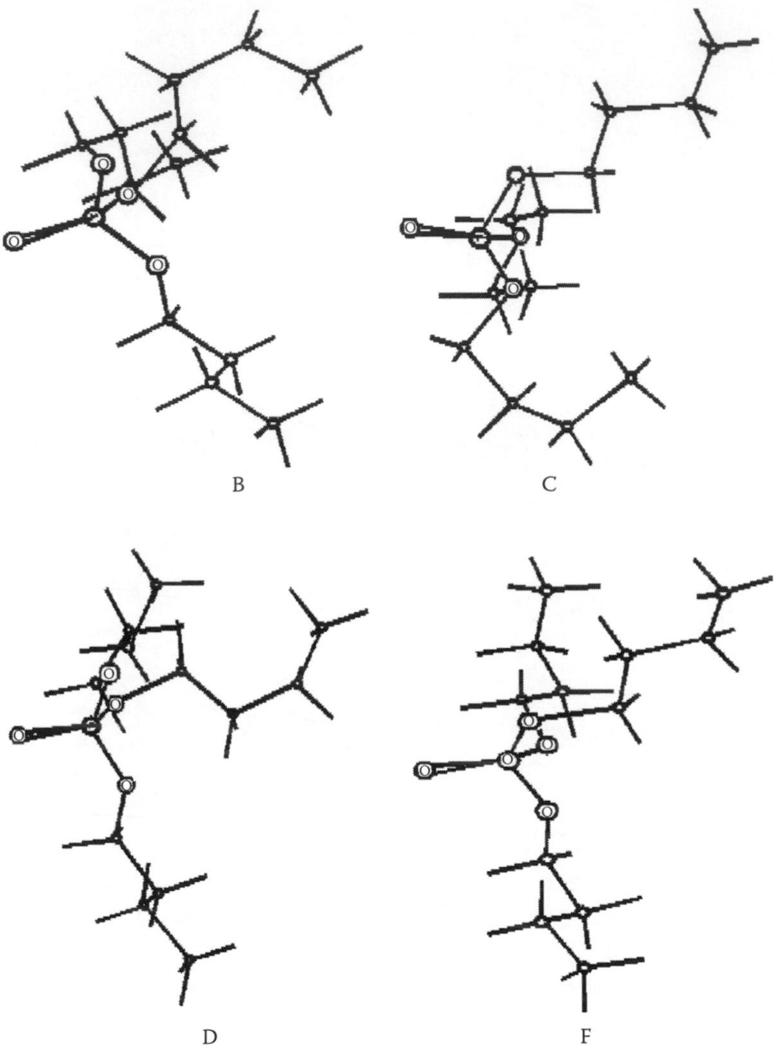

FIGURE 8.13 Projections of the conformers B, C, D, F on x–y plane.

the conformation equilibrium. Dipole moment of TBP conformers is associated with the different mutual orientations of O-C and P = O bonds.

Since donor properties of TBP are connected with the distribution of electron density on phosphoryl oxygen, the higher reaction ability will be associated to the high-polar TBP conformers as compared with low-polar ones. Since surface interaction stabilizes high-polar TBP conformers, the surface activation of TBP molecules is driven by the shift of conformation equilibrium.

TABLE 8.5
Extraction Enthalpy for TVEX-TBP and TVEX-DIOMP

	Aqueous Phase Acidity		
	8 mol/L HCl	8 mol/L HNO_3	6 mol/L H_2SO_4
TVEX-TBP	10.7 kJ/mol	–5.26 kJ/mol	–19.51 kJ/mol
TVEX-DIOMP	23.9 kJ/mol	–8.13 kJ/mol	–5.35 kJ/mol

8.5 MINERAL ACIDS EXTRACTION BY SOLID POLYMERIC EXTRACTANTS: TVEX CONTAINING NEUTRAL ORGANOPHOSPHORUS COMPOUNDS (TBP, DIOMP)

Since most of the hydrometallurgical metal extraction processes take place in highly acidic media the extraction of acids to the resin phase may be material limitation for solid polymeric extractions due to the reduction on metal capacity of the extractant. It is known that inorganic acids are extracted by solvating extractants (S) as TBP, DIOMP, TOPO, and Cyanex 923.

For these types of extractants, a hydration-solvation mechanism including the formation of complexes in organic phase that contain different numbers of acid and water molecules is postulated [17]:

$$H^+X^- + nH_2O + mS_o = (S_m{\cdot}H^+X^-(H_2O)_n)_o \qquad (8.2)$$

where H^+X^- is a mineral acid (e.g. HNO_3, H_2SO_4, HCl) and S is a solvating extractant.

The extraction rate of common mineral acids (e.g., HNO_3, H_2SO_4, HCl) by TVEX resins containing two solvating extractants (TBP and DIOMP) is comparable to the extraction rates with solvent extraction homologues systems with contacting times of 5 to 10 minutes for hydrochloric and nitric acids and 15 to 20 minutes for sulfuric acid to achieve equilibrium (Figure 8.14). Diffusion coefficient into TVEX granule is $1.18{\cdot}10^{-10}$ m^2/s for HNO_3, $9.78{\cdot}10^{-12}$ m^2/s for H_2SO_4 and $4.16{\cdot}10^{-10}$ m^2/s for HCl.

Extraction isotherms of nitric, sulphuric, and hydrochloric acids in comparison with solvent extraction data (Figure 8.15) show that extraction capacity of TBP in TVEX resins. An increase of TBP capacity in TVEX-TBP is caused by the extractant state in the polymer matrix. The extraction capacity of mineral acids by TVEX-TBP decreases in the sequence $HNO_3 > H_2SO_4 >$ HCl (Figure 8.15a,b,c), and when compared with TBP solvent extraction performance, the capacity of the impregnated polymers exceeds 1.5 times the capacity of HNO_3, 3.5 times that of HCl, and several times that of H_2SO_4.

8.6 SOLID POLYMERIC EXTRACTANT PROPERTIES ON METAL IONS EXTRACTION: MECHANISM OF RARE METALS EXTRACTION

Although impregnated materials have been postulated as "homologues of solvent extractions systems," the influence of polymer properties on the extractant activity

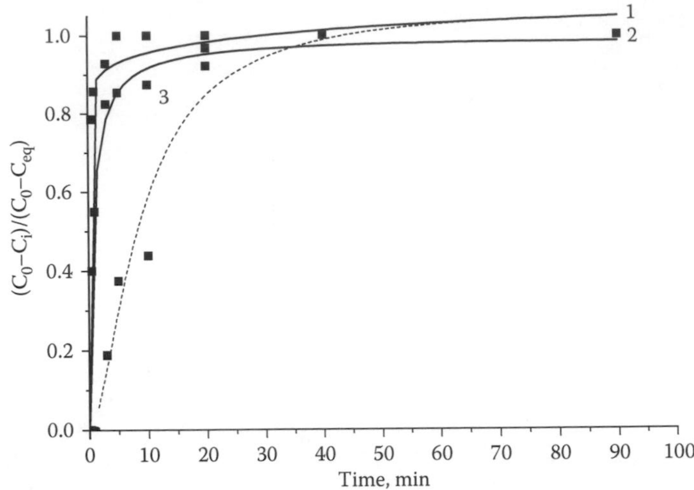

FIGURE 8.14 Extraction rates of mineral acids extraction by TVEX-TBP: 1–HCl, 2–HNO_3, 3–H_2SO_4.

and state is promoting changes on the extraction reactions when compared with the solvent extraction homologue.

The influence of TVEX polymer matrix on the extractant state causes differences in extraction and complex formation for some metals depending on extraction mechanism. These differences in extraction by TVEX when compared with solvent extraction systems were measured under equal organic to aqueous phase ratios.

8.6.1 Characterization of Sc Extraction with TVEX

Scandium separation from lanthanides and other accompanying elements has been described in the literature [22–25]. Typically effective scandium purification involves the separation from lanthanides and zirconium from concentrated hydrochloric solutions using TBP. Sc is nowadays extracted from mineral ores—uranium, wolframites, titanium—and zirconium-containing ones by hydroxide precipitation in the form of scandium oxide of 99% purity. Many scandium-containing minerals are industrial raw source for elements U, Al, Fe, Cr, Ni, Zr, Ti, W, Be, Sn, and Nb and rare-earth elements. Some of them are objects of mining, and it is possible to recover scandium simultaneously with extraction of the main metal (e.g., U, Al, Fe, Cr, Ni, Ti, Zr).

Scandium process recovery includes leaching of solid chloride wastes by hydrochloric acid, settling of the pulps formed during leaching, reduction of Fe(III) ions to Fe(II) by magnesium powder, and extraction with TBP in kerosene (70%) and Sc reextracted by diluted HCl (40 to 60 g/L). This technology applied at the industrial level in a titanium-magnesium plant revealed a number of considerable drawbacks, such as the formation of significant amounts of third phase due to extraction of titanium, zirconium, and silicon. The scandium recovery is 15 to 22% for one-step extraction, and TBP losses in the raffinate were up to 0.6 g/L. Analysis of numerous

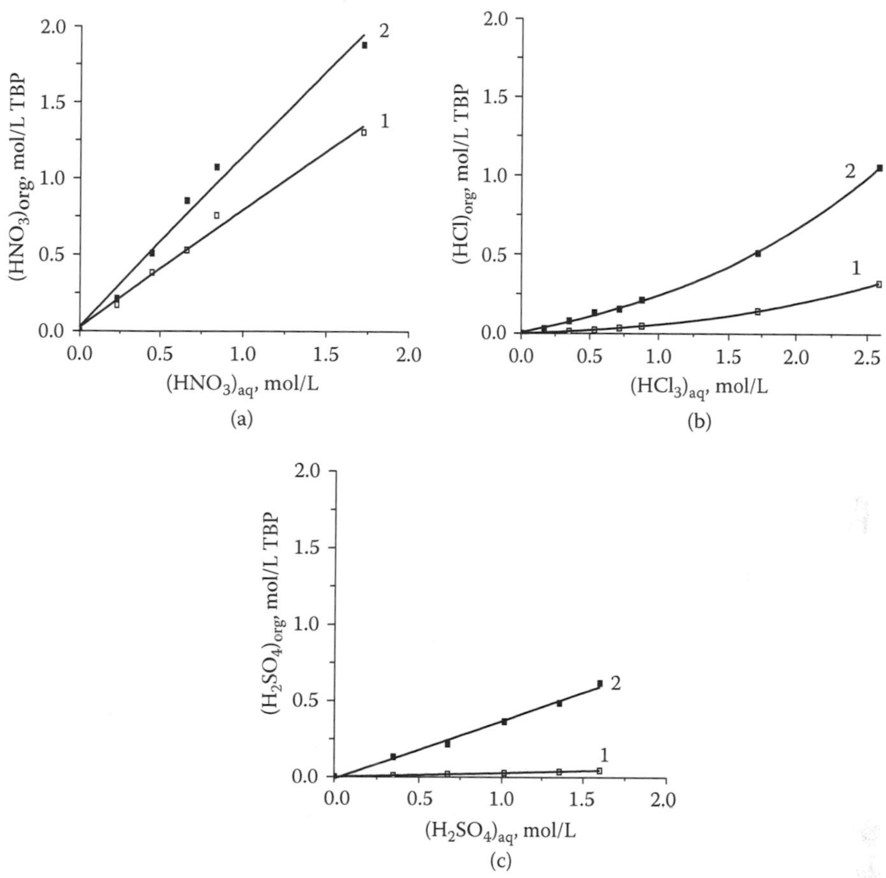

FIGURE 8.15 Extraction isotherms of nitric (a), hydrochloric (b), and sulfuric (c) acids by tributylphosphate liquid (1) and supported on TVEX matrix (2).

technological techniques for scandium recovery from different mineral raw shows that ion exchange and TVEX extraction could provide high efficiency both at first technological steps of Sc extraction from raw sources and during its fine purification from many impurities. Then, Sc extraction from acidic media by organophosphorus extractants supported on TVEX porous carrier containing TBP and DIOMP was evaluated in detail.

The Sc extraction process by TVEX-TBP and TVEX-DIOMP from hydrochloric, nitric, and sulfuric is endothermic in hydrochloric solutions and exothermic for sulfuric and nitric solutions (Table 8.6). Additionally, the extraction rates showed diffusion coefficients 1 to 2 orders of value higher (K_{dif} of Sc^{3+} ions into TVEX granule 10^{-11}–10^{-12} m^2/s) than for phosphorus-containing ion-exchange resins.

Scandium extraction capacity of TBP and DIOMP in TVEX porous matrix is higher as compared with corresponding solvent extractant capacity [10,19–22]. Isotherms of Sc extraction from hydrochloric solutions by TBP, TBP 50% solution in CCl_4 and TVEX-TBP almost coincide (Figure 8.16) at low Sc concentration;

TABLE 8.6
Composition of Sc-Containing Mineral Pulps

Component	Content, g/L	Component	Content, g/L
$FeCl_2$	100–200	$AlCl_3$	8–20
$FeCl_3$	28–45	$CrCl^3$	15–20
KCl	120–135	$MnCl_2$	8–20
NaCl	80–85	TiO_2	0.5–0.8
$MgCl_2$	110–115	ZrO_2	35–0.4
$CaCl_2$	8–12	HCl	15–20

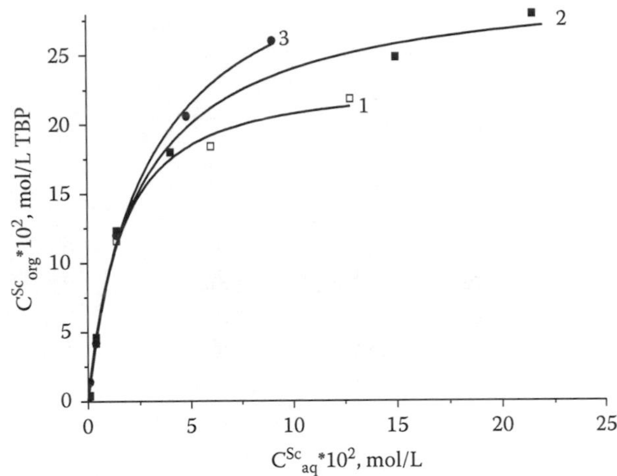

FIGURE 8.16 Isotherms of scandium extraction by TBP (1), TVEX-TBP (2), and TBP solution 50% in CCl4 (3) from 8 mol/L hydrochloric solution.

however, impregnated TBP in TVEX resins has a higher capacity as compared with 100% TBP at high Sc content. Considerably higher (almost 2 times), the equilibrium capacity of TVEX-TBP on Sc extraction from sulfuric medium is obtained (Figure 8.17) [20]. Similarly, the extraction capacity of DIOMP in TVEX matrix is close to 1.5 times higher [22,23] in comparison with solvent extraction (Figure 8.18). The observed difference for scandium distribution between liquid–liquid and TVEX extraction is caused by the extractant state in the TVEX matrix. The influence of the extractant state due to the porous matrix effect is transduced on a change in the chemical composition of the scadium complexes.

8.6.1.1 Peculiarities of Scandium Extraction by TVEX Materials from HCl Media

^{31}P NMR spectra of scandium extracts by 100% TBP, by a TBP solution in CCl_4, and by TVEX-TBP from hydrochloric medium contain the three peaks. The first

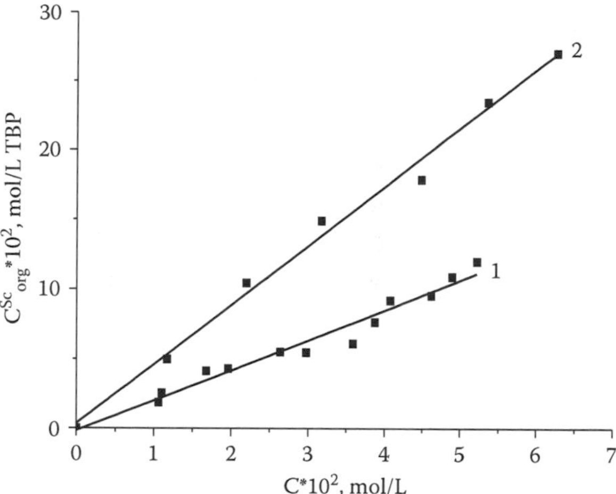

FIGURE 8.17 Isotherms of scandium extraction by TBP (1) and TVEX-TBP (2) from 6 mol/L sulfuric solution.

(1 ppm shift to the 85% H_3PO_4 phosphorous reference) corresponds to TBP noncoordinated with Sc, the second (with 4 ppm shift) corresponds to tri-butylphosphate coordinated to metal ion, and the third (–10 to –14 ppm shift) related to the products of tributylphosphate hydrolysis—dibutylphosphoric and monobutylphosphoric acids bonded with metal. These byproducts of TBP hydrolysis are bonded with scandium at the same time by ionic and coordination bonds, forming various multinuclear complexes in extracts; the presence of several signals in this peal indicates the variety of polymer forms in extract [10,19,21].

The average solvate number (ASN) defining the number of extractant molecules bonded with metal—calculated by the TBP/Sc mole ratio and signal intensities of TBP free and bonded with Sc—differs essentially for Sc extraction by undiluted

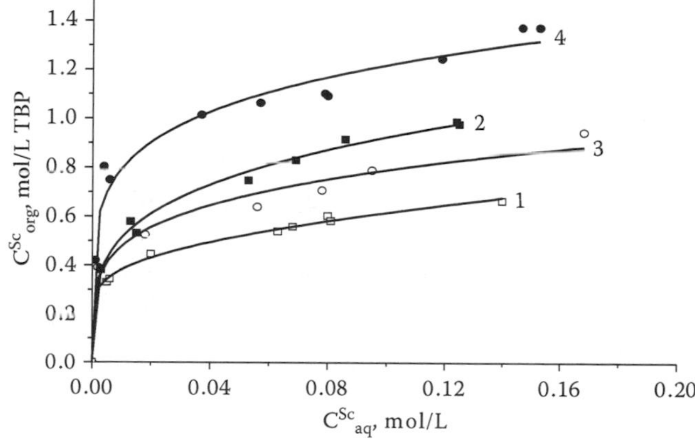

FIGURE 8.18 Isotherms of scandium extraction by DIOMP (1, 3) and TVEX-DIOMP (2, 4) from 6 mol/L (1, 2) and 8 mol/L (3, 4) hydrochloric solutions.

TBP (3.4 to 2.9), 50% TBP solution in CCl_4 (2.4 to 2.8), and TVEX-TBP (2.8–1.9) and has fractional value indicating simultaneous formation of several scandium complexes in organic phase and different ratio between them. In addition, the ASN value depends on the metal concentration in the resin phase and decreases for TVEX-TBP resins and increases for 50% TBP solution in CCl_4 with scandium content increase.

^{45}Sc NMR spectra show formation of several different complexes in extracts containing water molecules (Figure 8.19). The ratio between extracted forms is different for TBP, TBP solution in CCl_4, and TVEX-TBP with a large number of signals in spectra due to the high ligand lability in scandium coordination sphere.

^{45}Sc NMR signals in spectra of TBP, TBP solution in CCl_4, and TVEX-TBP spectra were assigned (Figure 8.19) to the following complexes: $Sc(TBP)_3Cl_3$, $[Sc(H_2O)_2(TBP)_2Cl_2]^+$, $[Sc(H_2O)(TBP)_3Cl_2]^+$, $[Sc(TBP)_4Cl_2]^+$, $[ScCl(H_2O)(TBP)_4]^{2+}{}_{cis}$, $[ScCl(H_2O)(TBP)_4]^{2+}{}_{trans}$, $[Sc(H_2O)_3(TBP)_3]^{3+}$. Then the Sc extraction process may be described in the following way:

$$[Sc(H_2O)_3Cl_3]_{aq} + 3TBP_r = [Sc(TBP)_3Cl_3]_r + 3H_2O_{aq} \tag{8.3}$$

$$[Sc(H_2O)_4Cl_2]^+{}_{aq} + 2TBP_r = [Sc(H_2O)_2(TBP)_2Cl_2]^+{}_r + 2H_2O_{aq} \tag{8.4}$$

$$[Sc(H_2O)_4Cl_2]^+{}_{aq} + 3TBP_r = [Sc(H_2O)(TBP)_3Cl_2]^+{}_r + 3H_2O_{aq} \tag{8.5}$$

$$[Sc(H_2O)_4Cl_2]^+{}_{aq} + 4TBP_r = [Sc(TBP)_4Cl_2]^+{}_r + 4H_2O_{aq} \tag{8.6}$$

$$[Sc(H_2O)_4Cl_2]^+{}_{aq} + 4TBP_r = [ScCl(H_2O)(TBP)_4]^{2+}{}_r + 3H_2O_{aq} + Cl^-{}_{aq} \tag{8.7}$$

$$[Sc(H_2O)_4Cl_2]^+{}_{aq} + 3TBP_r = [Sc(H_2O)_3(TBP)_3]^{3+}{}_r + H_2O_{aq} + 2Cl^-{}_{aq} \tag{8.8}$$

Scandium forms in a TBP solution in CCl_4 neutral complex $[ScCl_3(TBP)_3]$ and other complexes with low charge and a minimal amount of water molecules in the Sc coordination sphere. Extraction equilibrium for liquid TBP is shifted to formation of complexes with charge from +1 to +3 and presence of H_2O molecules in the Sc coordination sphere. For Sc extraction by TVEX-TBP hydrophobic polymer matrix influences scandium extraction shifting the equilibrium ratio of Sc complexes to the compounds with the smaller ionic charge and lower number of H_2O molecules in the Sc coordination sphere. The presence of hydrophobic polymer matrix reduces the migration of H_2O molecules to the organic phase together with Sc^{3+} ions.

Additionally, ^{31}P and ^{45}Sc NMR spectra for TVEX- 50%TBP with cross-linking degree from 10 to 45% DVB (Figure 8.20a–f) show the dependence of composition and relative amount of extracted Sc complexes on DVB content [24]. Complexes $[ScCl_3TBP_3]$ and $[ScCl_2(H_2O)_2TBP_2]^+$ prevail during Sc extraction by TVEX-TBP that is also typical for Sc extraction by TBP solution in CCl_4. The complex formation

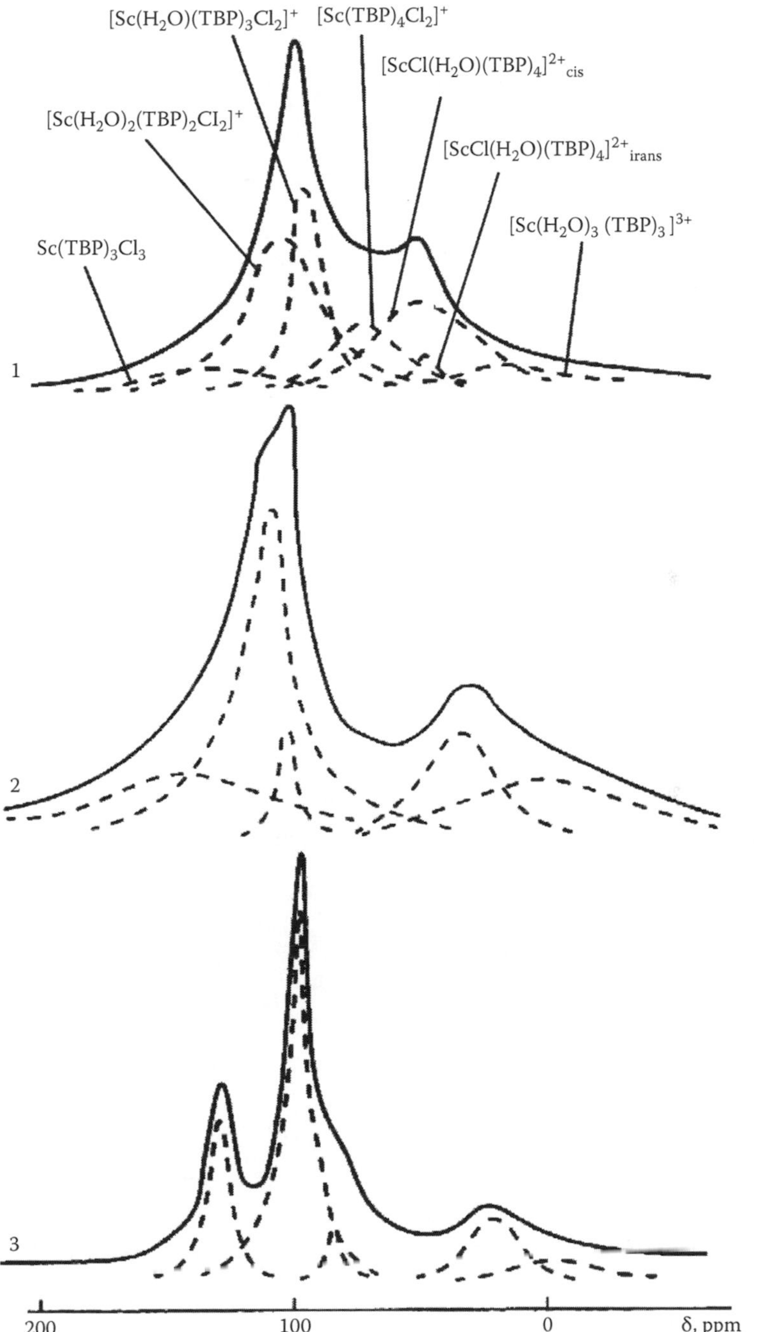

FIGURE 8.19 45Sc NMR spectra of scandium extracts from hydrochloric media by liquid TBP (1), TVEX-TBP (2), and 50 % TBP solution in CCl4 (3).

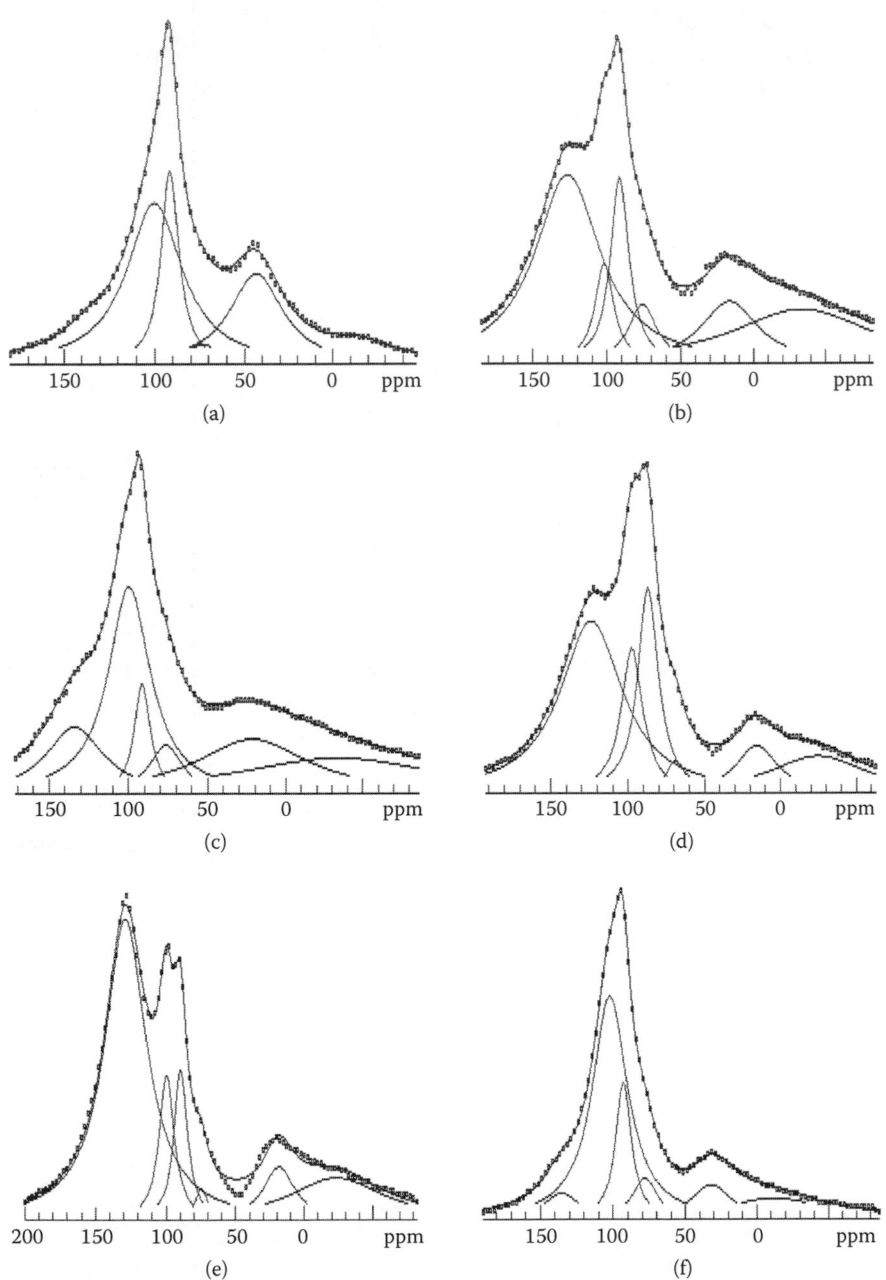

FIGURE 8.20 45Sc NMR spectra of the extracts (a) TBP, TVEX-TBP with different cross-linking degree, (b) 10% DVB, (c) 15% DVB, (d) 20% DVB, (e) 35% DVB, and (f) 45% DVB.

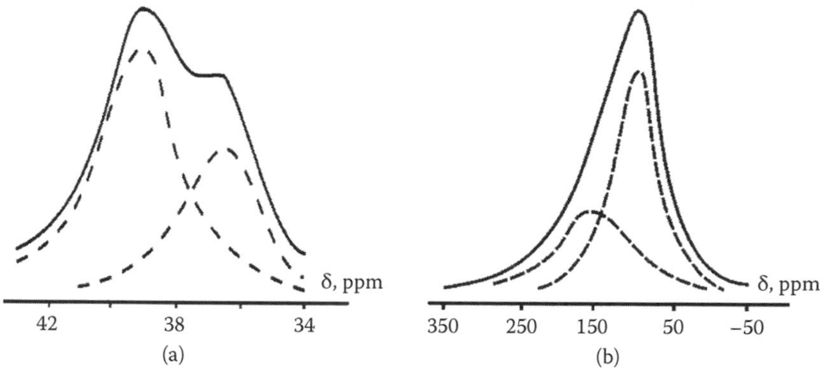

FIGURE 8.21 ^{31}P(P) and ^{45}Sc (b) NMR spectra of scandium extract by TVEX-TBP from 6 mol/L HCl.

increases with the cross-linking degree increase up to 35%. The further increase of divinylbenzene content to 45% leads to the fact that composition of Sc extracted complexes becomes similar to the extracted using 100% TBP.

Similar studies on Sc extraction with TVEX-DIOMP resins from 6 mol/L hydrochloric solution showed two signals in ^{31}PNMR spectra (at 36 and 39 ppm) (Figure 8.21a) attributed to free extractant and coordinated to scandium [23]. The ASN of DIOMP of 3 corresponds to a scandium complexes with three DIOMP molecules directly coordinated to the metal ion. This is corroborated by the presence of two peaks on the ^{45}Sc NMR spectra of Sc-containing TVEX-DIOMP resins (Figure 8.21b) at 102 and 152 ppm, respectively, with a number of chloride ions in the Sc complexes between 2 and 3, while the other coordination sites are occupied by water molecules. Thus, the extraction of Sc by TVEX-DIOMP from 6 mol/L HCl solutions proceeds with the formation of $[ScCl_3(DIOMP)_3]$ and $[ScCl_2(DIOMP)_3(H_2O)]^+$ complexes, and the extraction process may be described by the following reactions:

$$[ScCl_3(H_2O)_3]_{aq} + 3DIOMP_r = [Sc(DIOMP)_3Cl_3]_r + 3H_2O_{aq} \tag{8.11}$$

$$[ScCl_2(H_2O)_4]^+_{aq} + 3DIOMP_r = [ScCl_2(DIOMP)_3(H_2O)]^+_4 + 3H_2O_{aq} \tag{8.12}$$

During scandium extraction by TVEX-DIOMP resins an increase of DIOMP capacity in TVEX matrix is observed at lower Sc concentrations as compared with TVEX-TBP due to the higher DIOMP donor abilities in comparison with TBP.

From the evaluation of Sc extraction from hydrochloric solutions, it was established that TVEX capacity gradually decreases with the number of extraction—reextraction cycles are caused by tri-butylphosphate leakage from TVEX porous matrix. However, phosphorus content in TVEX resins (3.9 to 4.1%) after 100 extraction cycles is 90% of initial TBP content.

^{45}Sc NMR spectra of TVEX-TBP resins after two extraction cycles present a broad signal that corresponds to Sc extracted complexes $[ScCl_n(H_2O)_m(TBP)_{6-n-m}]^{3-m}$

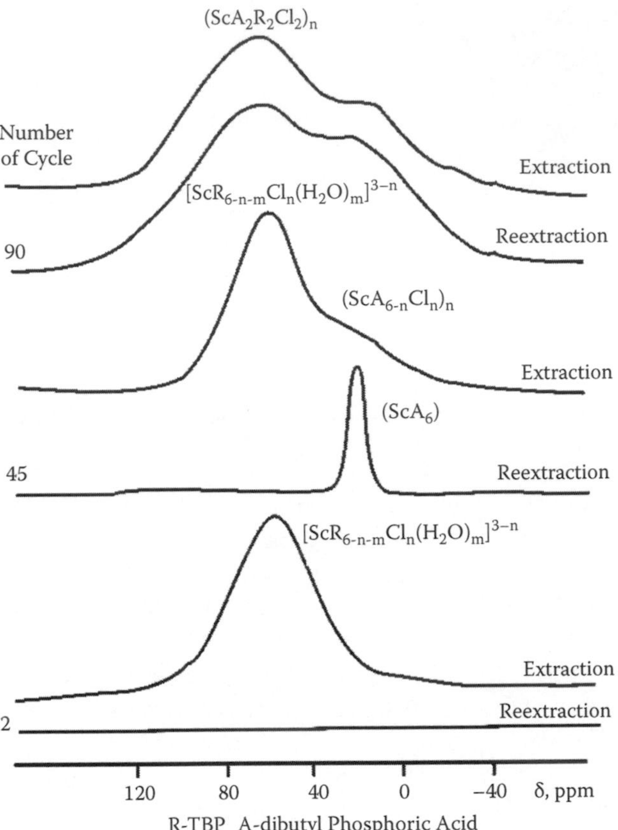

FIGURE 8.22 45Sc NMR spectra of TVEX-TBP after different number of extraction–reextraction cycles.

(Figure 8.22). The absence of signals in the spectrum of TVEX sample after scandium reextraction indicates total Sc removal from TVEX beads.

After 45 extraction cycles, NMR spectrum of TVEX-TBP contains signals related to Sc compounds with TBP and di-butylphosphoric acid (DBP), which is a product of TBP hydrolysis. The narrow signal remaining in TVEX spectrum after Sc reextraction with a chemical shift indicates the absence of Cl-ions in Sc coordination sphere. The small width of the signal indicates high symmetry of the central ion environment as it corresponds to the Sc complex $[Sc(DBP)_6]$.

After 90 cycles of extraction, the reextraction TVEX-TBP capacity significantly decreases, and the ^{45}Sc NMR spectra before and after reextraction do not differ. The nature of the spectrum suggests the formation of linear Sc polymers with TBP hydrolysis products. These compounds with a general composition of $[ScCl_2(DBP)_2(TBP)_2]^+$ are not extracted and accumulate in the resin matrix.

The nature of changes led to TVEX-TBP "aging" in a cyclic process of scandium extraction–reextraction from hydrochloric solutions. A decrease in extraction ability is caused by a decrease in TBP content due to extractant gradual hydrolysis and formation of strong scandium compounds with hydrolysis products that do not wash out TVEX during reextraction by water.

8.6.1.2 Peculiarities of Scandium Extraction by TVEX Materials from H_2SO_4 Media

^{31}P NMR spectra of TBP and TVEX-TBP both after Sc extraction from 6 mol/L H_2SO_4 solutions present singlet signals shifted 1 ppm of the initial TBP and TVEX-TBP, due to TBP solvation by acid molecules [20]. Such a minor solvate shift indicates that the coordination bond P = O-Sc is not taking place. ^{45}Sc NMR spectra of TBP and TVEX-TBP samples after Sc extraction contain one signal –16.0 ± 1.0 ppm that does not depend on a concentration of scandium and H_2SO_4 in the organic phase. Similarly, aqueous Sc-containing sulfuric solutions (0.05 mol/L Sc and mole ratio SO_4^{2-}/Sc from 10 to 120) showed that the complex $[Sc(SO_4)_3]^{3-}$ is not formed in aqueous phase, and the complexes $[Sc(H_2O)_4SO_4]^+$ and $[Sc(H_2O)_2(SO_4)_2]^-$ are the predominant. However, Sc is present as $[Sc(SO_4)_3]^{3-}$ complex in the organic/resin phases where the ratio SO_4^{2-}/H_2O is considerably higher as compared with aqueous phase, and its composition does not depend on Sc and acid concentration in the organic phase. Taking into account that monosolvates and proton hydration is favored in an organic phase, extracted scandium complex may be presented as $[TBP][H_5O_2(H_2O)_{m-2}]_3^+[Sc(SO_4)_3]^{3-}$, where m depends on the reagent concentration.

Then, scandium extraction from H_2SO_4 media by both TBP and TVEX-TBP may be presented by the following reactions:

$$[ScSO_4(H_2O)_4]^+_{aq} + 3H^+_{aq} + 2SO_4^{2-}{}_{aq} + (3m-4)H_2O_{aq} + TBP_r \rightarrow [TBP][H_5O_2(H_2O)_{m-2}]_3^+[Sc(SO_4)_3]^{3-}{}_r \qquad (8.9)$$

$$[Sc(SO_4)_2(H_2O)_2]^-_{aq} + 3H^+_{aq} + SO_4^{2-}{}_{aq} + (3m-2)H_2O_{aq} + TBP_r \rightarrow [TBP][H_5O_2(H_2O)_{m-2}]^{3+}[Sc(SO_4)_3]^{3-}{}_r \qquad (8.10)$$

8.6.2 Characterization of Nb Extraction with TVEX

Many scandium-containing minerals are industrial raw source for elements U, Al, Fe, Cr, Ni, Zr, Ti, W, Be, Sn, and Nb and rare-earth elements. Some of them are objects of mining, and it is possible to recover niobium simultaneously with extraction of the main metal (e.g., U, Al, Fe, Cr, Ni, Ti, Zr).

The extraction capacity of liquid TBP and TVEX-TBP increases with increase of aqueous phase acidity for 7, 9, and 11 mol/l HCl solutions (Figure 8.23). TBP capacity in TVEX matrix is higher as compared with liquid tri-butylphosphate; this difference depends on aqueous phase acidity and becomes maximal (1.5 to 2 times) at 9 mol/L HCl. TVEX-TBP capacity is also higher for Nb extraction for sulfuric media; however, dependence of extractant capacity on aqueous phase acidity is dependent on the redox unstability of Nb in sulfuric media [25,26].

The process is temperature dependent, and a temperature increase leads to increase of TVEX-TBP capacity for extraction from HCl media and negligibly

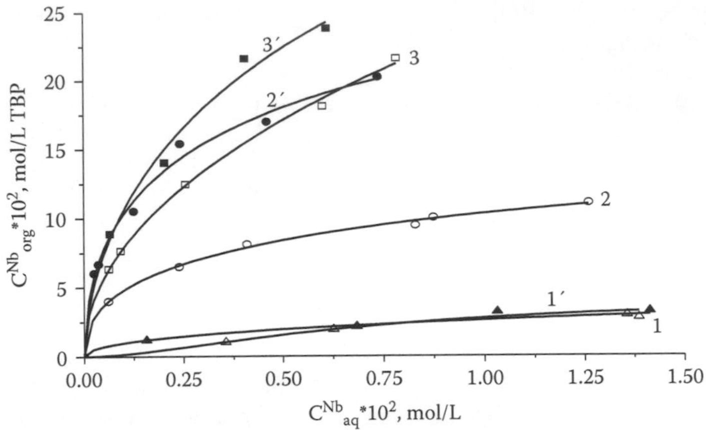

FIGURE 8.23 Isotherms of Nb extraction by TBP (1–3) and TVEX-TBP (1'–3') from hydrochloric solutions: 1 and 1'–7mol/L; 2 and 2'–9 mol/L, 3 and 3'–11 mol/L.

influences Nb extraction from H_2SO_4 media. The enthalpy of Nb extraction values are 35.87 kJ/mol for TVEX-TBP and 50.44 kJ/mol for TBP. The extraction rate of niobium by TVEX-TBP from HCl solutions is characterized with a Nb diffusion coefficient on TVEX granules of $4.3 \times 10^{-12} - 1.9 \times 10^{-11}$ m^2/s in 6 mol/L HCl solutions and $1.1 - 4.9 \times 10^{-10}$ m^2/s from 10 mol/L H_2SO_4 solutions.

Analysis of ^{93}Nb spectra of $NbCl_5$ aqueous solutions with the absence of signals from 7 to 11 mol/L HCl solutions indicates the absence of high-symmetric niobium complexes like $NbCl_6^-$, and then $[NbOCl_4]^-$ complex is the predominant Nb species. However, ^{93}Nb NMR spectra of $NbCl_5$ solution in TBP contains one signal (at 680 ppm), whereas two separated signals 1.26 and –2.5 ppm in ^{31}P NMR spectra (Figure 8.22b) attributed to extractant free and coordinated to niobium are found. The ASN equal to 1 attributes the ^{93}Nb spectra signal to the adduct $NbCl_5{\cdot}TBP$.

^{31}P NMR spectra of Nb containing TBP and TVEX-TBP from 11 mol/L HCl contain two separated signals 1.26 and –2.5 ppm in ^{31}P NMR spectra (Figure 8.24c,d) attributed to extractant free and coordinated to niobium. The calculated ASN values are 0.4 and 1 for pure tri-butylphosphate and TVEX-TBP resins, respectively. These data indicate formation of Nb:TBP = 1:1 solvate in TVEX at Nb extraction from 11 mol/L HCl. The similar complex and compound, in which the metal is not bonded with TBP directly, are formed in organic phase during niobium solvent extraction. Nb-containing TBP and TVEX-TBP samples do not contain signals in ^{93}Nb spectra similar to the system $NbCl_5$ solution, and then the $[NbOCl_4{\cdot}TBP]^-$ complex is extracted species on the resin phase.

^{31}P NMR spectra of TBP and TVEX-TBP extracts after contact with 7 and 9 mol/L HCl show singlet signals shifted 0.5 to 1.5 to strong field relatively pure TBP signal. Unlike extracts from 11 mol/L HCl, the spectra do not contain signals corresponded to TBP bonded with niobium. There are no signals in ^{93}Nb NMR spectra of TBP and TVEX-TBP indicating Nb extraction as complexes with low symmetry.

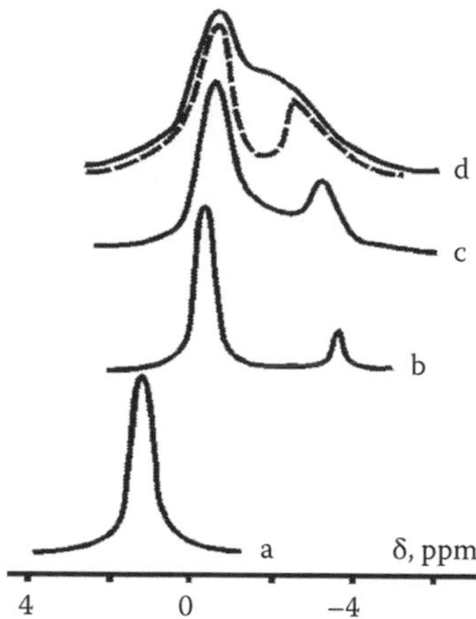

FIGURE 8.24 ^{31}P NMR spectra of liquid TBP (a), system NbCl5-TBP (b), Nb extracts by TBP (c), and by TVEX-TBP (d) from 11M HCl.

Thus, the extraction of Nb from 7 to 9 mol/L HCl solutions is explained as anion exchange reactions where anionic forms of Nb complexes are extracted to the resin phase:

$$[NbOCl_4]_{aq}^{-} + nTBP_r + H_{aq}^{+} + mH_2O = \{H_3O.(m-1)H_2O.nTBP\}^{+}[NbOCl_4]^{-}_{r} \quad (8.13)$$

whereas Nb extraction from 11 mol/L HCl could be explained by the following reaction:

$$[NbOCl_5]^{2-}{}_{aq} + TBP_o = [NbOCl_4{\cdot}TBP]^{-}{}_{o} + Cl^{-}{}_{aq} \quad (8.14)$$

An ASN value of 0.4 indicates that Nb is extracted from an 11mol/l HCl solution by formation of complexes described by reactions (Equation 8.13) and (Equation 8.14). The change to mechanism (Equation 8.14) in 11 mol/L HCl is connected to the transition of $[NbOCl_4]^{-}$ complex to $[NbOCl_5]^{2-}$ in an aqueous phase caused by an increase of chloride-ions concentration.

8.6.2.1 Characterization of Ga Extraction with TVEX

Isotherms of Ga extraction by TVEX-TBP, TVEX-DIOMP as well as by liquid TBP and DIOMP from 3, 5, and 8 mol/L HCl solutions show (Figure 8.25 and Figure 8.26) that extractant equilibrium capacity increases with aqueous phase acidity increase, extractant capacity in TVEX matrix is 1.3 to 1.5 times higher as compared with solvent extractant. Enthalpy of gallium extraction from 3 mol/L HCl is –11.7 kJ/mol for TBP and –12.0 kJ/mol for TVEX-TBP. Diffusion coefficient of gallium extraction is $4.2{\cdot}10^{-11}$ m²/s for TVEX-TBP (3mol/L HCl) and $5.4{\cdot}10^{-11}$ m²/s for TVEX-DIOMP (6 mol/L HCl) [27].

Ga(III) is present in aqueous media in the form to hexaaquacomplex $[Ga(H_2O)_6]^{3+}$, and it is complexed by chloride anions in HCl media to form aquo-chloride species

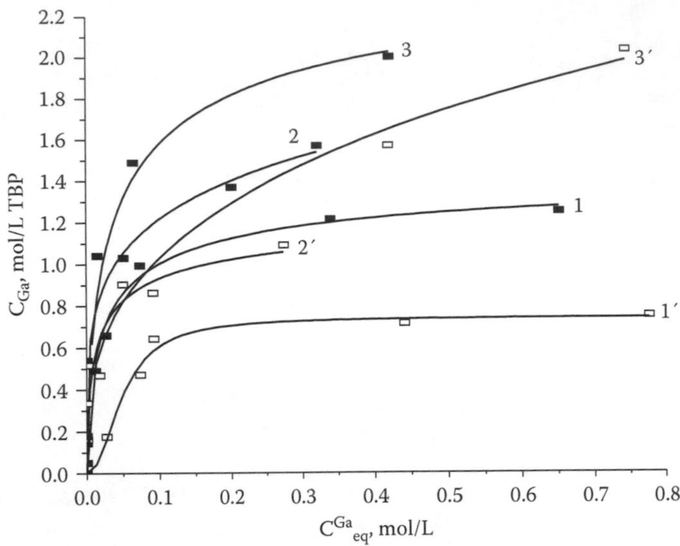

FIGURE 8.25 Isotherms of gallium extraction by TBP (1', 2', 3') and by TVEX-TBP (1, 2, 3) from hydrochloric solutions: 1 and 1'–3 mol/L HCl; 2 and 2' 5 mol/L HCl; 3 and 3'–8 mol/L HCl.

$[GaCl_{4-n}(H_2O)_n]^{n-1}$ in the range 0 to 7 mol/ HCl and to $[GaCl_4]^-$ in 7 to 8 mol/L HCl, as has been determined by ^{71}Ga NMR spectroscopy. ^{31}P spectra of tri-butylphosphate and TVEX are singlet signals shifted to strong field relatively signals of initial TBP and TVEX-TBP in the whole studied acidity range. The similar shift has for the

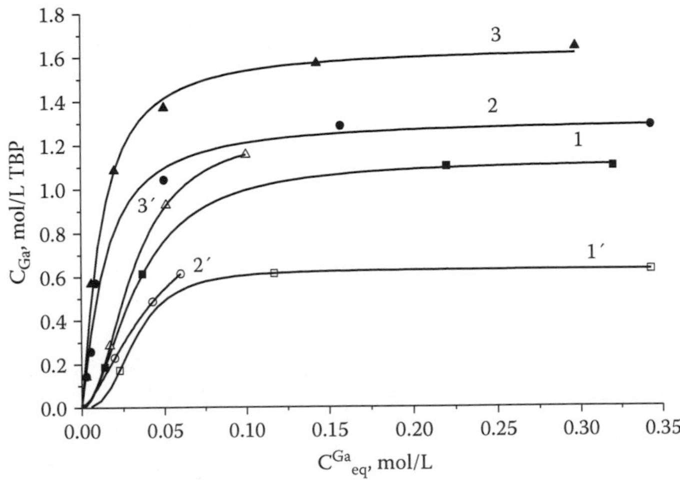

FIGURE 8.26 Isotherms of gallium extraction by DIOMP (1', 2', 3') and by TVEX-DIOMP (1, 2, 3) from hydrochloric solutions: 1 and 1'–3 mol/L HCl; 2 and 2' 5 mol/L HCl; 3 and 3'–8 mol/L HCl.

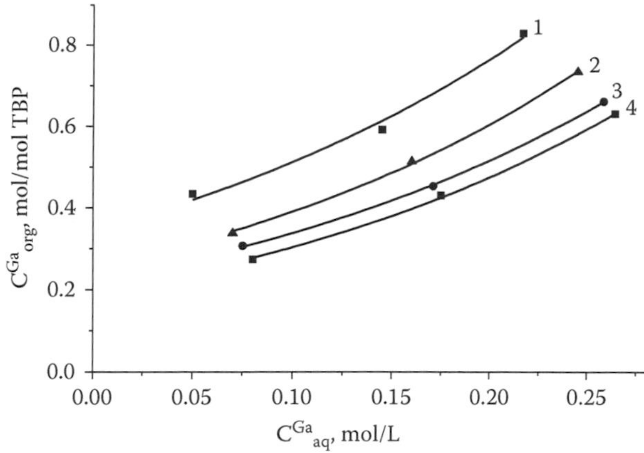

FIGURE 8.27 Equilibrium gallium extraction by TVEX containing 11% (1), 21% (2), 35% (3), and 50% (4) tri-butylphosphate.

phosphorus signal in spectra of TBP and TVEX-TBP saturated by hydrochloric acid indicates the absence of TBP coordination to gallium in extracts.

Then, the extraction of Ga(III) with TVEX-TBP resins could be explained by the following solvation reaction:

$$[GaCl_4]^-_{aq} + [H_3O{\cdot}nH_2O]^+_{aq} + 3TBP_r \rightarrow [H_3O{\cdot}nH_2O{\cdot}3TBP]^+[GaCl_4]^-_r \quad (8.15)$$

The model was consistent with Ga extraction with TVEX-TBP resins containing TBP contents of 50%, 35%, 21%, and 11% where the decrease of TBP (Figure 8.27) increased resin capacity in terms of mole ratio Ga/TBP, probably due to contribution of absorbed extractant to total TBP capacity.

Increase of extractant capacity in porous TVEX resins is associated with the activity of the TPB portion adsorbed on the matrix surface at the expense of butyl radicals. Polymer matrix changes the ratio between TBP high- and low-polar conformers stabilizing on the matrix surface more polar conformers with dipole moments higher than 4D. As a result, on inner surfaces of TVEX matrix, a layer of adsorbed extractant molecules is formed having higher dipole moments as compared with liquid TBP with dipole moment values of 3D. This extractant state in TVEX leads to a certain benefit during the recovery of metal species as compared with solvent extraction. This benefit is minor, and it does not influence extraction in case of solvate mechanism, when extractant is included into metal first coordination sphere. Indeed, there were no differences in capacity of TBP liquid and supported on TVEX carrier for extraction of some metals by solvate mechanism. However, if several metal complexes, easily transformed into each other, are formed in the organic phase at extraction by solvate mechanism (e.g., scandium extraction by TBP from hydrochloric media), extractant state in TVEX may result in change between the complexes. TBP in TVEX leads to a considerable increase of extractant equilibrium capacity in case of hydrate-solvate mechanism, when metal ion is not directly coordinated by extractant, and extraction takes place due to electrostatic forces.

The described extractant state and complex formation peculiarities in polymer carriers should be taken into consideration in the materials preparation steps as well as during the equilibrium and kinetics studies to be undertaken as preliminary steps of the process design applications with impregnated membranes and spherical carriers.

8.7 HYDROMETALLURGICAL APPLICATIONS OF SOLID EXTRACTANTS (TVEX) OF RARE AND RARE-EARTH ELEMENTS

Several metal separation chemistry studies based on the use of solid polymeric extractants have been developed; for a long time their application in different fields has been on small and industrial scales. In particular, the actual applications cover the following fields: (1) analytical applications on analysis schemes of preconcentration and separation of interferences; and (2) recovery of valuable metals from secondary sources. The promising metal recovery processes-based TVEX systems appear to be in the following cases: (1) recovery of metals from dilute solutions; (2) separation of metals from concentrated solutions obtained by hydrometallurgical processing of complex ores, concentrates, mattes and scraps, and purification of process solutions; and (3) separation and purification of metals that are economically or strategically important (e.g., gold and platinoids, rare-earth metals, gallium, niobium, hafnium, lithium).

8.7.1 Scandium Selective Extraction from Hydrochloric Media by TVEX-TBP

Due to absence of rich scandium raw minerals, scandium extraction is a complex process closely connected with the separation from impurities present on the leaching solutions. Chloride wastes after titanium-magnesium production (TMP) may be considered as a resource for scandium recovery. These wastes are highly mineralized hydrochloric pulps containing 0.08–0.1 g/L of Sc, more than 8 mol/L of HCl, and a large quantity of solid particles.

The high extraction and kinetic properties of TVEX-TBP allowed for the development of and introduction to technology of selective scandium recovery ("TVEX-Sc-process" [19,27]) from hydrochloric pulps obtained by leaching of accumulated mine wastes and salt chlorinators during titanium production.

The initial step (Figure 8.28) includes waste leaching with HCl to obtain a hydrochloric pulp. After phase separation, scandium is extracted by TVEX-50% TBP from the clarified pulp in several cycles with continuous agitation. TVEX-TBP is washed by concentrated hydrochloric acid to achieve a preliminary scandium purification from a number of impurities.

Reextraction is carried out by water (Figure 8.29) with further precipitation of scandium oxalate by oxalic acid. Technical scandium oxide is produced after filtration and oxalates decomposition containing 60 to 70% of Sc_2O_3. To produce high-pure scandium oxide (99.9% purity) technical oxide is dissolved in hydrochloric acid, scandium is extracted by tri-butylphosphate with subsequent Sc reextraction, precipitation as oxalate, and decomposition to oxide. Additional purification may

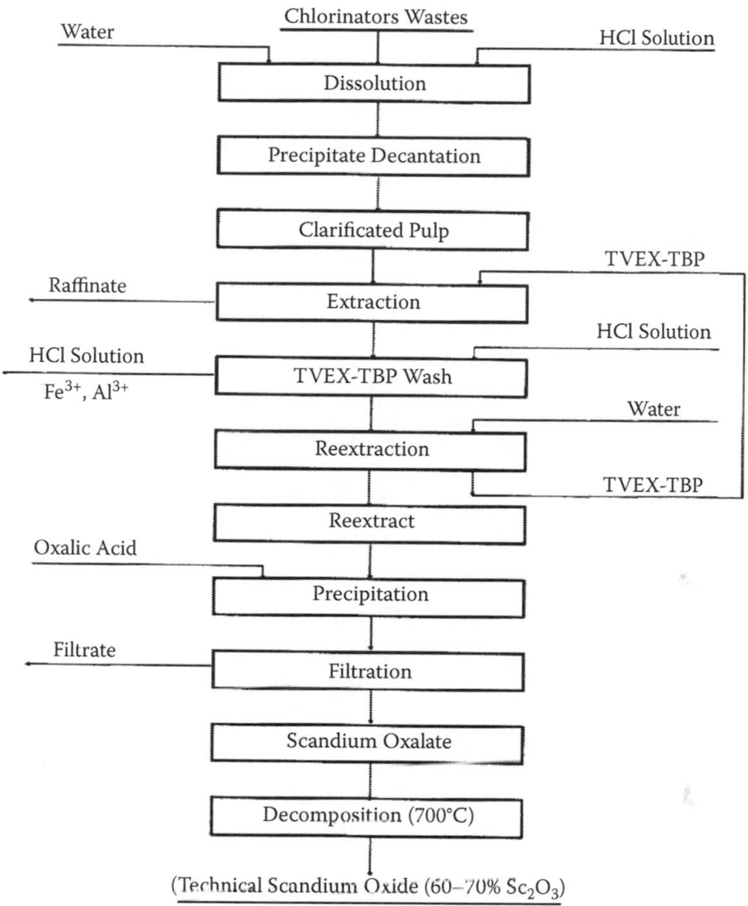

FIGURE 8.28 Basic flowsheet of Scandium extraction with solid polymeric extractants ("TVEX-Ss process").

be performed to remove thorium, zirconium, and titanium. A Sc-extraction process by DIOMP containing TVEX is characterized by a high Sc extraction capacity that is accompanied, however, by a selectivity decrease and then an additional expense during final purification.

The TVEX-TBP scandium extraction process from highly concentrated hydrochloric pulps with high content of solid particles and high silica content allowed for the exclusion of pulp filtration, reduction of tri-butylphosphate consumption and extractant loss with emulsions, and increase of extraction of aimed component. The introduction of "TVEX-Sc-process" at Ust'-Kamenogorsk titanium-magnesium production union (Kazakhstan) made it possible to reduce completely the impact on the Irtysh River due to reduction of TBP contents on plant effluents with a reduction on tons of TBP per year in comparison with the established TBP solvent extraction process.

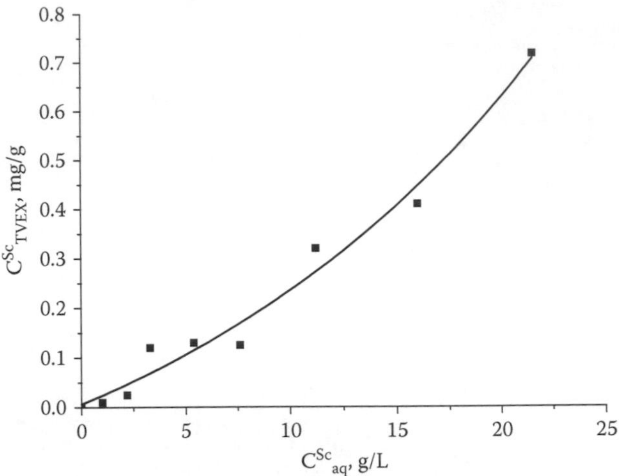

FIGURE 8.29 Isotherm of scandium desorption from TVEX-TBP.

8.7.2 Zirconium and Hafnium Selective Extraction from Hydrochloric Media by TVEX-TBP

Zirconium-containing raw is a promising scandium source [19–23]. It was established that Sc is accumulated in carbon-containing residue after chlorination [21] and in manifold after separation of base zirconium sulfate [22–23]. The extraction technology used for Sc recovery is based on leach neutralization, sludge separation, sludge leaching by nitric or hydrochloric acid, and scandium extraction by TBP solution in kerosene. The treated solution free of Sc is a valuable source of Zr and Hf.

TVEX-TBP resins also have shown high capacitive and kinetic properties for Zr and Hf extraction from HNO_3 media [28,29], a high separation factor for the couple Zr/Hf. The isotherms of Zr and Hf extraction by TVEX-65% TBP under different aqueous phase acidity showed Zr and Hf capacities up to 0.6 mol Zr/mol TBP and 0.3 mol Hf/mol TBP, respectively (Figure 8.30 and Figure 8.31). The particle diffusion coefficients measured were $1.5{\cdot}10^{-12}$ m^2/s for Zr and $2.2{\cdot}10^{-11}$ m^2/s for Hf, respectively, in 6 mol/L nitric solutions.

Semiindustrial tests were carried out on zirconium and hafnium extraction by TVEX-65% TBP from nitric silica-containing pulps with the following composition: Zr 17.0–44.0 g/L, Hf 0.2–0.5 g/L, Al 0.2–0.37 g/L, Fe 1.7–4.0 g/L, Si 0.8–8.2 g/L, HNO_3 260.0–420.0 g/L. The pulp was received as a result of HNO_3 leaching of cake of zircon with soda.

The tests were performed using a facility composed of two sequentially connected pulsation columns with KRIMZ packing (10 m of packing height) on the extraction step and a KNSPR column (12 m of height) for the reextraction step (Figure 8.32). Application of this equipment allows for the performance of metal extraction from media with high content of silica and solid particles. A limit of silica content of 3 to 4 g/L provides acceptable rate of phase division in countercurrent for

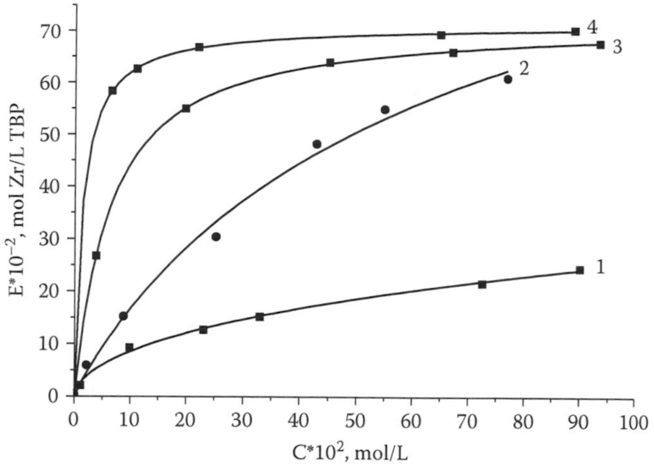

FIGURE 8.30 Isotherms of Zr extraction by TVEX-65% TBP depending on nitric acid content: 1–2 mol/L, 2–4 mol/L, 3–6 mol/L, 4–8 mol/L.

application of KRIMZ pulsation column (Figure 8.33). Typical assays treated 250 m^3 of pulp.

Zirconium and hafnium extraction was carried out in countercurrent flow mode using TVEX-65%TBP in pulsation column, with KRIMZ packing. Acidified TVEX was loaded from the bottom of column whereas the nitrate pulp (at 40 ^{0}C) was guided to the top with a phase ratio (solid /liquid) = (2.5)/1.0. Zirconium concentration in TVEX was 25 to 32 mg/L depending on the Zr content in pulp. TVEX was washed by 300 to 500 g/L nitric acid in upper part of column and guided to metal reextraction with a 50 to 70 g/L HNO_3 solution (Figure 8.34). Reextraction was carried out in a countercurrent mode with fixed TVEX layer in KNSPR column by 30 g/L nitric acid (taking into account HNO_3 captured by TVEX after nitric acid rinsing). As a result a

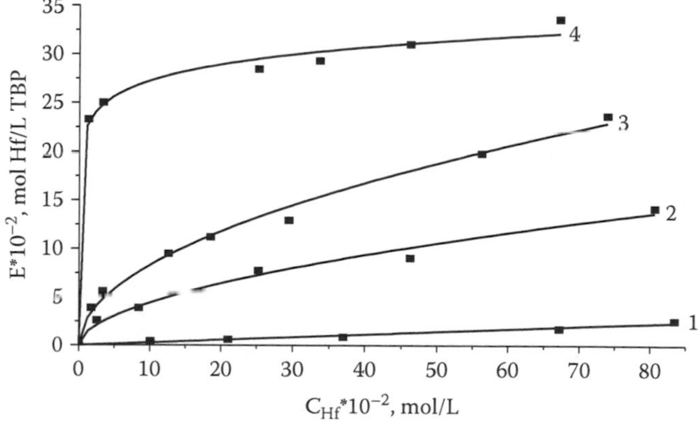

FIGURE 8.31 Isotherms of Hf extraction by TVEX-65% TBP depending on nitric acid content: 1–2 mol/L, 2–4 mol/L, 3–6 mol/L, 4–8 mol/L.

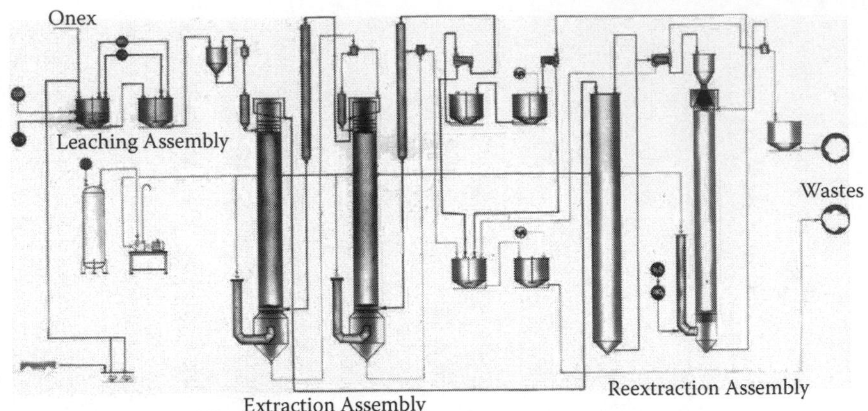

FIGURE 8.32 Technological scheme of zirconium and hafnium extraction by TVEX-65% TBP.

concentrate of Zr and Hf with a total metal content about 20 g/L is obtained. Zirconium concentration in overflow pulp was 0.1 g/L and corresponds to waste norms.

The chromatographic separation of zirconium and hafnium from 4 to 6 M HNO_3 solutions using TVEX-65% TBP in a 10 m column height was evaluated (Figure 8.35 and Figure 8.36). The height of TVEX layer was 8 m. Treated solutions contain 1.8 g/L Hf (85% from sum of metals) and 0.3 g/L Zr, and 5.4 M nitric was heated to 40°C and fed downward through acidified TVEX-65% TBP with a flow rate of 16.5 L/h. Column dynamic TVEX capacity on Zr was 2.9 mg/L and 1.2 g/L for Hf with a dynamic separation factor of 4.3. The reextraction of Zr and Hf from TVEX with 50 g/L nitric acid produced an enriched solution of 2.7 g/L of Zr and 1.2 g/L of Hf, providing a preconcentration factor of 9.

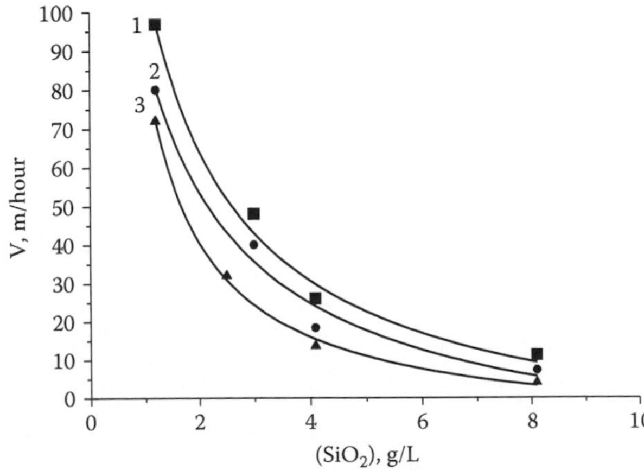

FIGURE 8.33 Effect of Si content on TVEX emergence rate depending on temperature: 1–60°C, 2–40°C, 3–23°C.

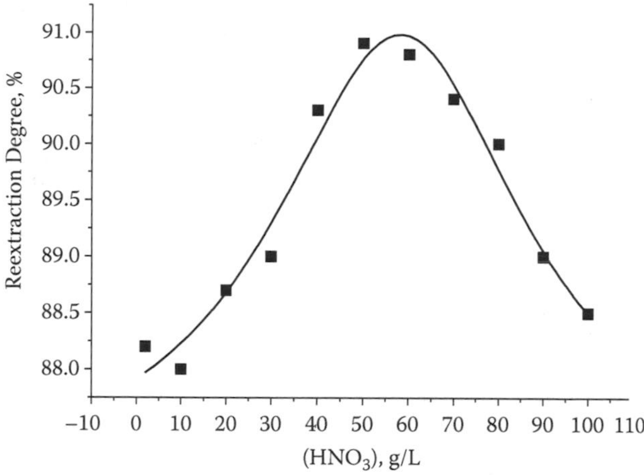

FIGURE 8.34 Dependence of reextraction degree of total Zr and Hf on acidity of reextracted solution.

8.7.3 TVEX-TBP Uranium Extraction from Dense Pulps by Using Solid Extractants

TBP is a well-known extractant on many solvent extraction processes in the nuclear fuel processing industry. TBP has been an excellent all-around solvent for the reprocessing of irradiated nuclear fuel. It possesses all major requirements that extractants must have to be successfully applied in an industrial solvent extraction separation process. These properties have been also transferred to the solid polymeric extractant TVEX-TBP.

The application of TVEX-TBP resins on uranium extraction (Figure 8.37 and Figure 8.38) show an uranium capacity dependence on TBP and DVB when extracted from 3 M HNO_3 solutions solution containing 5 g/L uranium. The optimal component ratio was 50% TBP and 20 to 25% DVB for uranium extraction.

TVEX-TBP properties were used to develop the biggest full-scale application of TVEX-TBP in the former USSR for uranium extraction at uranium ore-processing

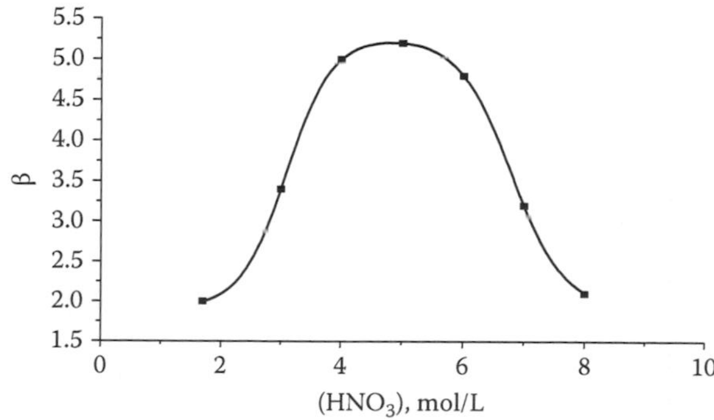

FIGURE 8.35 Dependence of separation coefficient Zr/Hf on nitric acid concentration.

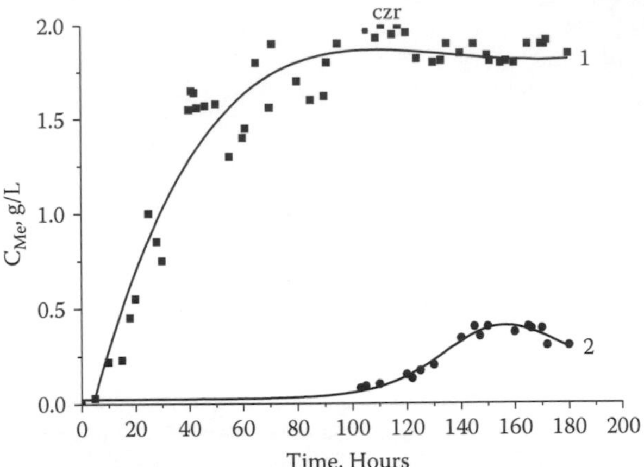

FIGURE 8.36 Breakthrough curves of hafnium (1) and zirconium (2) on column chromatographic separation with TVEX-65% TBP resins.

plants. The process was designed using pulsating sorption column PSK [30] (Figure 8.39) in a countercurrent mode from acid nitric pulps with density up to 1.8 g/cm^3 and solids content up to 40%. This technology was applied on an industrial scale from 1974 through 1988.

Application of solid extractants on hydrometallurgical applications has the benefits of increasing extraction capacity and reducing labor expenses and improving the working conditions of personnel at the expense of elimination of highly filtrated pulps operation. Exclusion of flammable solvents from production process allows for a decrease in the associated fire hazard. Moreover, tri-butyl-phosphate consumption decreases significantly by reduction of losses with nonseparated emulsions.

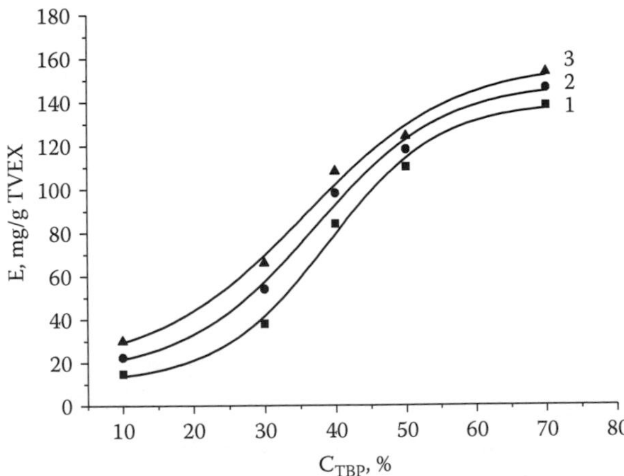

FIGURE 8.37 Dependence of TVEX capacity on TBP content as a function of cross-linking degree: 1–10% DVB, 2–20% DVB, 3–30% DVB.

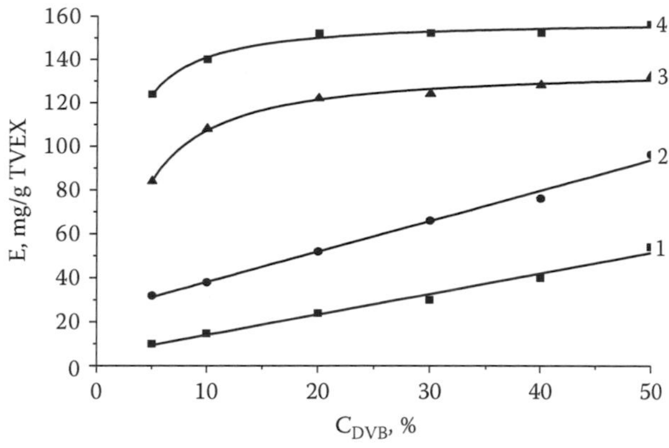

FIGURE 8.38 Dependence of TVEX capacity on DVB content in matrix: 1–10% TBP, 2–30% TBP, 3–50% TBP, 4–70% TBP.

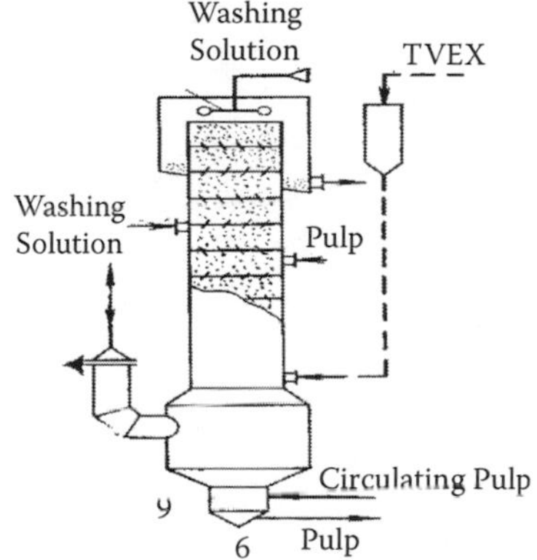

FIGURE 8.39 General scheme for a PSK pulsation column for mineral pulps processing with TVEX resins.

ACKNOWLEDGMENTS

The authors express their gratitude to the International Scientific Foundation (Grant K2F100) and the Science and Technology Center in the Ukraine (Project #2487) for partial financial support of research.

REFERENCES

1. A. Warshawsky, Extraction with solvent-impregnated resins, *Ion Exch. and Solv. Extr.*, 8, 229–310 (1981).
2. B. Kroebel and A. Meyer, West Germany Patent Application 2,162,951 (1971).
3. B. Kroebel, A. Meyer, and B. Bayer, Application of newly developed materials for extraction chromatography of inorganic salts in columns. *Proc. ISEC'74*, 3, 2095–2107 (1974).

4. A. Bolotov, Y. Korovin et al., Patent USSR 476279, Publication 05.07.75, Bulletin 25.
5. Y. Kuzovov, Y. Korovin, L. Kodubenko, B. Gromov, V. Savel'eva, and G. Kireeeva, Synthesis of new sorbents based on styrene-divinylbenzene co-polymer and tributylphosphate, *Proc. Moscow Mendeleev Chem.-Tech. Inst.*, 97, 43–48 (1977).
6. V. Korovin, S. Randarevich, and Y. Kuzovov, Regularities of formation of porous structure and exractant state in TVEX-TBP matrix, *Ukrainian Chem. J.*, 56, 1042–1046 (1990).
7. V. Korovin and S. Randarevich, Synthesis, properties and application of solid extractants (review), *Khimischeskaya Technologiya [Chemical Technology]*, 5, 3–13 (1991).
8. B. Smirnov, *Physics of Fractal Clusters*, Moscow: Nauka Publisher (1991).
9. D. M. Smith, G. P. Johnston, and A. G. Hurd, Structural studies of vapor-phase aggregates via mercury porosimetry, *J. Coll. Interface Sci.*, 135, 227–237 (1990).
10. Y. Kuzovov, V. Korovin, V. Savel'eva, and A. Komarov. 31P, 45Sc NMR Comparison of mechanisms of rare metals extraction by solvent extractants and ones introduced in TVEX porous matrix, *Proc. ISEC'88*, 3, 159–162 (1988).
11. V. Korovin, Y. Pogorelov, and I. Plastun. Tri-butylphosphate state in TVEX polymer matrix based on 1H and 13C NMR data, *Rus. J. Inorg. Chem.*, 38, 1866–1869 (1993).
12. A. Washman and I. Pronin, *Nuclear Magnetic Relaxation Spectroscopy*, Moscow: Energoatomizdat Publisher (1986).
13. V. Mank and N. Lebovka, *NMR Spectroscopy of Water in Heterogoneous Systems*, Kiev: Naukova Dumka Publisher (1988).
14. V. Korovin, A. Varnek, A. Kuznetsov O. Petrukhin, and O. Sinegribova, Electron structure and electrostatic potential of some neutral organo-phosphorus extractants and their complexes, *Proc. ISEC'88*, 1, 74–77 (1988).
15. W. Burkett and N. Ellinger, *Molecular Mechanics*, Moscow: Mir Publisher (1986).
16. L. Mazalov and V. Yumatov, *Electron Structure of Extractants*, Novosibirsk: Nauka Publisher (1984).
17. A. Rozen, *Extraction by Neutral Organic Compounds*. Moscow: Atomizdat Publisher (1976).
18. V. Korovin, Y. Shestak, and E. Valuaeva, Extraction of acids by TVEX based on tributylphosphate, *Ukrainian Chem. J.*, 62, 16–20 (1996).
19. V. Korovin, Y. Pogorelov, and A. Chikodanov, Scandium extraction by TVEX-TBP from titanium-magnesium production wastes, *Proc. ISEC'93*, 2, 173–178 (1993).
20. V. Korovin, S. Randarevich, S. Bodaratsky, and V. Trachevsky, Scandium extraction from sulfuric solutions by TBP and TVEX-TBP based on data of 31 and 45Sc NMR spectroscopy, *Rus. J. Inorg. Chem.*, 34, 2404 (1990).
21. V. Korovin, S. Randarevich, Y. Pogorelov, and S. Bodaratsky, Scandium extraction by tri-butylphosphate solvent and supported on TVEX polymer matrix from hydrochloric solutions based on 31P and 45Sc NMR data, *Rus. J. Coord. Chem.*, 22, 633–640 (1996).
22. V. Korovin, Y. Shestak, and Y. Pogorelov, Scandium extraction by organo-phosphorus compounds supported on a porous carrier, *Hydrometallurgy*, 52, 1–8 (1999).
23. V. Korovin and Y. Shestak, Scandium extraction by TVEX-DIOMP from hydrochloric media, *Ukrainian Chem. J.*, 62, 22–26 (1996).
24. V. Korovin and Y. Pogorelov, Effect of TVEX-TBP polymer matrix on scandium extraction based on 31P and 45Sc NMR data, *Ukrainian Chem. J.*, 60, 695–698 (1994).
25. V. Korovin, S. Randarevich, and Y. Shestak, Niobium extraction by TBP and TVEX-TBP from hydrochloric media based on 31P, 93Nb NMR data, *Rus. J. Inorg. Chem.*, 36, 3167–3171 (1991).
26. V. Korovin and Y.Shestak, Niobium extraction by TVEX—polymer resin containing tri-butilphosphate, *Reac. Polym.*, 40, 107–113 (1999).

27. V. Korovin, Y. Pogorelov, A. Chikodanov, and A. Komarov, Scandium extraction by TVEX-TBP from the wastes of titanium-magnesium production, *Rus. J. Appl. Chem.*, 8, 1744–1750 (1993).
28. V. Korovin, G. Yagodin, and V. Savel'eva, Extraction of zirconium and hafnium by TVEX-TBP from nitric solutions, *Rus. J. Appl. Chem.*, 67, 753–757 (1994).
29. V. Korovin, G. Yagodin, and V. Savel'eva, Division of zirconium and hafnium by solid extractant on tri-butylphosphate basis, *Rus. J. Appl. Chem.*, 67, 758–761 (1994).
30. S. Karpacheva and B. Ryabchikov, *Pulsation Equipment in Chemical Technology*, Moscow: Atomizdat Publisher (1983).

9 Solvent Impregnated Resin Applications on Metal Separation Processes

Abraham Warshawsky, José Luis Cortina, and Karel Jeřábek

CONTENTS

9.1 INTRODUCTION

The separation of metal ions is of fundamental importance in research and technology. It is widely used in analytical chemistry and radiochemistry and also has industrial applications in the chemical industry, in the nuclear industry, in the mineral and hydrometallurgical processes, and in environmentally related applications. The selective removal of toxic components from waste effluents and the separation and preconcentration of the different metal species from aqueous solutions can be achieved by solvent extraction (SX), ion exchange (IX), and solvent-impregnated resin technique.

Solvent extraction involves the distribution of different components between two immiscible phases (i.e., aqueous and organic) [1]. The organic phase contains a lipophilic extraction reagent (extractant), which forms complexes with metal ions extracted from the aqueous phase. Then, in the elution step, the metal ions are transferred to the aqueous phase to allow using the organic solution again for metal extraction (cyclic operation). The specific complexation between the extractant and different metal ions provides the difference in extractability that permits the selective separation of the metals.

Ion exchange technique consists in the exchange of ions in solution with the counterions of insoluble polymers containing fixed anionic or cationic groups [2]. Coordinating and chelating ion exchange resins are polymers with covalently bound side chains containing donor atoms that are able to form complexes with selected metal ions or with their complexes [3]. It is now widely accepted that the use of polymeric resins in metal recovery offers many advantages over the use of the solvent extraction technique. The most important of these advantages are the simplicity of equipment and operation and the possibility of using the polymeric adsorbent for many extraction cycles without losses in the metal extraction capacity.

The preparation of coordinating polymers is usually quite complex, time consuming, and costly. The properties of the resulted polymers are not very desirable; many times hydrophobic materials are obtained. Other times, the selectivity properties of the chelating ligand are not retained. Therefore, the concept of solvent-impregnated resins (SIRs) was developed [4]; it is very simple and is, in many cases, the only way to prepare ion exchange resins containing reactive groups with special properties that cannot be immobilized by chemical bonding. This concept includes the incorporation of the metal extraction reagent (ligand) by a physical impregnation technique into a polymer support.

In the 1970s, almost simultaneously, Warshawsky [4] and Grinstead [5] investigated SIRs in hydrometallurgical and effluent treatment applications. SIRs can be envisaged as a liquid-complexing agent dispersed homogeneously in a solid polymeric medium. The impregnated extractant should behave as a liquid ligand but should exhibit strong affinity for the matrix.

Limitations in the preparation of chelating and ion exchange resins together with the complexity and time-consuming procedures needed for covalently linking the functional group to the backbone of the resin contribute to their high cost. All this together with the recent advances in the synthesis of new organic reagents for solvent extraction establishes the development of impregnated resins as a link between solvent extraction processes and ion exchange processes in this important and developing segment of separation science. Macroporous polymeric supports with a number of the desired properties (i.e., high surface area, high porosity, and high chemical and mechanical resistance) were selected as the extractant carriers, and since that time the research and applications of SIRs has grown steadily as reflected in the large number of scientific contributions during the last three decades [6,7]. Basically, the impetus has been directed to the development of new metal separation processes, particularly for liquid wastes and effluents.

This chapter provides fundamental concepts of the development of solvent-impregnated resins useful for rational selections of solutions for target applications. This includes basis for the selection of the SIR components, reagents and polymeric support, the tools for assessment of the properties of used materials needed for the determination of equilibrium, and kinetic parameters useful in the development of metal separation, recovery, and removal processes.

9.2 PREPARATION OF IMPREGNATED RESINS

Sorbents containing immobilized extractants can be classified into two basic types: (1) solvent-impregnated resins prepared by adsorption of an liquid extractant on polymer supports; and (2) resins with liquid extractant encarcerated within polymer matrix during polymerization (TVEX and Levextrel resins) [6,7].

9.2.1 Solvent-Impregnated Resins (SIRs)

The preparation methods to produce efficient SIRs have been classified in the following ways [8–11]:

1. The "dry" impregnation method, in which the extractant is adsorbed directly into the polymeric support after contacting a solution of the extractant with the polymer and subsequent evaporation of the organic solvent.
2. The "wet" impregnation method, in which a solution of the extractant is imbibed into the polymeric support, and the obtained impregnated resin is then a three-component system: polymeric support–extractant-organic solvent.
3. Modifications of these two possibilities, in which the extractant is mixed with a phase modifier to promote water wetability. This approach was proposed to reduce high hydrophobicity of impregnated resins obtained by the dry impregnation method.

Without the need of adding a third component into the system the hydrophobicity could be reduced by direct adsorption of the extractant from a mixture of an organic solvent and water. Mixtures of acetone, methanol, and ethanol with water have been used. Subsequently, the polymer beads are removed from the solution by filtering, and the excess of solvent retained on the polymer is removed by washing with water. Before further use the product is stored in water. In comparison with the classical impregnation methods, this procedure has the advantage of providing a polymeric material with improved hydrophilicity.

The impregnation of the extractant on the polymeric support sorption process is mainly a sorption process [9–11]. Integrating a number of interactions as attraction forces between alkyl chains or aromatic rings of most ligands and those of the resin backbone, and subsequent physical trapping of these ligands within the pores of the resin beads. However, it is difficult to predict accurately just which extractants will be well adsorbed by a given polymeric substrate. In effect, nonionic resins, such as Amberlite XAD2 and XAD-4, Lewatit OC, are excellent supports for retaining molecules of a given organic reagent and also ionic compounds. Adsorption of these reagents occurred due to physical attraction, namely π-π dispersion forces, arising from the aromaticity of the styrene-divinylbenzene-type resin and the benzene rings in the reagent molecule. In most cases, structures with a permanent porosity with aliphatic hydrocarbons in their backbone are more suitable for sorption of large anions. Moreover, acrylic components have a favorable influence on the retention of organic reagents [11].

The impregnation isotherms of different organophosphorous extractants (i.e., TBP, DEHPA, Cyanex 272, Cyanex 301, Cyanex 302 and Cyanex 471), tertiary amines (Alamine 336), and quaternary amines (Aliquat 336) into Amberlite XAD2, Amberlite XAD4, and Amberlite XAD7 supports have shown a general trend the increase of the quantity of extractant incorporated into the polymer increases with the extractant concentration in the impregnation solutions. However, a plateau value for extractant loading into the support is reached for each extractant.

9.2.2 TVEX and Levextrel Resins

TVEX and Levextrel resins are macroporous styrene/divinylbenzene copolymers containing a metal selective extractant entrapped within the polymer matrix. They are achieved by adding the extractant directly to the mixture of the monomers during the bead polymerization process [12–14]. The content of the extractant and degree of cross-linking of the final bead product may be varied at will. The conditions of polymerization can be varied, thus compensating for any differences in the properties of the individual extractants. Generally, using this approach can be within the polymer nework incarcerated various active components, such as organophosphorous-based extractants, aliphatic amines, and aliphatic and aromatic oximes. The polymerization conditions depend on the extractant properties as its acidity, viscosity, and water solubility [16,17]. Information on on such materials commercialized by the Production Union Pridneprovski Chemical Plant (Ukraine) under the trade name of TVEX are shown in Table 9.1. Bayer Ag (nowadays Lanxess) produced impregnated resins under the trade name Levextrel resins, containing phosphoryl compounds: tributylphosphate (Lewatit VP1023OC), di-(2-ethylhexyl)phosphoric

TABLE 9.1
List of Solid Extractants Produced by Production Union Pridneprovski Chemical Plant

	Liquid Extractant Introduced to Polymer Styrene-Divinylbenzene Matrix	Extractant Content, % Mass
1	tri-butylphosphate (TBP)[a]	50
2	tri-isobutylphosphate	50
3	tri-isoamylphosphate (TIAF)	50
4	tri-oktylphosphate	50
5	tri-2(ethyl)hexylphosphate	50
6	di-2(ethyl)hexylphosphoric acid (D2EHPA)[b]	50
7	di-2(ethyl)decylphosphoric acid	50
8	isododecyl phosphetane acid (IDDPA)[b]	50
9	poly-2(ethyl)hexylphosphonitrile acid	50
10	phosphoric acid hexabutyltris amide[b]	50
11	dioctylmethylphosphonate	50
12	diisooctylmethylphosphonate (DIOMP)[b]	50
13	octyldioctylphosphonate	50
14	tri-isoamyl phosphine oxide[b]	50
15	tri-octyl phosphine oxide	50
16	octyl, heptyl, pentyl phosphine oxide (PHOR)[b]	50
17	TBP/TIAF	25/25
18	TBP/D2EHPA[a]	25/25
19	TBP/PHOR	25/25
20	TBP/IDDPA	25/25
21	DIOMP/D2EHPA	25/25
22	DIOMP/kerosene	50/10
23	DIOMP/PHOR[a]	25/25
24	PHOR/D2EHPA[b]	30/20
25	D2EHPA/kerosene	35/15

[a] Industrial production.
[b] Pilot production.

acid (Lewatit VPOC 1026), and di(2,4,4-trimethylpentyl)phosphinic acid (Lewatit TP807´84) [18].

9.3 INTRODUCTION OF NEW POLYMERIC SUPPORTS AND IMPREGNATION PROCEDURES

Improvements in SIR preparation were primarily aimed to solve the major problems to the implementation of SIR columns, which is the leakage of the ligand into the aqueous phase. This leakage can be multiplied by use of the SIR column in several extraction–elution cycles. The problem remained as to how to ensure the highest possible efficiency for removing low levels of metal ions and to increase the kinetics of the extraction process by improving the hydrophilic nature of the polymer matrix. To solve these problems different approaches were evaluated.

Recently, preparation of impregnated resins by immobilization of the reagent onto conventional ion exchange resins has been introduced [10]. The use of these polymeric supports should serve two different purposes:

1. Impregnation of the reagent onto the support bearing ionically active groups could produce a synergistic metal extraction reaction [10,19–23].
2. Increase in the hydrophilic character of the impregnated resin throughout the favored interaction of the active groups of the resin with water [24].

Selection of the physical properties of the polymeric support and the chemical properties in terms of the complexing properties of the active groups of the resin is an important step in determining the behavior of participation of complex systems. It could be possible that active groups of the ion exchange resin may produce a loss in the selectivity patterns of the loaded reagent and thus lose one of the advantages of impregnated resins.

9.3.1 New Polymeric Supports

Typical nonreactive macroporous polymeric supports based on polystyrene-divinylbenzene were substituted by reactive polymeric supports including polyvinylpiride (PVP) and weak or strong ion exchange resins. All these new types of polymeric supports for SIR are characterized by the presence of either acid or base groups. The selection of a suitable reactive polymeric support should be controlled by the following rules.

First, when impregnating an acidic ligand (HL) (e.g., organophosphoric [DEHPA] and organothiophosphoric [DEHPTA]) and neutral (TBP, TOPO) derivatives, a polymeric support containing a basic functionality should be selected (e.g., PVP or a weak base ion exchange resins). The acid-base interaction between the extractant and the polymeric support could be described as follows:

$$\text{PS-R}_2\text{N}_{\text{res}} + (\text{RO})_2\text{P(O)-OH} \Leftrightarrow (\text{PS-R}_2\text{N}\cdot\cdot\ \text{OH P(O)(RO)}_2)_{\text{res}} \tag{9.1}$$

for example, for a tertiary anion-exchange resin (**$\text{PS-R}_2\text{N}_{\text{res}}$**) and an organophosphoric derivative (**$(\text{RO})_2\text{P(O)-OH}$**), and where PS indicated the polymeric support network.

Second, when impregnating a basic extractant (B), (e.g., an amine type extractants as [Alamine 336] or LIX79), a polymer containing an acidic functionality could be selected (e.g., strong and weak acid ion exchange resins). The acid-base interaction could be described as follows:

$$\text{PS-S(O)OH}_{\text{res}} + \text{R}_3\text{N} \Leftrightarrow \text{PS-S(O)OH}\cdots\text{NR}_{3\ \text{res}} \tag{9.2}$$

where ($\mathbf{PS\text{-}XOOH_{res}}$) represents a sulphonic cation-exchange resin and $\mathbf{R_3N}$ a tertiary amine derivative.

In the case of solvating extractants as TPB, TOPO, Cyanex 923, the interaction proces with the acidic groups of the ion exchange resins could be described by the following reaction:

$$PS\text{-}XOOH_{res} + B \Leftrightarrow (PSXOOH...B)_{res} \tag{9.3}$$

where ($\mathbf{PS\text{-}XOOH_{res}}$) represents a sulphonic cation-exchange resin and **B** a solvating derivative.

The interactions described in the previous equations may be detected by spectroscopic measurements (FTIR, ^{31}P, ^{1}H, ^{13}C), and in general, these interactions are much stronger than the physical adsorption of the extractant on nonfunctional (inert) polymeric supports.

9.3.2 Modifying Impregnation Procedures

Improvements in the kinetic properties of the materials were also achieved by modification of the impregnation procedures in several different ways. When using the direct adsorption of the extractant onto the polymeric support, a mixture of a solution of the reagent in a water-miscible organic solvent is imbibed in the polymer. The solvent should also be a "good" solvent for the polymer, meaning a solvent with a high solubility in the polymer phase. It promotes efficient absorption and good distribution of the extractant in the polymer phase, and in combination with presence of water in the impregnation mixture it may produce materials with improved wetting properties. Mixtures of methanol, ethanol, and acetone with water allowed preparations of SIRs containing both acidic and basic extractants showing mass transfer data similar to those shown typically by hydrophilic ion exchange resins.

9.3.3 Increasing Extractant Loading

For making SIRs competitive as metal removal materials, there is a need to increase the extractant loadings. Two different options were evaluated: (1) the use of high-surface hypercross-linked polymeric supports (such as the Hypersol-Macronet™ [Purolite]); and (2) the use of reactive supports (ion exchange resins) [10,19]. The use of polymeric supports with high surface areas (around 1000 m^2/g) did not produce a substantial increase in extractant loading or in subsequent metal loadings [10]. But the use of ion exchange materials as supports for SIRs allowed extractant loadings to reach close to 0.8 to 1.0 mol/Kg with metal ion exchange or extraction capacities very similar to capacities of ion exchange resins (between 2 to 5 mol/Kg). However, although those values of capacity are achievable, the activity of the reagent adsorbed on the SIRs needs to be carefully studied considering the fact that the increase of the extractant loading is very likely to affect some other important properties of the SIRs when applied on a technological scale. First, the increase of loading implies an increase in the extractant losses. Second, the increase in the loadings also causes increase in the hydrophobic nature of the SIR resin. Hence, it is very important to define what the range of optimum loadings is and how the loadings compromise a different number of objectives that need to be accomplished: (1) maximum capacity; (2) low hydrophobic nature (affecting kinetics); and (3) low extraction losses.

9.4 DESCRIPTION OF METAL EXTRACTION PROCESSES WITH IMPREGNATED RESINS

When comparing SIRs to conventional ion exchange resins, the SIRs provide increased selectivity for a metal without reduction in the kinetics and capacity shown by chelating ion exchange resins. On the other hand, when comparing SIRs with liquid–liquid extraction systems, the SIR systems show a more efficient extractant use with minimal use of organic solvents in the SIR preparation step—but no organic solvent is used in the metal extraction step. The early papers did not discern any differences in the behavior of these SIRs [4–6]. However, recently it has been shown in a number of cases that the polymer matrix influences not only the extraction kinetics but also the formation of metal complexes in the organic phase and then the extraction reactions. Description of the metal extraction reactions and the influence of the solid carrier have been studied. Knowledge and understanding of the metal extraction processes was a useful tool for designing new separation and preconcentration schemes based on SIRs [7]. Then, extensive efforts were made to develop a unified approach to describe the metal extraction and separation properties of solid supported ligands and to compare their behavior with those found when these supported ligands are used in liquid–liquid extraction. A concerted effort was made to describe simultaneously the chemical and physical processes involved in the preparation and application of SIRs in metal extraction coupled with the characterization of their surface and morphological properties. This was brought about by the combination of the expertise vested in three fields: solvent extraction chemistry, SIRs and polymer chemistry, and physico-chemistry of surfaces. At the starting point the thermodynamic approaches developed to describe liquid–liquid and solid ion exchange extraction systems were used as models [7,8,11]. The thermodynamic evaluation was accomplished by the following:

1. A description of the support-extractant system to determine the extractant distribution between the aqueous and the polymer phases as well as the aggregation state of the extractant in the polymer phase.
2. A description of the extractant–metal reactions to identify exactly the metal extraction species responsible for the metal extraction process.

Typically, different families of acidic organophosphorous extractants (DEHPA, Cyanex 272, DEHPTA), and basic ligands (Alamine 336 and LIX79) in the removal of heavy and precious metals have been thermodynamically characterized. As a general trend, behavior of these extractants retained on the solid matrix and the extracted metal–ligand species follow similar trends to those found when dissolved in organic solvents. Only in a few specific cases some changes in the pH dependence order were found. This supports usefulness of the use of liquid–liquid distribution data as the initial step required in the design of new SIRs helping to economize characterization of the metal extraction properties of any new SIRs.

The differences observed in metal removal by liquid extractant and by SIRs obtained by the immobilization of the same extractant onto a porous polymeric carrier are the result of the effects of the polymeric carrier matrix on the state of the extractant. We have only recently observed and reported such interactions [3–7]. The elucidation of the mechanism by which the supporting matrix influences metal complex formation may enable new and more efficient methods of metal separation and recovery to be devised.

Determination of the morphology of the original polymers and the SIRs obtained from them by impregnation is very helpful in elucidation of the adsorption mechanism of the extractant. The results of specific surface area determinations, mercury intrusion porosimetry, and inverse steric exclusion chromatography (ISEC) were used to determine the different mechanism involved in extractant retention in the polymer phase as a function of the polymer structure and the type of mechanism involved in the retention: (1) physical adsorption or (2) chemical interaction.

A description of the methodologies developed in the preparation and application of SIRs on metal extraction and separation is provided in the next sections of this chapter.

9.5 SOLVENT-IMPREGNATED RESINS BASED ON REACTIVE POLYMERS

SIRs represent a simple way for implementation of the popular idea to immobilize well-proven chemical reagents on a solid support and, hence, to simplify their application. In any extraction process of target species (e.g., metallic species) from aqueous solutions, SIRs provide the possibility to replace relatively complicated liquid–liquid extraction methods with a simple stationary bed apparatus [4]. Transfer of the reagent from freely mobile fluid phase onto or into a solid carrier substantially changes the conditions for its functioning.

For easy contact of the polymer-supported ligand with the treated aqueous solution it would be advantageous if the ligand could form an adsorbed layer on the pore walls of the support. However, experimental evidence suggests that in the aforementioned SIR-type materials the ligand is not spread on the surface but rather fills the pore volume [8,9]. If one considers the hydrophobic character of both the supported liquid reagent and the polymer support, such morphology does not look too promising for practical applications for treatment of metal-bearing aqueous solutions. During treatment of aqueous metal ion solutions by SIRs, all the mass transfer must travel trough the relatively small area of the ligand menisci in the resin pore orifices. Another substantial disadvantage of these conventional SIRs is a leakage of the ligand into the treated liquid. To diminish the leakage of the ligand there was a proposition to support acidic ligands on polymeric carriers bearing basic active groups where the strong chemical interaction between the basic groups of the support and acidic groups of the ligand molecules could hold the ligand more firmly than a mere physical adsorption [8]. Another possible advantage of ion exchange resins as supports for SIRs lies in their hydrophilic nature, facilitating the contact of the resin with the treated aqueous solutions.

TABLE 9.2
Properties of the Polymer Supports

Designation	Basicity	Resin Type[a]	Effective Exchange Capacity, meq/g	Content of DVB, mol. %	Trade Name
PVP-G	weak	G	8.8[b]	2	Reillex 402
PVP-M	weak	M	5.5[b]	25	Reillex HP
AW-M	weak	M	6.1	n.s	Lewatit MP-62
AS-M	strong	M	4.4	n.s.	Lewatit MP-600
AS-G	strong	G	4.9	n.s.	Lewatit M 600

[a] G, gel-type resin; M, macroreticular resin.
[b] Manufacturer's information.

9.5.1 Preparation of Solvent-Impregnated Resins Based on Reactive Polymers

For preparation of the novel type of SIRs, two types of resins were proposed: (1) divinylbenzene (DVB)-cross-linked polyvinylpyridine resins, and (2) conventional anion exchange resins with cross-linked polystyrene skeleton bearing strong basic quaternary amine or weak basic tertiary amine groups. Their properties are summarized in Table 9.2.

In the preparation of these new materials three alkyl phosphoric acids were used that differed in degree of substitution of sulfur for oxygen: (1) Di-(2-ethylhexyl) dithiophosphoric acid (D2EHDTPA); (2) Di-(2-ethylhexyl) monothiophosphoric acid (D2EHMTPA); and (3) di-(2-ethylhexyl) phosphoric acid (D2EHPA). Impregnation was performed by suspending a weighed amount of the resin in a known amount of acetonitrile containing various concentrations of the ligands. The samples of the ligand-impregnated resins were drained on a glass frit and washed with distilled water.

Adsorption isotherms of all three alkylphosphoric acid ligands are shown in Figure 9.1. Shapes of the sorption isotherms on reactive supports with steep initial ascend and flat plateau are typical for a chemisorption mechanism. The surface area of the pores (about 14 and 60 m^2/g for AW-M and PVP-M, respectively) can accommodate an insignificant fraction of the basic group only. Chemical interaction of ligand molecules with the active groups of the supports requires penetration of the ligand inside the polymer mass. The ligand is obviously rather absorbed then adsorbed, which is evidenced by the fact that even the nonporous, gel-type resin AS-G was able to sorb 1.35 mmol D2EHDTPA/g. The changes of the surface area of the resins resulting from the ligand absorption shown in Table 9.3 can be explained by changes in the shape of pores caused by the swelling of the polymer mass rather than by filling of the support pores by the liquid ligand, which was observed in a single case when the sorbed ligand filled the pores. This is the case of sorption of D2EHPA on PVP-M at equilibrium concentrations of the ligand in the acetonitrile solution approaching 6 wt.%, which is the saturation concentration of this compound.

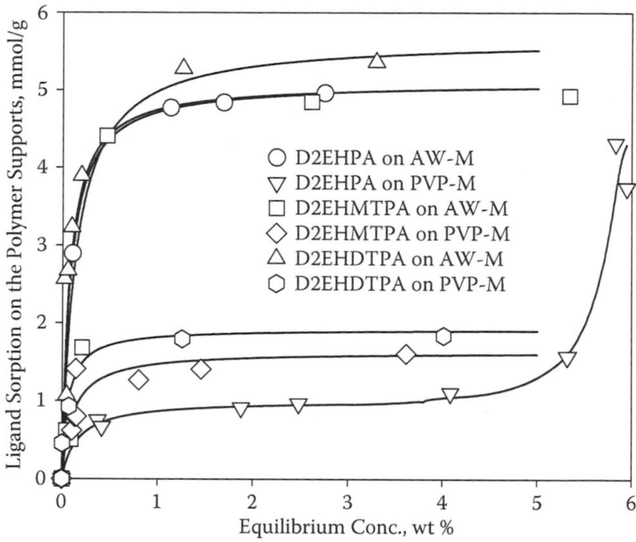

FIGURE 9.1 Comparison of sorption isotherms of the three ligands from their acetonitrile solutions on polyvinylpiridine (PVP-M) and conventional anion exchange resin (AW-M) supports.

TABLE 9.3
Separation Factors(S) for Synthetic and Mineral Leaching Solutions on the Processing of Gold Mineral Ores

Solution	pH_{eq}	$S_{Au/Ag}$	$S_{Au/Zn}$	$S_{Au/Fe}$	$S_{Au/Cu}$
Synthetic	10,2	3	6	40	20
Synthetic	10,4	4	8	45	23
Leaching 1	10,4	2	4	20	15
Leaching 2	10,6	2	5	30	15

As shown in Figure 9.1, the adsorption isotherm of D2EHPA under these conditions starts to rise sharply. This effect of sharp rise of the isotherm near the saturation point is an analogy to the capillary condensation that is well known from the adsorption measurements in solid–gas systems.

The degree of chemisorption saturation of the supports with the ligand as shown in Figure 9.1 is in all cases lower than the exchange capacity of the supports (Table 9.2). The saturation degree (i.e., the fraction of the basic groups usable for chemisorption) depends on the polymeric support morphology, its ability to swell and accommodate the ligand, and strength of the chemisorption bond.

9.5.2 Morphological Characterization of Solvent-Impregnated Resins Based on Reactive Polymers

The swelling and the swollen-state morphology of the vinylpyridine polymers was studied by (ISEC) in tetrahydrofuran (THF) [25,26]. Figure 9.2 compares swollen-

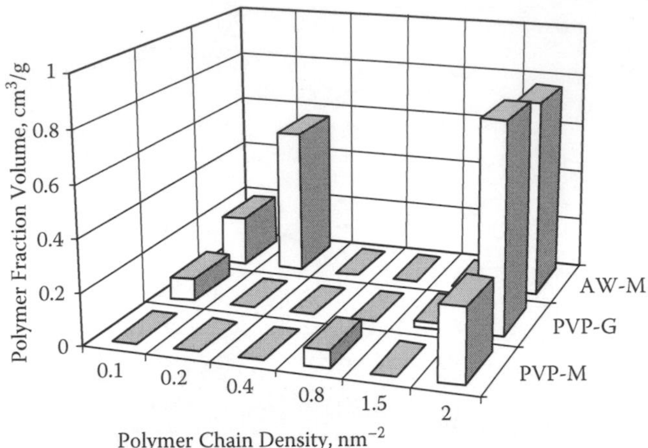

FIGURE 9.2 Swollen-state morphologies of selected polymer supports as determined by ISEC in THF (for polymer designations see Table 9.1).

state morphologies of polymer mass of three investigated polymers, depicted as a histogram of volume distribution of polymer fractions of different polymer chain densities expressed in units of chain length per unit of volume, nm^{-2}.

The scale of the polymer chain densities spans from 0.1 nm^{-2}—corresponding to the density of an extremely expanded, nearly uncross-linked polymer network—up to the density 2 nm^{-2}, which is characteristic for very dense polymer network into which even the smallest probe molecules during the ISEC measurements penetrate with difficulties. Both polyvinylpyridine resins PVP-G and PVP-M are predominantly composed of a very dense polymer mass characterized by polymer chain concentration 2 nm^{-2}. The swelling ability of PVP-G containing only 2% DVB is not too different from that of PVP-M containing 25% DVB. The great difference in the nominal degree of crosslinking results only in small difference in swelling ability. It indicates very high significance of the physical cross-linking for these polymers, which seems to be a common feature for polymers prepared from reactive monomers [25,26]. On the other hand, the swelling ability of the polymer mass of AW-M is much better. Beside the dense polymer mass domains it contains also quite a significant volume of low-density, highly swollen fractions characterized by polymer chain concentration 0.1 and 0.2 nm^{-2}. This better swelling ability explains why the sorption capacity for all the ligands on AW-M corresponds to about 80% of its exchange capacity, whereas in PVP-M less than 20% of the basic groups are accessible for the ligands.

The importance of a presence of well-expandable, low-density fractions (chain concentration 0.1 to 0.2 nm^{-2}) in the polymer mass of the support for effectiveness of ligand sorption is illustrated in the comparison of the differences between macroreticular resins AW-M and AS-M and their gel-type analog AS-G. ISEC characterization of these resins shows the presence of these low-density fractions only in the macroreticular resins, whereas the gel-type AS-G is composed of more dense fractions only (Figure 9.3).

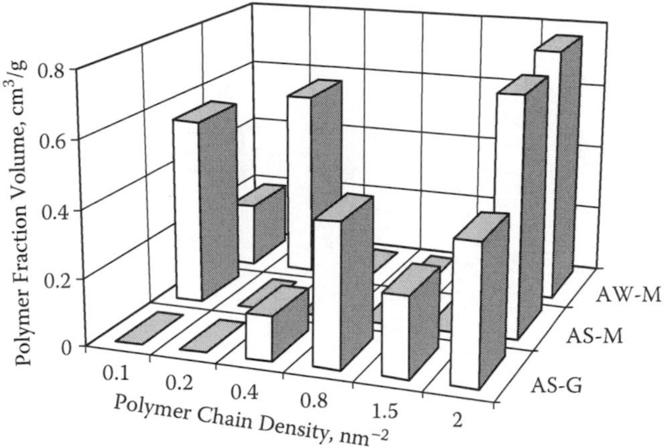

FIGURE 9.3 Swollen-state morphologies of anion exchange resin supports as determined by ISEC in THF (for polymer designations see Table 9.2).

The strikingly bidisperse morphology of polymer skeleton of AW-M and AS-M is typical for polymer skeletons of macroreticular resins [25,27]. Porogenic solvents used during the preparation of macroreticular polymers not only create the macropores but also modify morphology of the polymer mass. Figure 9.4 compares adsorption isotherms of the D2EHDTPA ligand on three resins; Figure 9.3 depicts its morphologies.

The polymer skeleton morphology has a much stronger influence on the ligand sorption than the strength of the chemical interaction between the ligand and the basic groups of the support. The difference between resins AS-G and AS-M differing in morphology but containing the same strongly basic active groups is quite substantial, whereas the weakly basic AW-M and strongly basic AS-M having similar morphology exhibit also similar sorption behavior.

However, as similar as the morphologies of AW-M and AS-M are, they are not exactly identical. Hence, on this basis the effects of the strength of the interaction

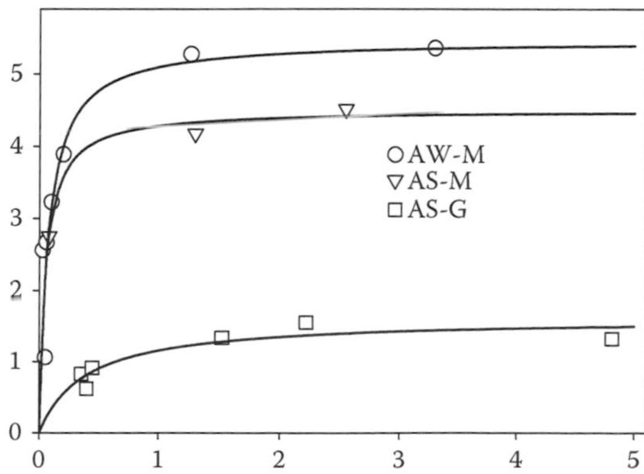

FIGURE 9.4 Sorption isotherms of D2EHDTPA from acetonitrile on anion exchange resins.

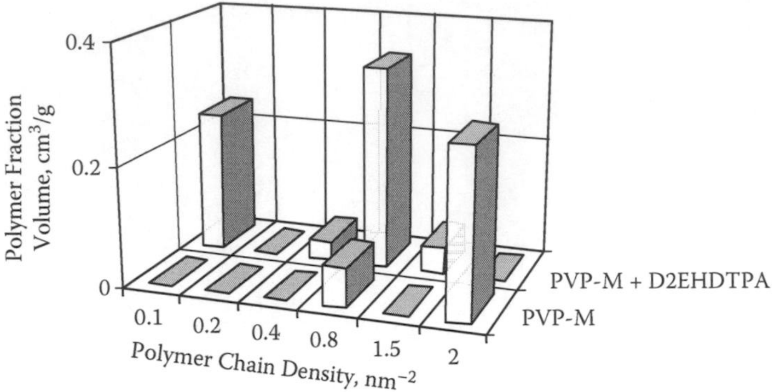

FIGURE 9.5 Swollen-state morphologies of PVP-M as determined by ISEC in THF before and after the ligand deposition.

of the acid ligand with the basic groups of the supports cannot be safely assessed. For this purpose it is better to compare the sorption isotherms of different ligands on the same support (Figure 9.1). For the conventional anion exchanger AW-M the observed differences are not very significant. The polymer mass of this resin is much more flexible than that of PVP-M, and almost all the exchange groups are accessible for the ligands. Polymer skeleton of PV-M is much more rigid, and on this support the accessibility of the basic centers depends quite significantly on the strength of the interaction between the ligand and basic groups. Out of the examined ligands the highest affinity toward the polymer support is shown by D2EHDTPA, whereas the lowest affinity was observed for D2EHPA. This does not correspond to the acid strength order of this type of acid in aqueous environment, where thiophosphoric acids appear to be weaker than their oxo-analogs. However, the sorption isotherms were measured in acetonitrile, and it is known that the order of acidities in organic solvents may be reversed [28].

The chemisorption bond between the ligand and support is rather strong and cannot be easily broken by a simple solvent extraction. ISEC-determined swollen-state morphologies of the Reillex HP in THF resin without and with the ligand are shown in Figure 9.5. It is evident that the chemisorption of the ligand on the polyvinylpyridine chains induces substantial expansion of the polymer network.

9.5.3 Metal Extraction Properties of Solvent-Impregnated Resins Based on Reactive Polymers

It must be expected that metal ion sorption influences the bond between the ligands and the basic groups of the polymer supports. It was found that the interaction of metal ions by the ligand breaks the bond between chemisorbed ligand and the support. The free ligand is not a swelling solvent for the polymer mass and is expelled from the polymer gel phase. This effect was most apparent during contact of metal ion containing aqueous solutions with SIRs prepared on the base of gel-type resin AS-G with D2EHDTPA, when from the beginning of the interaction the metal saturated water insoluble ligand starts to appear outside the polymer beads. Porous

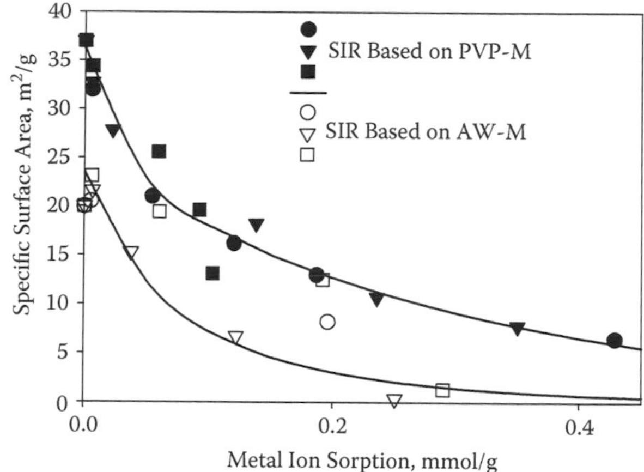

FIGURE 9.6 Surface area dependence on adsorbed amount of metal for PVP-M-based SIRs (filled marks) and AW-M-based SIRs (open marks). D2EHDTPA loadings 0.4 g/g of SIR and 0.65 g/g of SIR, respectively.

system of the macroreticular resins is able to accommodate the expelled ligand. The gradual filling of the pore volume with the metal-saturated ligand is possible to observe as changes of the surface area with the increase of the amount of sorbed metal ions (Figure 9.6).

The pore volume of the PVP-M resin determined by ISEC is 0.87 cm^3/g. The volume equivalent of the highest amount of ligand that PVP-M is able to sorb is 0.65 cm^3/g; hence, pore system of this support is evidently able to accommodate all the ligand expelled from the polymer mass. The ligand leaks from the beads only when the expelled ligand amount exceeds the volume of the macropores. Volume of the pore system of Lewatit MP-62 is rather small: In the water-swollen state, as determined by ISEC, it is only about 0.3 ml/g of the dry resin. This volume can be completely filled by only about 0.85 mmol of a ligand, which is substantially less than is the ligand sorption capacity of this resin. Therefore, during the metal ion sorption on AW-M fully saturated with any of the three used ligands, there were observed "maps" (i.e., differently colored areas) on the surface of the resin beads, or drops of liquid ligand even appeared outside of the resin beads. Dried samples of such resins exhibited practically zero specific surface area, which is proof of complete filling of their pore system with the metal saturated ligand.

D2EHDTPA forms very strong complexes with metal ions [29], and SIRs based on this ligand are hard to regenerate. Complexes of metal ions with D2EHMTPA or D2EHPA are possible to be freed from the metal ions by washing, for example, with mineral acids and so regenerating the ligand. SIRs prepared by impregnation of AW-M with D2EHMTPA (0.54 mmol/g) adsorbed in a column experiment from 0.013 M solution of cadmium sulfate 0,27 mmol Cd/mmol of supported ligand. After regeneration by washing with 3 M hydrochloric acid the sorption capacity was 0.10 mmol Cd/mmol of supported ligand; after second regeneration the capacity was found to be 0.13 mmol Cd/mmol of supported ligand.

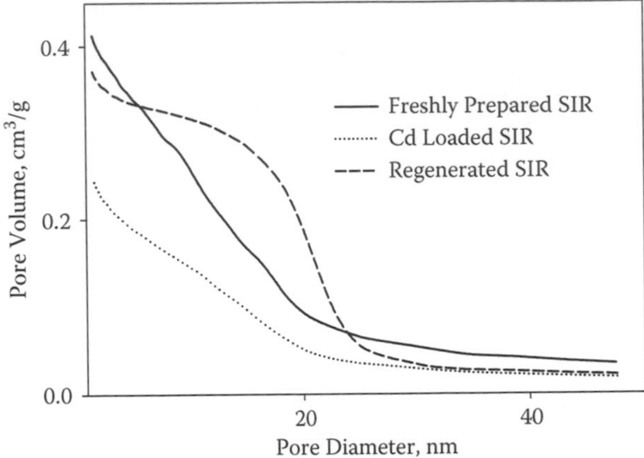

FIGURE 9.7 Culumlative pore volume distribution of AW-M with D2EHMTPA loading 0.257 g/g of support in different stages of the working cycle.

Figure 9.7 shows cumulative pore volume distributions obtained from mercury porosimetry characterization of SIRs in different stages of the working cycle (AW-M loaded with D2EHMTPA used in the previously described column experiments). Results in Figure 9.7 show practical disappearance of the porosity after saturation of the SIR with cadmium cations due to filling of the pore system with the ligand–metal complex expelled from the polymer mass. Stripping of cadmium from the resin by thorough washing with 3 M hydrochloric acid followed by washing with water regenerated the ligand. The regenerated resin shows again the pore volume similar to the freshly prepared SIR, as the metal-free ligand is able to be associated again with the basic groups of the support and, hence, to be reabsorbed into the polymer mass.

The extractant migration from the polymeric gel to the polymer pore space and back may be followed using the difference in the chemical shifts in the ^{31}P solid state NMR spectra of the extractant molecule inside the SIR resin. In the gel phase the extractant is chemically associated with the polymeric amine groups, whereas in the macropores the extractant exists as a "liquid" occupying the pore volume. The acidic extractant, forming a salt with the polymeric amine group in the polymer gel phase, is part of the solid polymeric matrix so that the corresponding phosphorous peak must be accompanied by sidebands and has a quite substantial line width, HHFW (half height full width, ca. 500Hz, "solid"-type peak). The solid-type peak is shifted slightly to a strong field, similarly to the peaks of salts. On the other hand, the liquid extractant in the macropores shows a narrow peak without sidebands ("liquid"-type peak). The migration phenomena under pH changes in the SIR based on reactive supports is shown in Figure 9.8.

The acidic feed solution containing mineral acid ions is fed through an SIR column. This forces accumulation of the organic extractant in the macropores (Step 2), and the gel phase has become extractant deficient. In the next step, Step 3, the water rinse of the SIR column forces the mineral acid out of the resin pores with concurrent migration of the extractant from the macropores to the inner gel phase,

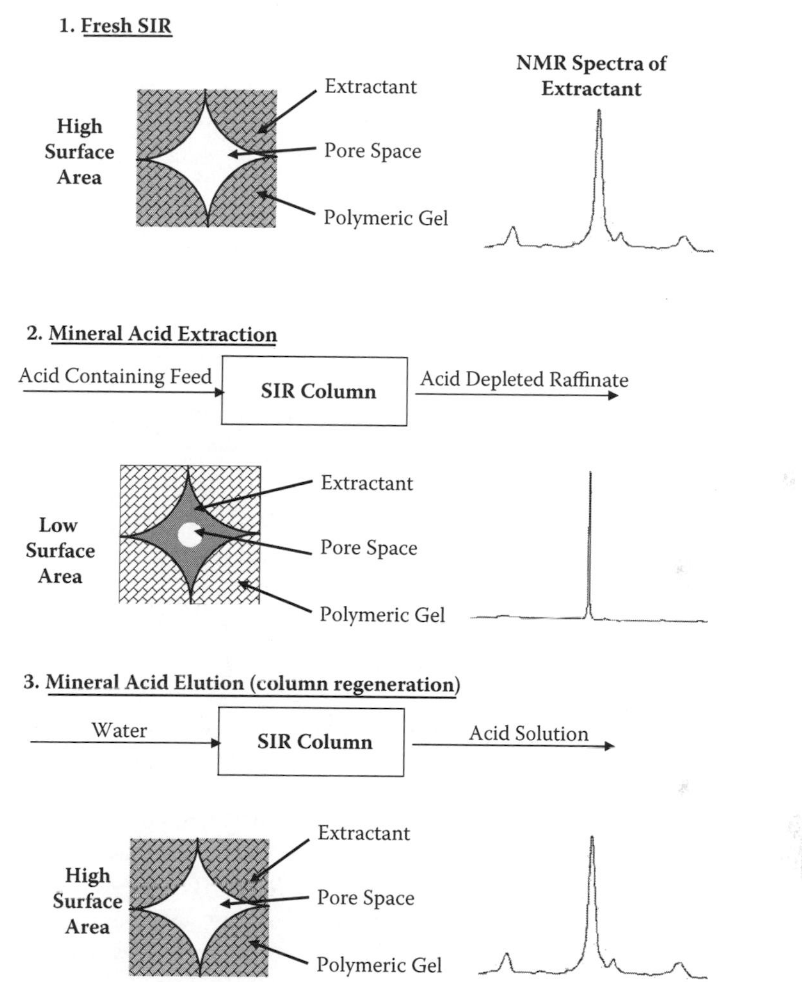

FIGURE 9.8 Migration phenomena under pH changes in the SIRs based on reactive supports.

rendering the macropores free of the extractant. In the case of creation of metal–extractant complex, the extractant is not bound to the functional groups of the resin anymore but creates its own phase in the macropores.

Morphology of SIRs resulting from supporting of liquid ligands on reactive supports seems to have distinctive advantages. On the reactive supports the intrinsically hydrophobic ligand is absorbed in the hydrophilic matrix, and it should facilitate its contact with the treated aqueous phase. In a preliminary experiment, sorbent prepared by supporting of Di-(2-ethylhexyl)dithiophosphoric acid (D2EHDTPA) on the macroreticular basic ion exchanger Lewatit MP-62 was able to diminish cadmium concentration in water from 100 ppb to less than 5 ppb at flow rate 4 bed volumes/min (Figure 9.9).

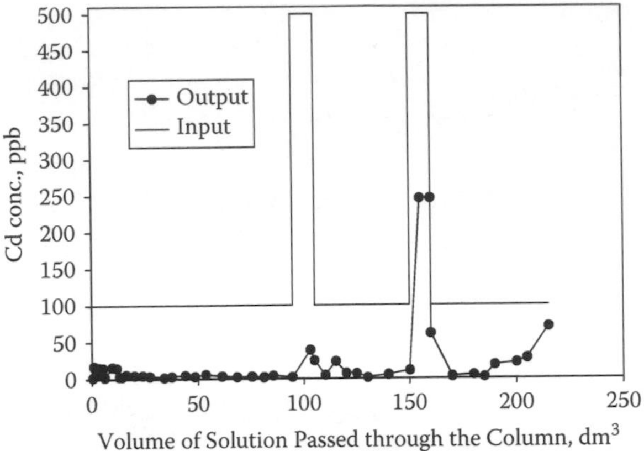

FIGURE 9.9 Diminishing of cadmium concentration in water after treatment on column with 7.2 cm³ (1.82 g) SIR prepared by impregnation of Lewatit MP-62 with D2EHDTPA.

Quite exceptional is the mobility of the ligand during the working cycle of the sorbent. Only the ligand anion can "dissolve" in the polymer phase, because it is held in it as the counterion to the polymer-bound cationic groups. The ligand itself is incompatible with the polymeric phase, and when the ionic bond is interrupted without this strong interaction the freed ligand is immediately expelled outside. It means that after the "chemical" separation (i.e., formation of the metal–ligand complex) follows a "physical" separation, where the metal/ligand complex migrates outside the polymer mass while the unspent ligand remains in the polymer. Physical separation of the product (i.e., the metal–ligand complex) makes it possible to recover it in a concentrated form. In another preliminary experiment, SIRs were prepared from Lewatit MP-62 and Di-(2-ethylhexyl)thiophosphoric acid (D2EHMTPA) and used for sorption of lead from 0.001 M solution of lead perchlorate. Material balance based on the differences between input and output lead concentration show that after passing of 7.5 dm³ of the solution through column filled with 10 g of SIR there was sorbed 870 mg Pb. After drying the resin was extracted with toluene. Analysis of the extract (determination of dry residuum and Pb content) showed that the extract contained pure complex $Pb(D2EHMTPA)_2$ in the amount almost exactly corresponding to the weight of lead separated from the water solution. With toluene was extracted only the metal–ligand complex, and moreover, the extraction was very easy; more than 90% of the metal was contained in the first 25 cm³ of the extract. The unreacted ligand (about 75% of the originally supported amount) remained absorbed in the resin.

The previously described two illustrative examples show very promising exceptional features of the novel solvent-impregnated resins. Especially the unique combination of chemical and physical separation mechanisms can bring completely novel tools for the design of separation processes. Reclaiming the separated metal in the form of concentrated organic solution could be in some cases more attractive than its elution with aqueous regenerating solvent, which could be sometimes rather difficult (e.g., in applications of powerful sulfur-containing ligands). On the other hand, these resins represent very complicated systems, because their morphology in the course of the working cycle varies and can influence both their working capacity and the kinetics of the processes. Information on these effects is still very scarce, and that is why we would like to investigate in detail properties of the solvent-impregnated resins based on reactive supports, especially from the chemical engineering point of view.

9.6 DEVELOPMENT OF SOLVENT-IMPREGNATED RESINS FOR GOLD HYDROMETALLURGICAL APPLICATIONS

Dilute solution of goldcyanide and other metal cyanide complexes (Ag, Fe, Cu, Ni, Co) are currently purified and concentrated by activated carbon [30]. However, in the last few years the following innovations have been attempted: substitution of activated carbon by ion-exchange systems and implementation of processes combining leaching and adsorption steps [31,32]. The main goal of this effort was solving existing problems associated with activated carbon—for example, the need to improve the selective recovery of goldcyanide, the need to prepare ion exchange materials with high loadings and stripping efficiencies, and the development of integrated process of leaching and extraction as resin-in-leach and resin-in-pulp processes. In this last case, the ion exchange material to be used should fit the experimental conditions of the system and should be able to extract gold from the leaching solutions containing $Au(CN)_2^-$ at about pH 10 [43]. The objective might be accomplished if it is possible to develop a polymeric support with a functional group (R), which would operate on the hydrogen ion cycle shown in the following reaction:

$$PS\text{--}R_{res} + H^+X^- \Leftrightarrow P\text{--}R\ H^+X^-_{res} \tag{9.4}$$

where the equilibrium lies far to the right at pH < 10 and far to the left a pH > 12.5. The functional group in the protonated form, (RH+), at pH < 10 would be an active anion extractant; the functional group in the neutral form R at pH > 12.5 would not be an anion extractant. Then, goldcyanide extraction would be accomplished by the following reaction:

$$P\text{--}RH^+X^-_{res} + Au(CN)_2^- \Leftrightarrow PS\text{--}RH^+Au(CN)_2^-{}_{res} \tag{9.5}$$

Available single extractants tested do not meet the aforementioned objectives on pH cycle, with the exception for recently developed family of reagents from Cognis Co containing a guanidine functionality with pKa values around 12 or greater, depending on the R groups attached to the various nitrogen atoms [34–35]. The reagent known as LIX79 contains a guanidine ion-pairing functionality (Figure 9.10).

In previous studies [34,36] the extraction of gold(I) from aurocyanide aqueous solutions using LIX 79 was evaluated and showed that the aurocyanide complex

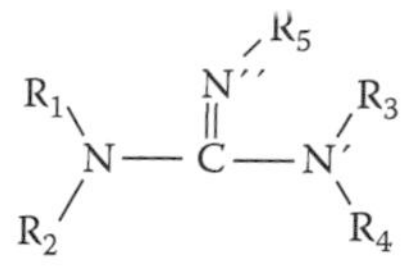

FIGURE 9.10 Basic structure of the guanidine functional group (R_1 to R_4 represent alkyl groups).

is extracted preferentially over other metal–cyano complexes at alkaline pH and is extracted into the organic phase by formation of the $RHAu(CN)_2$ species.

9.6.1 Preparation of Impregnated Resins Containing LIX 79 for Goldcyanide Extraction

Solid adsorbents can be prepared also by immobilization of liquid extractant during the polymer synthesis in the presence of the extractant [37]. Synthesis of solid extractant (TVEX) containing 40 to 50% LIX79 was carried out in a glass reactor fitted with a mixer, thermometer, and backflow condenser [38]. Starch solution (1%), used as emulsion stabilizer, was first added to the reactor. Then reaction mixture containing styrene, divinylbenzene, LIX79, and benzoyl peroxide as initiator of radical copolymerization was fed into the reactor at 55°C with agitation. Polymerization was carried out for 5 hours at 85 to 90°C. The TVEX granules were washed with water and dried with nitrogen.

9.6.2 Extraction Performance of LIX79 Impregnated Resins

The variation in gold extraction against pH for LIX79 resins showed an increase with pH decrease, and as the extractant concentration increased the corresponding gold extraction curve shifted to more alkaline pH values (Figure 9.11). This behavior is also extensive to impregnated resins prepared with pure LIX79 (TVEX/LIX 79, XAD2/LIX79).

The expected general equilibrium from which the aurocyanide complex is extracted by LIX 79 can be represented by the reaction

$$R_{res} + H^+ + Au(CN)_2^- \Leftrightarrow RH^+Au(CN)_2^-{}_{res} \tag{9.6}$$

where R denotes the extractant molecule and the subscript res represents a species in the resin phase. Assuming ideal behavior for the reaction in Equation (9.3) in resin and aqueous phases, the stoichiometric equilibrium constant (K) can be written as

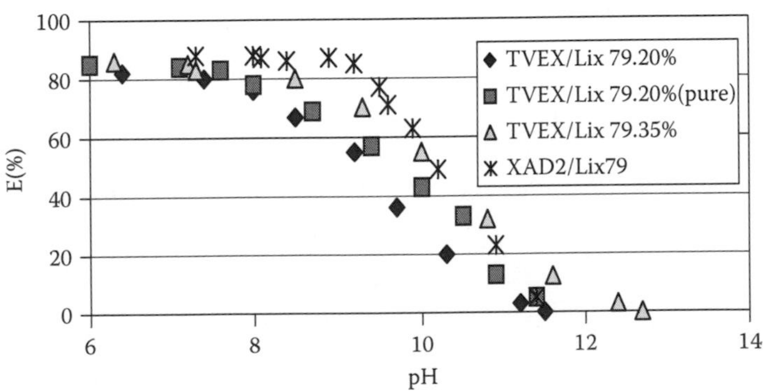

FIGURE 9.11 The influence of the LIX79 contents on impregnated resins on gold(I) extraction from cyanide solutions.

$$K = \frac{\left[RH^{+}Au(CN)_2^{-}\right]_{res}}{[R]_r\left[H^{+}\right]\left[Au(CN)_2^{-}\right]} \tag{9.7}$$

and, taking into account the distribution coefficient D_{Au},

$$D_{Au} = \frac{[Au(I)]_{r,\ tot}}{[Au(I)]_{aq,\ tot}} \tag{9.8}$$

dependence of the distribution coefficient and operation parameters can be described by the following equation:

$$\log D_{Au} = \log K - pH + \log[R]_r \tag{9.9}$$

The $pH_{50,}$ pH where a 50% of extraction (D_{Au} = 1) is obtained and is defined by the following equation:

$$pH_{50} = \log K + \log[R]_r \tag{9.10}$$

Figure 9.12 shows that the variation of pH_{50} as a function of the different LIX79 impregnates has a linear relation with a slope of 1 and an intercept of 12.9 as the logK value. Then, goldcyanide extraction on LIX79 impregnates occurs via formation of a $RH^{+}Au(CN)_2^{-}$ complex onto the resin phase, similar to the described behavior for this reagent when used in liquid–liquid extraction systems.

The selectivity of the present LIX79 impregnated resins against the extraction of different metal–cyano complexes was studied with a 35% LIX 79 content resin and aqueous solutions with aqueous 10 mg/L metal concentration. Results are plotted in Figure 9.13, which represents the percentage of metal extraction by TVEX/LIX 79 resins against equilibrium pH, showed that the aurocyanide complex is extracted at the most alkaline pH value and, consequently, can be separated selectively from other metal–cyano complexes present in the aqueous phase in this pH range.

It can be seen that an extraction sequence can be tentatively established for these metal–cyano complexes: $M(CN)_2^{-} > M(CN)_4^{n-} \sim M(CN)_6^{n-}$; thus, the extraction of these complexes depends on the metal coordination number. In general, those complexes with lower coordination numbers were extracted preferentially over those

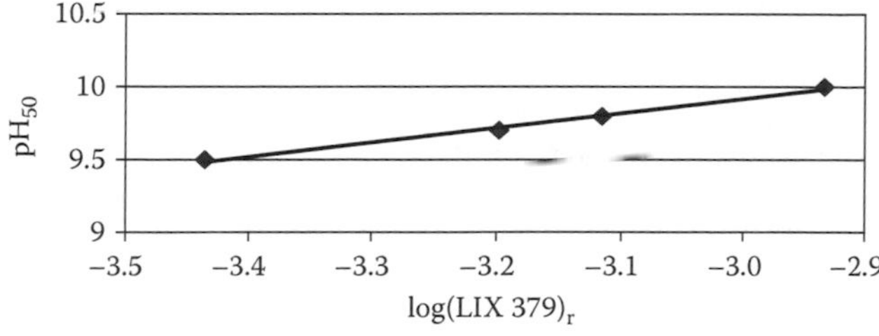

FIGURE 9.12 Dependence of pH_{50} on LIX79 content on the impregnated resins (XAD2-LIX79).

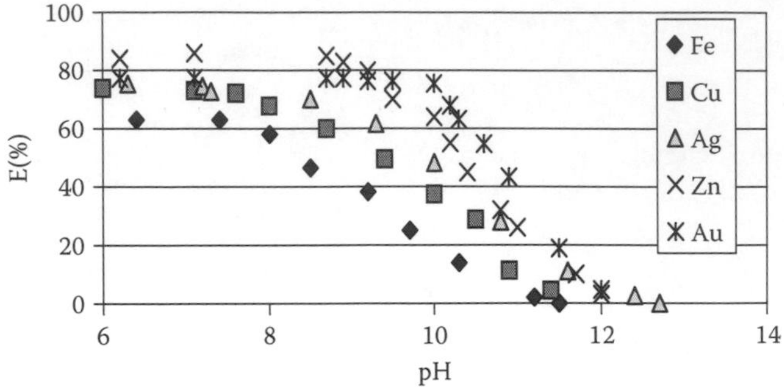

FIGURE 9.13 Extraction-pH dependence of $Au(CN)_2^-$, $Zn(CN)_4^{2-}$, $Ag(CN)_2^-$, $Fe(CN)_6^{4-}$, and $CU(CN)_3^{2-}$ with impregnated resins (XAD-LIX79 [35%]).

with higher numbers, and the lower charge complexes are extracted preferentially over higher charge complexes. Furthermore, extraction experiments were performed with the real leach solutions, and the results expressed in terms of the separation factors ($S_{Au/metal}$) are compared with those for synthetic liquor in Table 9.4.

It can be seen that, independent of the aqueous solution, the apparent order follows the series

$$S_{Au/Ag} > S_{Au/Zn} > S_{Au/Cu} > S_{Au/Fe}$$

This selectivity patern is in agreement with the fact that the hydrophilicity of the polymer matrix and the ionic density play important roles in determining the selectivity characteristics of an impregnated resin. Metal cyanide ions–water interactions are stronger in the aqueous phase than in the resin phase simply because more free water is available for solvation in the aqueous phase. Thus, on this basis alone the ions with the greatest ionic charge tend to have a lower extraction onto the resin phase. In the resin phase a low degree of hydrophylicity and a low ionic density increase the selectivity for gold and silver and favors those metals over the base metals, iron, and copper.

The stripping of Au(I) from loaded TVEX-LIX79 impregnates follows the order NaOH > NaCN and tends to increase with increasing reagent concentration, although good Au(I) stripping can be achieved using low NaOH or NaCN concentrations.

TABLE 9.4
Separation Factors(S) for Simulated and Mineral Leaching Solutions on the Processing of a Brazilian Gold Containing Mineral Ore

	pH_{eq}	$S_{Au/Ag}$	$S_{Au/Zn}$	$S_{Au/Fe}$	$S_{Au/Cu}$
Simulated	10,2	3	6	40	20
Simulated	10,4	4	8	45	22
Leaching 1	10,4	2	4	20	15
Leaching 2	10,6	2	5	30	15

When the stripping process is completed, the active component [R] of the impregnated resin is deprotonated (Equation (9.3) shifts to the left), and no metal cyanides are present in the resin phase.

9.7 SEPARATION AND PRECONCENTRATION OF HEAVY METALS WITH SIRs CONTAINING ORGANOPHOSPHOROUS EXTRACTANTS

Novel metal decontamination and water treatment technologies are required for a range of applications in the metallurgical and mining sectors [39]. Both types of industries are characterized by using huge volumes of water streams with high contents of strong acids and bases, high content of electrolytes, and moderate to low levels of metal ions.

Most nonferrous metal industries generate basically two types of metal-containing effluents: (1) highly concentrated liquors from spent baths of electroplating or tanning operations; and (2) diluted streams typically generated by rinsing operations. Whereas the first type is directed to recovery operations such as chemical precipitation, electrowinning, or cementation, the second type of liquor needs a preconcentration step to reduce volume and to achieve suitable concentrations to be treated like the spent baths. Additionally, as those diluted streams could contain mixtures of metal ions, a separation step is required to achieve suitable conditions for subsequent recovery. Metal ion mixtures are also found in acidic waters generated by dedusting off gases of many domestic and industrial waste incineration plants, thermal power stations, and metallurgical plants. These acidic solutions contain a wide range of nonferrous metals such as Zn, Cd, Pb, and Cu in the order of 0.5 to 0.01g/L and nonmetals such as As, Sb, and Bi, typically in the order of 0.01 to 0.001 g/L. The main components of such a stream are summarized in Table 9.5.

Although traditionally these solutions are treated by the physico-chemical process of precipitation with lime, current efforts are directed to the selective recovery and separation of the metals Zn, Cd, and Pb for subsequent treatment by electrolysis (for Zn, Pb, Cu) or cementation by powder metallic zinc (for Cd). Those effluents with relatively low levels of metal ions cannot be treated efficiently by SX but can be treated by transforming the SX system into the corresponding SIR system.

TABLE 9.5
Typical Composition on Metals, Nonmetals, and Salts of Acidic Metallurgical Waste Effluents

Metals (M): 0.1–0.5 g/L	Nonmetals (NM): < 0.01g/L	Common Anions (A) (0.1 to 100 g/l)	Common Cations (C) (0.05 to 5 g/l)
Zn	As	Cl^-	Ca^{+2}
Cu	Sb	SO_4^{-2}	Na^+
Cd	Bi	NO_3^-	Mg^{+2}
Pb	Se		H^+ (pH (0–5))

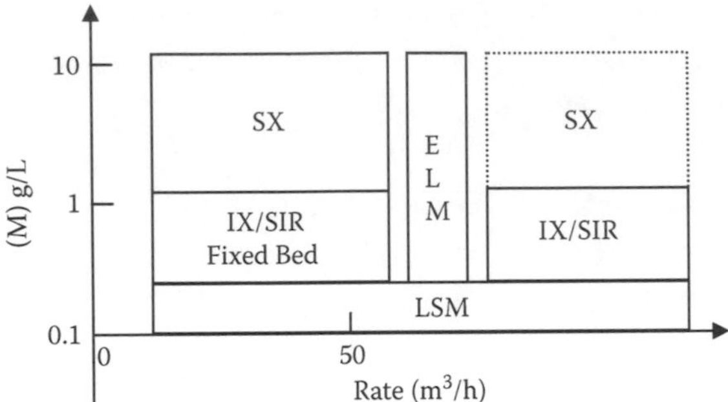

FIGURE 9.14 Application limits for selected metal separation processes. IX, ion exchange; SX, solvent extraction; SIRs, solvent-impregnated resins; SLM, supported liquid membranes; ELM, emulsified liquid membranes.

Many of the separation, transformation, and destruction processes required already exist, such as IX, SX, and membrane filtration processes. However, improvements are needed, notably in economic efficiency of the treatment and minimization (volume reduction), neutralization, and further processing and treatment of byproducts. This is reflected in the large number of scientific contributions during the last decades, and the impetus is in the development of new separation chemistry techniques, particularly for treatment of liquid wastes and effluents [40,41]. Taking into account these limitations in the preparation of chelating and IX resins and the recent advances in the synthesis of new organic reagents for SX, which are bound to significantly improve both separation efficiency and selectivity for a wide range of chemical species, the development of SIRs as a link between SX and IX processes has become an important field of development in separation science. The place and limits of these materials in the domain of metal recovery methods covering the IX and SX sectors are shown in Figure 9.14. As can be seen the role of SIR materials will be closely related to that played by IX covering the low-level range (below 1 g/l).

9.7.1 Preparation of Impregnated Resins for Transition Metal Extraction

The application of SIRs in the removal and separation of toxic and heavy metal ions (Zn, Cu, Cd) present in many industrial streams at low levels is presented. Acidic organophosphorous derivatives (DEHPA, Cyanex 272) have been used as active component due to their high affinity and complexing properties with heavy and toxic metal ions and also for their suitable properties in the stripping step.

A great effort has been directed to the separation of Zn from Cu and Cd, and problems were encountered when applying strong cation exchange (bearing sulphonic acid groups) or chelating (bearing carboxylic, diiminoacetic, or aminophosphinic groups) resins. Benefiting from the extensive extraction studies of Zn, Cu, and Cd using organophosphosphorus extractants in SX applications [42], significant pioneering work was initiated on the homologous SIR systems [8–11]. Impregnated resins containing acidic organophosphorous derivatives (Cyanex 272 and DEHPA) impregnated onto Amberlite XAD2 or Levextrel type resins (Lewatit 1026 OC, Lewatit TP807) were made and tested. The behavior of such materials (in terms of

TABLE 9.6
ΔpH_{50} of Zn(II), Cu(II), and Cd(II) with Impregnated Resins and Levextrel Resins

Resin	Zn/Cd	ΔpH[50] Zn/Cu	Cd/Cu	%M1/M2 Zn/Cd	(pHopt) Zn/Cu
Impregnated Resins					
DEHPA/XAD2	0.8	0.5	0.3	85/5 (2)	85/20 (2)
Lewatit 1026 (DHEPA)	1	0.6	0.4	85/10 (2)	85/10 (2)
Cyanex 272/XAD2	2.0	1.3	0.6	99/1 (3.5)	99/5 (3)
Lewatit TP807 (Cyanex 272)	2.0	1.5	0.5	99/1 (3.5)	99/5 (3)
Ion Exchange Resins					
Lewatit TP207[a]	1.2	0.8	0.4	NA[b]	NA[b]

[a] Lewatit 207, chelating resin containing aminophosphinic groups
[b] NA, no available data.

selectivity) was controlled modulating polymer properties or by modifying extraction properties or by using mixtures of extractants [43].

The extraction dependence of divalent metals on SIRs containing acidic organophosphorous extractants on the acidity of the aqueous solutions is described by the following general reaction:

$$M^{2+} + (2 + q)\, HL_{res} \Leftrightarrow ML_2\, (HL)_{q,res} + 2H^+ \tag{9.11}$$

The pH_{50} (pH for which a 50% extraction efficiency is achieved) has been used to measure the separation factors. A practical ΔpH_{50} $(M_1/M_2) \geq 1$ allows an effective separation efficiency of M1 (>95%) over M2 (<5%). The pH_{50} and ΔpH_{50} values for the set of selected metals for solvent extraction, ion exchange, and impregnated resins systems (Table 9.6) show that sulphonic, phosphoric, and carboxylic groups do not allow efficient separation factors, yet phosphinic acid extractants such as Cyanex 272 could offer an appropriate solution to solve this separation problem [44].

The pH dependence functions for the target group of transition metals are shown in Table 9.6. Cyanex 272 impregnated resins with the highest separation factors are shown in Figure 9.15. The extraction function of Zn(II) in relation to Cu(II) and especially to Cd(II) makes this resin suitable for use in the separation of this metal couple.

The extraction efficiency of SIRs prepared by direct adsorption of Cyanex 272 onto a macroporous polymeric support of styrene-divinylbenzene (Amberlite XAD2) and those prepared by polymerization of styrene-divinylbenzene in the presence of Cyanex 272 (Levextrel type) were evaluated in column operations. The breakthrough curves obtained for solutions containing Zn(II), Cu(II), and Cd(II) in the low concentration range 1–100 mg/L are shown in Figure 9.16.

As can be seen metal ion capacity of the impregnated resins is highly dependent on operating pH, and Cd(II) extracted at the higher pH values show lower resin loadings. Metal solutions were in sulfate media (1,000 mg/L), and pH was adjusted

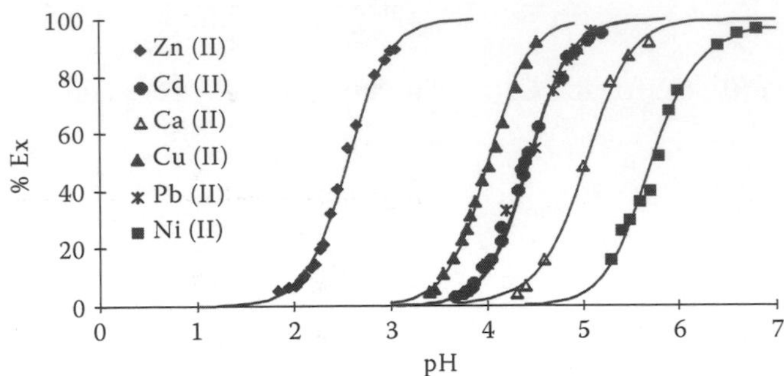

FIGURE 9.15 Extraction of Zn(II), Cu(II), Cd(II), Pb(II), Ca(II), and Ni(II) as a function of pH for Cyanex 272 impregnated resin.

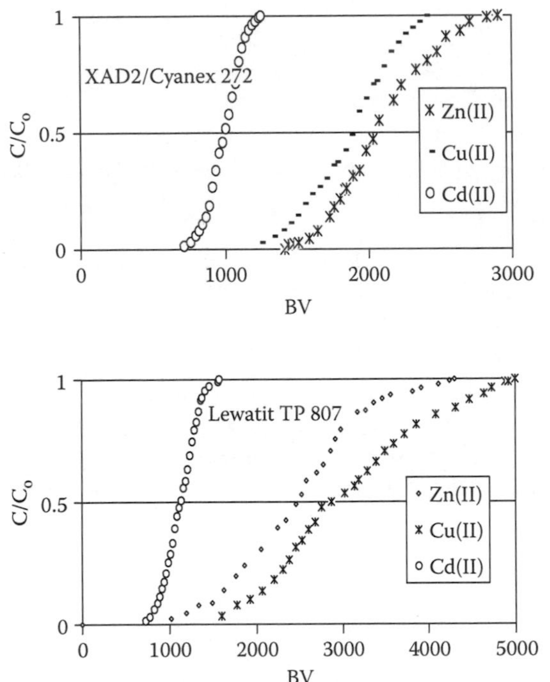

FIGURE 9.16 Breakthrough curves for Zn(II), Cu(II), and Cd(II for impregnated resins XAD2/Cyanex 272 and Levextrel TP807'84 resins.

to 5.5 by adding sodium hydroxide. As the metal extraction process progresses, the metal–proton exchange reaction takes place and the increased proton concentration causes decreased pH of the feed solution, which in turns results in favored extraction of Zn(II) and Cu(II) at lower pH values. Consequently, relative high metal loadings between 20 and 30 g metal/kg resin were obtained.

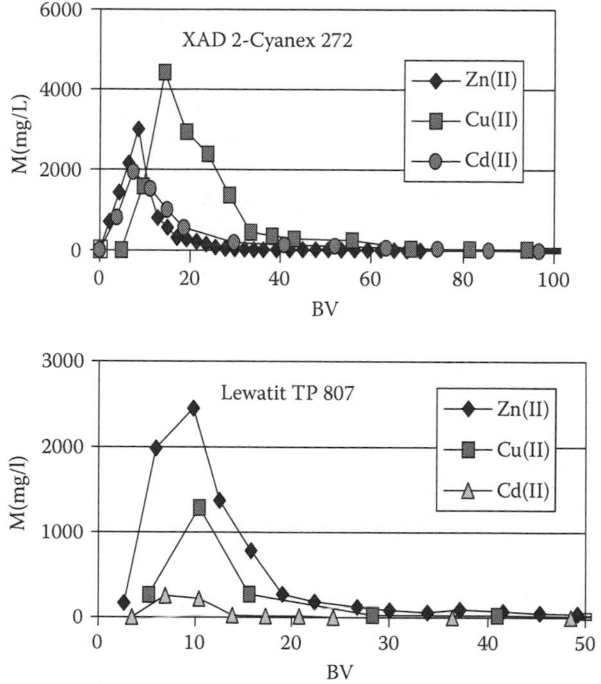

FIGURE 9.17 Elution curves of Zn(II), Cu(II), and Cd(II) for XAD2/Cyanex 272 and Levextrel TP807'84 resins.

Elution runs for single systems were performed using 1 mol/L HCl solutions, and results are shown in Figure 9.17. Concentrated solutions containing 4 g/L metals can be obtained.

Preconcentration factors (PFs) achieved are shown in Table 9.7. Higher PF factors obtained operating with lower flow rates allow the achievement of concentrated solutions (up to 10 to 20 g/L) suitable for subsequent cementation or electrowinning steps.

TABLE 9.7
Preconcentration Factors of Zn(II), Cu, and Cd(II) with Impregnated Resins Containing DEHPA and Cyanex 272

Resin	PF (Zn(II))	PF(Cu(II))	PF(Cd(II))
Reagent: DEHPA: XAD2/DEHPA	40	45	45
Lewatit 1026 OC	60	70	60
Reagent: Cyanex 272: XAD2/Cyanex	65	70	50
Lewatit TP807	60	55	80

9.7.2 Impregnated Resins for Transition Metal Extraction Separation by Parametric Pumping

More recently, new solutions to improve the separation of mixtures of dissolved heavy metals by means of parametric pumping with variation of pH have been presented for IX resins [45]. The most important property of this method is the fact that it is a regenerant-free separation or ion exchange process. Separation is possible if a parameter of substantial influence on the sorption equilibrium is varied.

The strong acid dependency observed during the column operation was evaluated as a possible separation method to be efficiently applied using parametric pumping. Only a few publications in the literature describe the variation of weak parameters (e.g., variation of ionic strength and the change of molecular sizes) as a factor in parametric pumping that will allow separation of heavy metals. The main current activity in this direction was only the use of pH as a controlling factor in parametric pumping. In this study parametric pumping assays were performed by dividing an original solution into two half volumes. Both half volumes were by turns contacted with a Cyanex 272 resins, and an exchange reaction occurred during each contact. As favorable parameter values were adjusted, Cu and Cd were transported from one solution to the other, whereas Zn was transported in the opposite direction.

Direct parametric pumping assays were performed by simply varying the pH of the solutions. Using more strongly acidic impregnated resins containing DEHPA, a significant effect was found when the hydrogen ion concentration of low pH solution exceeded the concentration of the heavy metals by some magnitude. On the other hand, the pH of the higher-pH solution is limited by the precipitation of heavy metal hydroxides and additionally increases the DEHPA molecules distribution to the aqueous phases. Direct efficient mode separations were not possible using resins containing phosphoric groups (e.g., DEHPA), and effort was directed to resins containing a weaker acidic organophosphorous function such as phosphinic groups (Cyanex 272). Parametric pumping experiments were carried out at laboratory scale using 1-liter containers. The pH was controlled and automatically adjusted by means of a pH-controlling unit with two dosage pumps for adding acid or base if required. In each half cycle, the stirrer with the resin material rotated for 120 minutes due to the kinetic behavior of those resins in each solution. Between the contact with the two solutions, the liquid in the resin packing was centrifuged off.

Separation experiments of the couples Zn/Cu and Zn/Cd with Lewatit TP807 using the direct mode were carried out, and results are shown in Figures 9.18 and 9.19.

Separations in the direct mode were achieved. The pH values were adjusted according to the pH-dependency functions of Cyanex 272 resins given in Figure 9.16. Figure 9.18 shows the separation of Zn from Cu. Zn is extracted at pH 3, whereas the extraction of Cu commenced only after reaching pH 4 or higher. Zinc therefore was transported into the pH 1.5 vessel. After ten cycles, the pH 4 vessel contained a Cu(II) solution with 2% Zn(II). Similarly, Figure 9.19 shows the separation of Zn from Cd, using in this case a vessel of pH 4.5 where Zn was transported into the pH 1.5 vessel. After ten cycles, the pH 4.5 vessel contained a Cd(II) solution with 1% Zn(II).

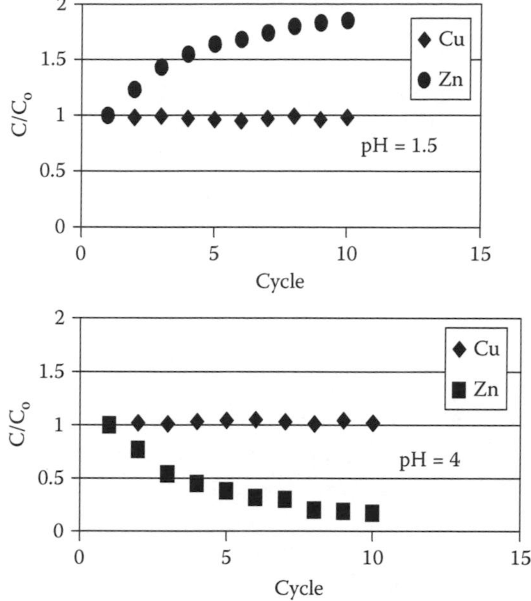

FIGURE 9.18 Parametric pumping runs for the system Zn(II)/Cu(II) Lewatit TP807 in the two half volumes system. Initial concentrations of Zn(II) and Cu(II) = 2 mmol/L.

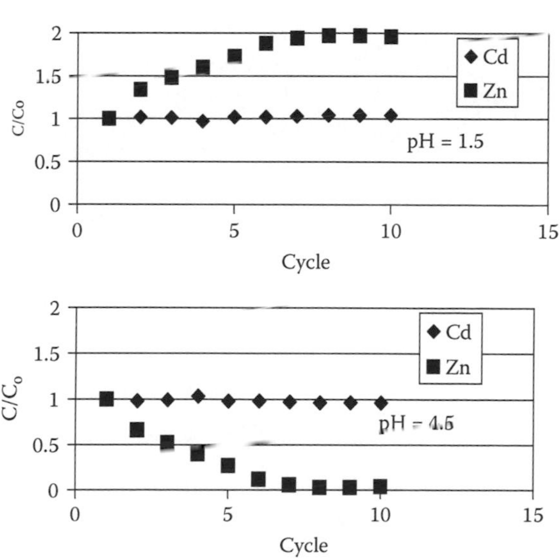

FIGURE 9.19 Parametric pumping runs for the system Zn(II)/Cd(II) Lewatit TP807 in the two half volumes system. Initial concentrations of Zn(II) and Cd(II) = 2 mmol/L.

9.8 DEVELOPMENT OF IMPREGNATED RESINS FOR RECOVERY AND SEPARATION OF PLATINUM GROUP METALS

Catalytic converters, which allow the purification of automotive exhaust emissions, contain significant amounts of platinum group metals (PGMs), specifically palladium (Pd), platinum (Pt), and rhodium (Rh). This resource is particularly significant because the PGM concentrations in catalysts are, in general, higher than those of the richest ore bodies. Since a substantial share of the world demand for PGMs is used for autocatalyst production, the quest for new processes focused on the recovery of these metals to be reused in the production of new catalysts is required [46,47]. When the PGMs are recovered by means of hydrometallurgical schemes, the ceramic matrix containing the metals is leached in acidic media, and subsequently Pd, Pt, and Rh would be separated using conventional techniques such as solvent extraction and ion exchange. Although processes based on ion-exchange and solvent extraction have been published or patented, the research for specific strategies for separation and recovery of PGMs based in these technologies is required [48]. Intensive research dealing with PGM recovery and separation using impregnated resins [26,27,49,50] was devised to develop straightforward PGM separation patterns in acidic media using impregnated resins prepared by adsorption onto Amberlite XAD2 of two different organic extractants: Alamine 336 and DEHTPA.

Organothiophosphoric extractants such as Di-(2ethylhexyl)thiphosphoric acid (DEHPTA) showed highly selective separation patters for PGM extraction when impregnated on XAD2 resins, as can be seen in Figure 9.20.

Pd(II) was completely extracted, whereas Pt(IV) and Rh(III) in practice exhibited nil sorption. Such a difference between Pd(II) and the other metals is mainly as a result of the Soft Lewis acid character of Pd(II); studies have unambiguously established that extractants containing sulfur as DEHTPA, Soft Lewis bases, are highly effective and selective for Soft Lewis acids [51,52]. Pd(II) distribution data with impregnated resins containing DEHTPA showed that the extraction of Pd(II) from HCl solutions with XAD2/DEHTPA was described through the following metal extraction reaction:

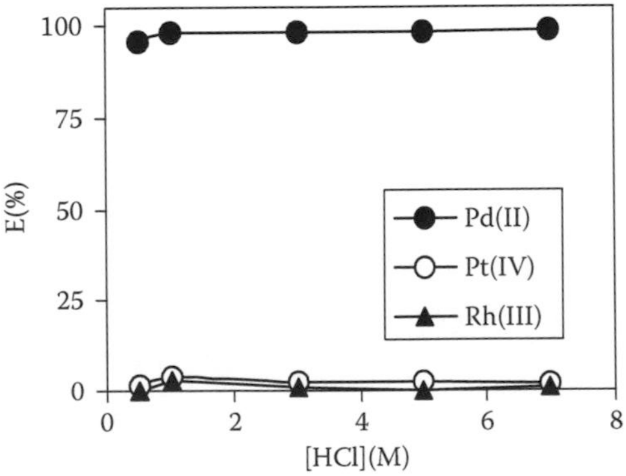

FIGURE 9.20 Pd(II), Pt(IV), and Rh(III) extraction from hydrochloric solutions using impregnated resins containing Di-(2ehtylhexyl)thiphosphoric acid (DEHPTA).

$$PdCl_4^{-2} + 4HL_r \Leftrightarrow PdL_2(HL)_{2\,res} + 4Cl^- + 2H^+ \quad (9.12)$$

The elution of Pd(II) from the resin could be achieved using hydrochloric thiourea (Tu) solutions due to the fact that thiourea ligands, which are a soft ligand, can reextract the Pd(II) from the resin phase as a water soluble complex ($PdTu_4^{+2}$) as is described by the following reaction:

$$PdL_2(HL)_{2\,r} + 4\,Tu + 2H^+ \Leftrightarrow PdT_4^{+2} + 4HL_r \quad (9.13)$$

On the other hand, when HCl was employed as a stripping agent varying the acid concentration from 0.5 to 10 mol/dm^3, a nil stripping ratio was found in all cases. Selective separation factors for Pd(II), Pt(II), and Rh(III) at different HCl concentrations was observed when using XAD2/Alamine 336 resins, as can be seen in Figure 9.21.

Almost complete extraction of Pd(II) is attained in the HCl concentration range studied. Pt(IV) extraction is complete at 0.5 M HCl but decreases by increasing HCl concentration. In contrast to Pd(II) and Pt(IV), Rh(III) is not extracted independently of the acid concentration.The sorption of Pd(II) and Pt(IV) on the impregnated sorbent can be expressed according the following reaction:

$$2\,R_3NH^+Cl^-_{org} + MCl_n^{-2}{}_{aq} \Leftrightarrow (R_3NH^+)_2MCl_n^{-2}{}_{org} + 2Cl^-_{aq} \quad (9.14)$$

with n = 4 for Pd(II) and n = 6 for Pt(IV).

HCl, $HClO_4$, and Tu in HCl were tested to strip Pd(II) and Pt(IV), and the results are summarized in Table 9.8. Either $HClO_4$ or Tu are able to effectively reextract Pd(II) and Pt(IV). $HClO_4$ acts through an anion-exchange mechanism, and Tu is effective since Pt(II)—Tu had previously reduced Pt(IV) to Pt(II)—and Pd(II) form very stable complexes with compounds containing sulphur as stated previously. Concerning HCl stripping ability, it was found that increasing HCl concentration, metal stripping increases and for instance, concentrated HCl can strip quantitatively Pd(II); nevertheless, only partial elution of Pt(IV) was found in all the HCl ranges investigated.

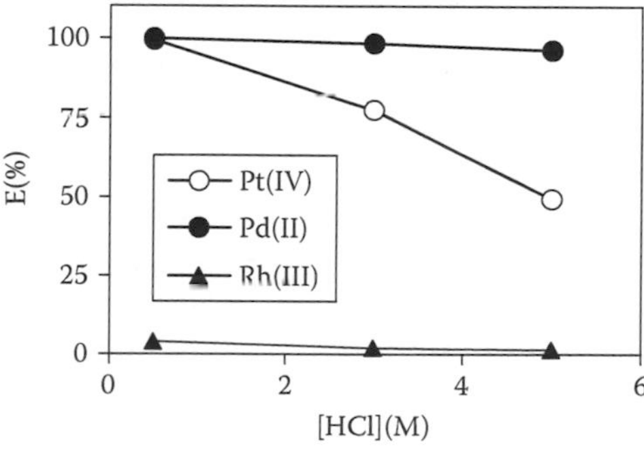

FIGURE 9.21 Effect of HCl concentration of Pt, Pd, and Rh extraction of Alamine 336 with XAD2/Alamine 336 resins.

TABLE 9.8
Stripping (S%) of Pd(II) and Pt(IV) from Impregnated Resins Containing Alamine 336 (XAD2/Alamine 336)

Tu($mol.L^{-1}$) in 1 $mol.L^{-1}$ HCl	0.05 $mol.L^{-1}$	0.1 $mol.L^{-1}$	0.2 $mol.L^{-1}$
S(%)-Pd	100 ± 2	93 ± 2	85 ± 1
S(%)-Pt	93 ± 1	91 ± 1	90 ± 2
$HClO_4$($mol.L^{-1}$)	**0.5 $mol.L^{-1}$**	**1 $mol.L^{-1}$**	**3 $mol.L^{-1}$**
S(%)-Pd	93 ± 2	100 ± 2	92 ± 2
S(%)-Pt	86 ± 1	91 ± 1	87 ± 1
HCl($mol.L^{-1}$)	**1 $mol.L^{-1}$**	**7 $mol.L^{-1}$**	**10 $mol.L^{-1}$**
S(%)-Pd	2 ± 1	60 ± 2	96 ± 2
S(%)-Pt	1 ± 1	12 ± 1	43 ± 1

9.8.1 Pd(II), Pt(IV), and Rh(III) Separation

In view of the results presented, a flow sheet for the separation of Pd(II), Pt(IV) as well as Rh(III) from HCl solutions using impregnated resins are envisaged and presented in Figure 9.22. A PGM solution containing Pd(II), Pt(IV), and Rh(III) in 1 M HCl is brought in contact with XAD2/DEHTPA impregnated resins, and selective Pd(II) extraction takes place; thus, the Pd(II) in the organic resin may be stripped with Tu in HCl and therefore separated from the other metals. Recovery of Pt(IV) can be accomplished by contacting the aqueous solution obtained in the previous step; rich in Pt(IV) and Rh(III); with XAD2/Alamine 336 resin, which will

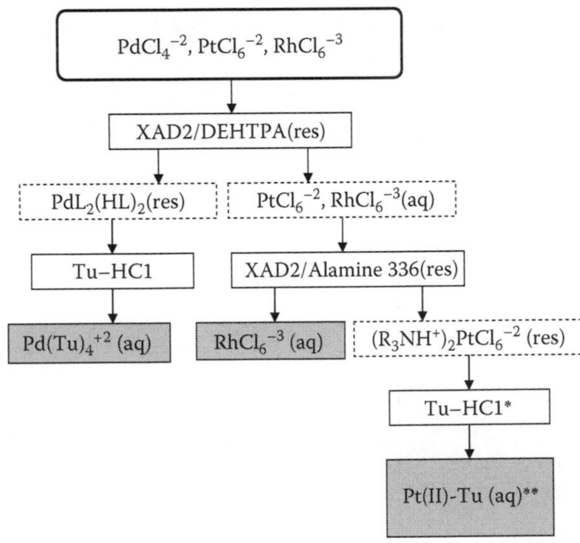

FIGURE 9.22 Pd(II)-Pt(IV)-Rh(III) separation from HCl media *HC104 can also be used for Pt(IV) stripping instead of Tu in HCl **Pt(II) forms stable complexes with thiourea, which had previously reduced Pt(IV) to Pt(II).

selectively extract Pt(IV) while Rh(III) will remain in the aqueous solution. Finally, Pt stripping may be conducted with either Tu or $HClO_4$.

ACKNOWLEDGMENT

Karel Jerabek and José Luis Cortina express their gratitude to Abraham Warshawsky for the opportunity to work together for more than one decade. The authors wish to dedicate this chapter to the memory of Abraham Warshawsky.

REFERENCES

1. G. M. Ritcey and A. W. Ashbrook, *Solvent Extraction Principles and Applications to Process Metallurgy,* Part 2, Elsevier Science Publishers B.V., Amsterdam, 1984.
2. F. G. Helfferich, *Ion-Exchange,* McGraw-Hill, New York, 1962.
3. D. K. Hale, Chelating Resins, *Research* 9, 104 (1956).
4. A. Warshawsky, Solvent impregnated resins in hydrometallurgical applications, Inst. Min. & Metal. (London) Trans. Sect. C, 83, (1974), 101.
5. R. R. Grinstead and K. C. Jones, *J. Nucl. Inorg. Chem.,* 36, 391 (1974).
6. A. Warshawsky, *Ion Exchange and Solvent Extraction,* vol. 8, ed. J. A. Marinsky and Y. Marcus, Marcel Dekker, New York, 229, 1981.
7. J. L. Cortina and A. Warshawsky, *Ion Exchange and Solvent Extraction,* vol. 13, ed. J. A. Marinsky and Y. Marcus, Marcel Dekker, New York, 209, 1996.
8. J. L. Cortina, N. Mirales, A. Sastre, M. Aguilar, A. Profumo, and M. Pesavento, *React. Polym.* 18, 67 (1992).
9. K. Jeřábek, L. Hankova, A. G. Strikovsky, and A. Warshawsky, *React. Funct. Polym.* 28, 201 (1996).
10. A. Warshawsky, A. G. Strikovsky, and K. Jerabek, *Israeli Pat. Appl.* 121369, *Czech Pat. Appl.* 3012-97, 1997.
11. A. G. Strikovsky, K. Jeřábek, J. L. Cortina, A. Sastre, and A. Warshawsky, *React. Polym.* 28, 149 (1996).
12. B. Kroebel and A. Meyer, West Germany Pat. Appl. 2, 162,951 (1971).
13. B. Kroebel and A. Meyer, *Proc. ISEC '74,* vol. 3, 2095–2107 (1974).
14. A. Bolotov, Yu. Korovin, et al., Pat. USSR 476279, Publication 05.07.75, Bulletin 25.
15. Yu. Kuzovov, Yu. Korovin, et al., *Proc. Moscow Mendeleev Chem.-Tech. Inst.* 97, 43 (1977).
16. V. Korovin, S. Randarevich, and Yu. Kuzovov, *Ukrainian Chemical Journal* 56, 1042 (1990).
17. V. Korovin and Yu. Shestak, *Reac. Pol.* 40, 107 (1999).
18. http://www.ion-exchange.com/products/specialty/index.html.
19. A. G. Strikovsky, A. Warshawsky, K. Jeřábek, and J. L. Cortina, *Solv. Extr. Ion Exch.* 15, 259 (1997).
20. J. L. Cortina, N. Miralles, M. Aguilar, and A. Sastre, *Solvent Extr. Ion Exch.,* 12, 371–391 (1994).
21. A. Warshawsky, A.G. Strikosky, J. Jerabek, and J. L. Cortina, *Solvent Extr. & Ion Exch.* 15, no. 2, 259–283 (1997).
22. V. I. A. Warshawsky, K. Jerabek, and A. G. Strikovsky, Method for Preparation of Sorbents Selective for Heavy Metals by Supporting of Acidic Organic Reagent on Polymer Carriers Containing Basic Active Groups, *Israeli Pat. Appl.* 121369 (1997).
23. K. Jeřábek, Inverse Steric Exclusion Chromatography as a Tool for Morphology Characterization, in *Strategies in Size Exclusion Chromatography,* eds. M. Potschka and P. L. Dubin, American Chemical Society, Washington, DC, 212, 1996.

24. A. G. Strikovsky, A. Warshawsky, L. Hankova, and K. Jeřábek, *Acta Polymerica* 49, 600–605 (1998).
25. K. Jeřábek, Inverse Steric Exclusion Chromatography as a Tool for Morphology Characterization, in *Cross-Evaluation of Strategies in Size-Exclusion Chromatography,* eds. M. Potschka and P. L. Dubin, American Chemical Society, Washington, DC, 211–224, 1996.
26. K. Jeřábek, H. Widdecke, and B. Fleischer, *React. Polym.* 19, 81 (1993).
27. K. Jeřábek, L. Hanková, and A. Revillon, *Ind. Eng. Chem. Res.* 34, 2598 (1995).
28. M. I. Kabachnik, S. T. Ioffe, and T. A. Mastriukova, *Zh. Obshch. Khim.* 25, 684 (1955).
29. I. P. Alimarin, T. V. Rodionova, and V. M. Ivanov, *Rus. Chem. Rev.* 58, 863 (1989).
30. J. Marsden and I. House, *The Chemistry of Gold Extraction,* Ellis Horwood, New York, 1992.
31. C. A. Fleming, *The Potential Role of Anion Exchange Resins in the Gold Industry,* The Minerals, Metals and Materials Society, 95–117, 1998.
32. B. R. Gren, M. H. Kotze, and J. P. Engelbretch, *Resin in Pulp: After Gold, Where Next?* EPD Congress 1998, The Minerals, Metals and Materials Society, 119–136, 1998.
33. R. Kautzmann, C. H. Sampaio, V. Korovin, Y. Shestak, M. Aguilar, and J.L. Cortina in *Solvent Extraction Applications, International Solvent Extraction Conference, Cape Town, South Africa*, Ed. P. A. South African Institute of Mininig and Metallurgy, Marshalltown, 506–511, 2002.
34. G. A. Kordosky, J. M. Sierakcski, J. Virnig, and P. L. Mattison, *Hydrometallurgy* 30, 291–305 (1992).
35. M. Virnig and G. A. Wolfe, *Proceedings International Solvent Extraction Conference (ISEC '96),* Melbourne, 311–316, 1996.
36. A. Sastre, A. Madi, J. L. Cortina, and F. J. Alguacil, *J. Chem. Tech. & Biochem.,* 74, 310–314 (1999).
37. V. Korovin and S. Randarevich, in *Proc. ISEC '88,* vol. 3, Moscow, USSR, 159–162, 1988.
38. V. Korovin, *Kchimicheskaya Technologiya (Chemical Technology)* 5, 3–13 (1991).
39. Environmental Signals 2000, Environmental Assesment Report 6, European Environment Agency, Copenhagen, 2000.
40. G. Tiravanti, D. Petruzzelli, and R. Passino, *Waste Mangement* 16, 597–605 (1996).
41. H. J. Bart and A. Schoneberger, *Chem. Eng. Technol.* 23, 653–660 (2000).
42. G. M. Ritcey and A. M. Ashbrook, *Solvent Extraction: Principles and Applications to Process Metallurgy,* Part 1, Elsevier, Amsterdam (1984).
43. J. L. Cortina, N. Miralles, M. Aguilar, and A. M. Sastre, *Reactive and Functional Polymers* 32, 221–229 (1997).
44. E. Castillo, M. Granados, M. D. Prat, and J. L. Cortina, *Solvent Extr. & Ion Exch.* 143, 243–245 (2000).
45. W. H. Holl, C. Stohr, R. Kiefer, and C. Bratosch, *Ion Exchange at the Millenium,* ed. J. A. Greig, Imperial College Press, 377, 2000.
46. J. E. Hoffmann, J. Met, June (1988) 40.
47. C. Hagelüken, *Recycling of Automotive Catalytic Converters,* 2d ed., European Precious Metals Conference, Lisbon, 1995.
48. R. K. Mishra, *Precious Metals '89*, ed. M. C. Jha and S. D. Hill, The Minerals, Metals & Materials Society, 483, 1988.
49. M. Rovira, L. Hurtado, J. L. Cortina, J. Arnaldos, and A. M. Sastre, *Reactive and Functional Polymers* 38, 279–287 (1999).
50. M. Rovira, J. L. Cortina, J. Arnaldos, and A. M. Sastre, *Solvent Extraction and Ion Exchange* 16, no. 2, 545–564 (1998).
51. T. Handley, *Anal. Chem.* 35, 991 (1963).
52. G. Cote and D. Bauer, *Rev. Inorg. Chem.* 10, 121 (1989).

Index

D

G

H

I

K

L

M

N

O

P

Q

R

S

T

U

V

W

Y

Z

Printed and bound by CPI Group (UK) Ltd, Croydon, CR0 4YY
24/10/2024
01778278-0015